Finite Elements in Solids and Stru

Finite Elements in Solids and Structures

An introduction

R. J. Astley

Reader in Mechanical Engineering
University of Canterbury
Christchurch
New Zealand

CHAPMAN & HALL
London · Glasgow · New York · Tokyo · Melbourne · Madras

Published by Chapman & Hall, 2-6 Boundary Row, London SE1 8HN

Chapman & Hall, 2-6 Boundary Row, London SE1 8HN, UK

Blackie Academic & Professional, Wester Cleddens Road, Bishopbriggs, Glasgow G64 2NZ, UK

Chapman & Hall, 29 West 35th Street, New York NY10001, USA

Chapman & Hall Japan, Thomson Publishing Japan, Hirakawacho Nemoto Building, 6F, 1-7-11 Hirakawa-cho, Chiyoda-ku, Tokyo 102, Japan

Chapman & Hall Australia, Thomas Nelson Australia, 102 Dodds Street, South Melbourne, Victoria 3205, Australia

Chapman & Hall India, R. Seshadri, 32 Second Main Road, CIT East, Madras 600 035, India

---

First edition 1992

Typeset in 10/12pt Times by Pure Tech Corporation, Pondicherry, India
Printed in Great Britain at The University Press, Cambridge

ISBN 0 412 44160 8 0 442 31629 1(USA)

A catalogue record for this book is available from the British Library
Library of Congress Cataloging-in-Publication Data available

Contents

Preface

This book has grown from a lecture course in computational solid mechanics given in one form or another over the past ten years to final year undergraduates at the University of Canterbury. Although given as part of a mechanical engineering degree the course is also appropriate for students of civil or aeronautical engineering. The objective of the course, and of the book, is to introduce finite elements for the analysis of stress and deflection, and to do so in a way that makes the method accessible to students with only a general interest in the topic. The book itself is very much an introductory text and has no pretensions to being a 'state-of-the-art' treatment.

Finite elements are applied to a broad range of engineering problems, but they are most widely used for solid and structural calculations, this being the area in which commercial finite element programs are well established in engineering practice. Such programs are available at all levels, often at quite a modest cost. As a consequence, the use of finite element models over quite a broad spectrum of design applications is now common, supplementing and refining the 'strength of materials' calculations which have dominated such analysis in the past and circumventing, in many instances, the provisions of conservative design codes. The requirement that engineering graduates be familiar at a basic level with the finite element concept, so that they can use such programs with an awareness of their capabilities and limitations, is becoming an important component of engineering training. Certainly, within our own mechanical degree programme, which has a strong emphasis both on design and on computer integrated engineering, we find that once students have access to finite element codes, as is increasingly the case given the integrated nature of the CAD environment, they are quick to use them in lieu of traditional methods. It is then imperative that they understand enough about the method, and about the appropriateness or otherwise of its use, to be able to produce reliable results. It is very much towards this end that the current text is directed. It is intended either as a prescribed text for a final year course in 'finite element methods', or 'computational solid mechanics', or as an accompanying text for more conventional 'solid mechanics/advanced strength of materials' courses which incorporate aspects of the finite element approach, or simply as a self-contained course of private study for students or graduate engineers who need to become familiar with the finite element method.

The subject matter is presented in a logical sequence which starts with a brief introduction to finite elements in general (Chapter 1) and the necessary relationships of linear elasticity and of beam and plate theory (Chapter 2). This is followed by a discussion of variational formulations and of the displacement energy principle (Chapter 3). The Rayleigh–Ritz method is then introduced (Chapter 4) and forms the basis for the remainder of the treatment. This precludes coverage of more advanced elements based on mixed and hybrid formulations, but provides a strong unifying thread for conventional 'displacement' elements. The remainder of the text then progresses through the different categories of element roughly in order of increasing complexity. One and two-dimensional bar elements are dealt with first (Chapter 4), followed by linear two and three-dimensional continuum elements (Chapter 6), more advanced continuum elements (Chapter 8), beam elements (Chapter 9) and plate elements (Chapter 10). On the way there are brief digressions into solution methods (Chapter 5) and the criteria for convergence (Chapter 7). The text concludes with a brief introduction to dynamic analysis (Chapter 11). All material is presented at a level which presupposes that the background of the reader extends no further than elementary coursework in calculus, linear algebra and strength of materials. Such material is common to most first and second year courses in mechanical, civil or aeronautical engineering. The book is therefore designed to supplement more advanced courses in applied elasticity and continuum/structural mechanics rather than requiring them as a prerequisite. It has been written very much as a self-contained text so that the reader can work sequentially through it without having to back-track or refer to material which has not yet been encountered. Exercises are included at the end of each chapter. These consolidate the material covered using simple manual examples, and also they provide a mechanism for introducing additional but related material. Many examples, therefore, start with phrases such as 'show that...', or 'confirm that...' and provide the reader with an opportunity to derive useful results not specifically included in the body of the text.

In teaching this material myself, I generally intersperse the presentation of theory with short computational assignments in which students are required to calculate the stresses and deflections in real (albeit simple) engineering components using the elements encountered in each chapter. This seems to me to be the most satisfactory way of teaching the subject at a practical level and can be done quite simply provided that students have access to one of the many finite element programs now available at a modest cost for educational use. At the University of Canterbury, we currently use ALGOR interfaced to Intergraph's MICROSTATION modelling system but other PC-based finite element programs such as MSC/PAL, ANSYS–PC, COSMOS/M are equally suitable as are more extensive mainframe programs. Whatever software is used, this type of activity replaces the exercises in programming or program modification which are to be found in many texts dealing with the finite element method. Although useful in the past, these are of limited value to most contemporary users who will never be required to write their own programs. The extensive program listings to be found in many texts are notably absent from the current one for the same reasons.

Finally, I would like to thank friends and colleagues who have been instrumental in the production of this book. These include Roger Jones and Clem Earle of Unwin–Hyman who provided the initial encouragement to get it off the ground, my colleagues in the Mechanical Engineering Department at the University of Canterbury who have assisted in various ways, notable among them are Philip Smith and Dr John Smaill whose assistance in all matters to do with computers and word-processors has been indispensable, and Dr Anne Ditcher for her help with ALGOR. I would also like to thank the many hundreds of final year students who have acted as unpaid proofreaders, and last but not least, my wife, Jane, who has had to put up with the sound of a keyboard from behind a closed door for rather more evenings than she had any right to expect.

Jeremy Astley
University of Canterbury

1

Introduction

1.1 HISTORICAL DEVELOPMENT

The term 'finite element' appears to have been coined by Clough in an article published in 1960 [1]. It was used to describe a computational approach to the analysis of elastic membranes in which the continuum was divided into a discrete number of small but finite subregions or 'elements'. The idea itself was not new. In fact, the notion of dividing a continuum into finite pieces had been suggested by Courant in 1943 [2]. The practical implementation of such an approach only materialized in the mid-1950s, however, with the advent of digital computation. It was Turner, Clough and others [3] who then combined the idea of discrete elements with the 'stiffness' approach to matrix structural analysis, to produce a systematic procedure which later became known as the **finite element method**. An interesting account of these early developments is to be found in Clough's own commentary on the period [4].

The popularity of the method grew rapidly in the years that followed. Within ten years of the invention of the term, more than one thousand articles dealing with finite elements had been published in the scientific literature [5]. Two decades later, the number of articles listed in the COMPENDEX-PLUS engineering index approaches 50 000. This phenomenal rate of growth is a reflection of the degree to which the finite element concept has complemented the emerging capabilities of the digital computer. Most importantly, it has lent itself to the development of multipurpose programs. In the late sixties, this gave rise to the first, commercial, finite element computer codes. These were programs, capable of solving different physical problems through changes to the input data rather than to the code itself. Such programs have expanded and proliferated in the years since and now include such industry standards as NASTRAN, ABAQUS, ADINA, ANSYS, PAFEC, SAP, MARC and EASE, to name but a few. More recently they have been joined by a new generation of PC or workstation orientated codes such as MSC/PAL, ANSYS–PC, SAP90, COSMOS/M, ALGOR and so on. Their application to practical problems of stress analysis constitutes one of the major advances in mechanical and structural design of the present era.

Also contributing to the rapid growth of the method from the late sixties onwards was the realization that it could be applied to problems other than those of solid and structural analysis. The first suggestion that this might be the case came with the work of Zienkiewicz and others [6] in demonstrating that the method could be used for field problems involving Laplace's and Poisson's equations (steady state thermal conduction, for example, or potential flow of an inviscid fluid). These developments were accompanied by the realization, also in the early sixties [7], that the Galerkin approach and other 'weighted residuals' techniques could be used as a theoretical basis for applying finite elements to virtually any problem which could be expressed in terms of partial differential equations. As a consequence, current areas of application now include fluid mechanics, aerodynamics, acoustics, lubrication theory, geomechanics, atmospheric dynamics, electromagnetic theory and so on. A survey of finite element applications in these and other areas is to be found in [8].

In addition to single applications, the method also lends itself to the solution of coupled problems involving two or more constituents: the seismic interaction of dams and reservoirs, for example, or the interaction of waves and offshore structures, acousto-structural coupling, elasto-hydrodynamic lubrication and so on. These lie well beyond the scope of this book but form the substance of much current research.

Confining our discussion now to solid and structural applications, the finite element method was well advanced by the late sixties in its capacity to solve practical two and three-dimensional problems in linear elasticity. However, it still required substantial computation – in the context of that commonly available at the time – and was accessible only to those engaged in large-scale industrial or institutional research. One of the most significant recent developments in the application of the method has arisen not so much from any specific advance in the methodology itself (new element formulations, solution algorithms etc.) but simply from the steady decrease in the effective cost of computating. This has brought finite element computation within the mainstream of engineering practice by making it accessible, for routine design analysis, to engineers with relatively modest computational facilities at their disposal. Many commercial programs now operate effectively on engineering workstations or personal computers [9] and perform substantial analyses for a small fraction of the cost associated with such calculations a decade or so ago. This is clearly a trend which has a long way to run and one which will make the method even more accessible to the engineering profession in years to come.

A side effect of the same phenomenon, the inexorable decrease in the real cost of computing, has been the use of computer graphics in the engineering environment. This in turn has driven the growth in computer aided design and manufacture (CAD/CAM). Practical implementation of the finite element method has been encouraged by such developments since the generation of finite element models can readily be integrated with the geometric modelling techniques which lie at the heart of the CAD/CAM revolution. There are consequently few CAD systems today which do not provide a facility for the interchange of geometrical information with compatible finite element programs. This integration between geometric design (using a CAD system) and quantitative analysis (using a finite element code) has rendered the preparation of finite element data and the display of the results of finite element analysis, if not trivial, at least an order of magnitude less time consuming than was the case with the finite element programs of the sixties and seventies.

1.2 SOLID AND STRUCTURAL ANALYSIS USING FINITE ELEMENTS

1.2.1 Types of analysis

Whether performed by hand or on a computer, the analysis required to determine stresses and deflections in a deformable body can be subdivided into several categories. The simplest, and by far the most common in routine application, is the analysis of **linearly elastic systems**. Here it is assumed that a linear relationship exists between force and deflection and that all deformation is recovered when the load is removed. For many engineering materials, the assumption of linearity implies also that the displacements of the structure are small compared to its overall dimensions. To fix ideas, a simple linearly elastic system–an elastic beam on simple supports subject to a point load at its centre–is shown in Fig. 1.1(a). The load is such that no yielding occurs, deflections are small, and when the load is removed, the beam springs back to its original shape.

A more complex problem is posed when the system is **materially nonlinear**, that is, when the force–displacement relationship of the material itself is no longer linear. In a ductile material, such as steel, this occurs when the yield stress is exceeded. The material then undergoes inelastic deformation which is no longer proportional to the load. Behaviour of this sort is often, though not always, accompanied by an irreversible 'permanent set'. An example of a materially nonlinear system is shown in Fig. 1.1(b). It is almost identical to that of Fig. 1.1(a), but the beam is now loaded past its point of yield. A materially nonlinear plastic region therefore forms in the highly stressed region near the centre and a plastic 'hinge' is created. This is less resistant to bending than the remainder of the beam and an elbow occurs as shown. Moreover, a portion of the deformation is irreversible so that when the load is removed a permanent 'kink' remains.

A different type of nonlinear behaviour occurs in cases where the internal forces are dependent upon the final deformation. Such systems are said to be **geometrically nonlinear**. A simple example is that of a wire hanging under its own weight which is loaded by a concentrated load at its centre (Fig. 1.1(c)). Provided that the tension in the wire does not exceed the elastic limit, the system remains 'materially' linear and the wire resumes its original length and shape when the load is removed. It is 'geometrically' nonlinear, however, since its ability to carry the load at all depends upon its final deformation, that

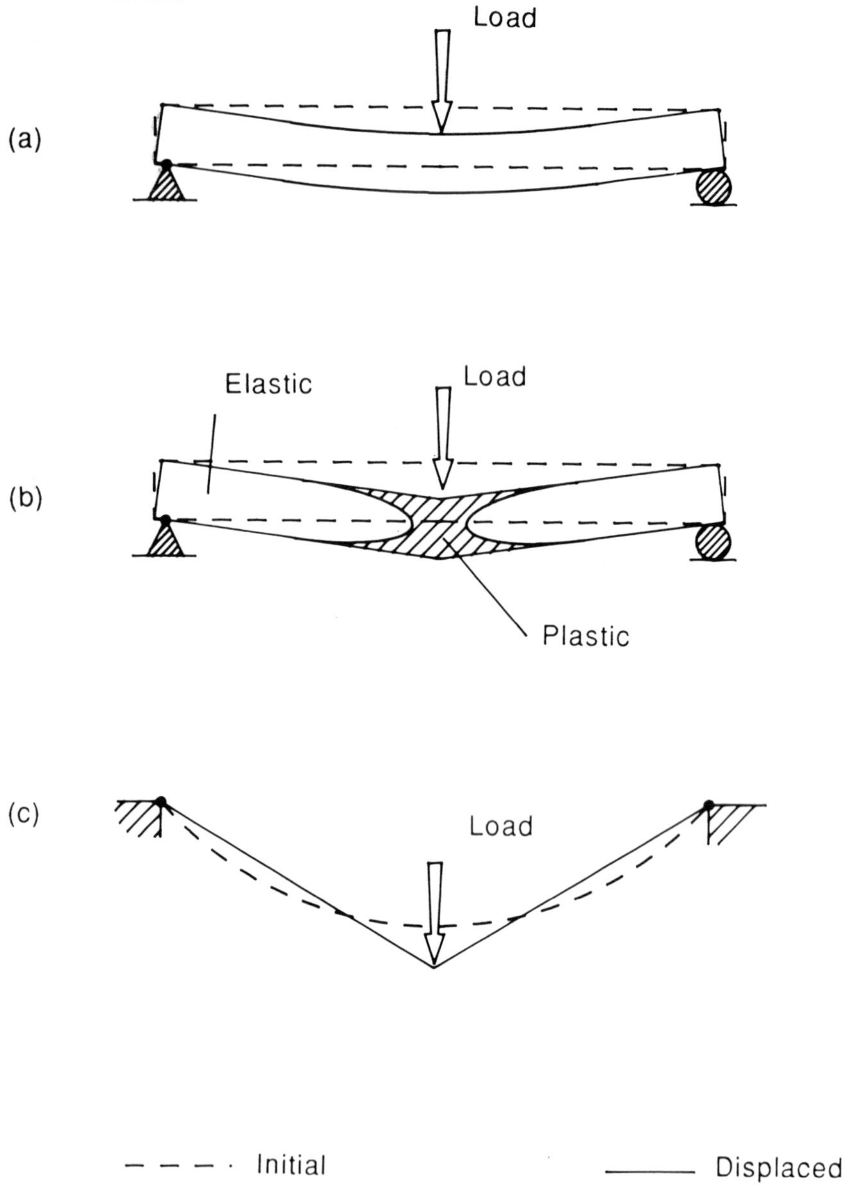

Fig. 1.1 *Types of analysis: (a) linearly elastic; (b) materially nonlinear; and (c) geometrically nonlinear.*

is the V-shape shown in Fig. 1.1(c). A less obvious example of geometrical nonlinearity, but one of importance in structural design, is the phenomenon of 'buckling', in which the lateral deflection of a beam or plate determines its in-plane load carrying capacity, or lack of it.

Linear and nonlinear problems may be subdivided further into **static** and **dynamic** cases. In static problems the loads are applied slowly in the sense that inertial forces are negligibly small throughout the loading process. This is an acceptable approximation provided that the time scale of the loading process is greater than the period of the lowest vibrational frequency of the system. In a dynamic problem, on the other hand, the inertial forces are significant compared to the applied loads, and the instantaneous forces and deflections experienced by the structure are then dependent upon the time history of the loading process. A preliminary to a full dynamic analysis, in the case of a linear system, is an analysis of the natural frequencies and associated modes of vibration. This is termed a **normal mode analysis**. Once it has been performed, a full dynamic response can be calculated relatively simply using mode superposition. When the system is materially or geometrically nonlinear, a more complex and time consuming step-by-step time integration of the dynamic equations is required.

All of these types of analysis (linear and nonlinear, static and dynamic) can be performed using finite element models. In order of increasing computational effort, they can be ranked (roughly) as follows:

1. linear static analysis;
2. normal mode analysis;
3. buckling analysis (linear);
4. linear dynamic analysis (mode superposition);
5. static/dynamic analysis (materially nonlinear);
6. static/dynamic analysis (geometrically nonlinear).

The capabilities of finite element codes tend to reflect this ranking. Smaller codes operating on workstations or personal computers often restrict themselves to categories 1 and 2, sometimes including 3 and 4 but rarely 5 or 6. More comprehensive codes are generally capable of performing all categories of analysis in one form or another.

1.2.2 Types of element

The first step in creating a finite element model for a solid object or a structure is to divide it into a finite number of discrete elements. These are usually selected from a library of element types available within a given program. Each element is characterized by its 'topology' (an ordered sequence of points or 'nodes', usually on its periphery) and by a number of material or structural properties (density, Young's modulus, thickness and so on).

Upon first acquaintance many finite element codes present the user with a bewildering choice of elements. Fifty or more distinct elements for solid and structural problems is not uncommon. To make matters worse, elements of similar physical appearance and the same topology can have quite different capabilities.

There are, however, broad categories into which individual elements can be placed. When these are clearly understood, they greatly simplify the identification of elements applicable to a particular physical problem. These categories are now reviewed, partly to facilitate immediate discussion of specific types of element but also to preface some general comments on the scope and structure of the remainder of the book.

The most straightforward subdivision of elements is into one, two and three-dimensional groupings. These do not relate directly to the function of the element but only to its physical form. This division is represented by the horizontal layers of Fig. 1.2. Bars and beams, for example, which idealize a prismatic member by a single one-dimensional line, qualify as 'one-dimensional' elements and occupy the upper level of the diagram. Membranes and plates, which idealize thin sheets of material as two-dimensional planes or

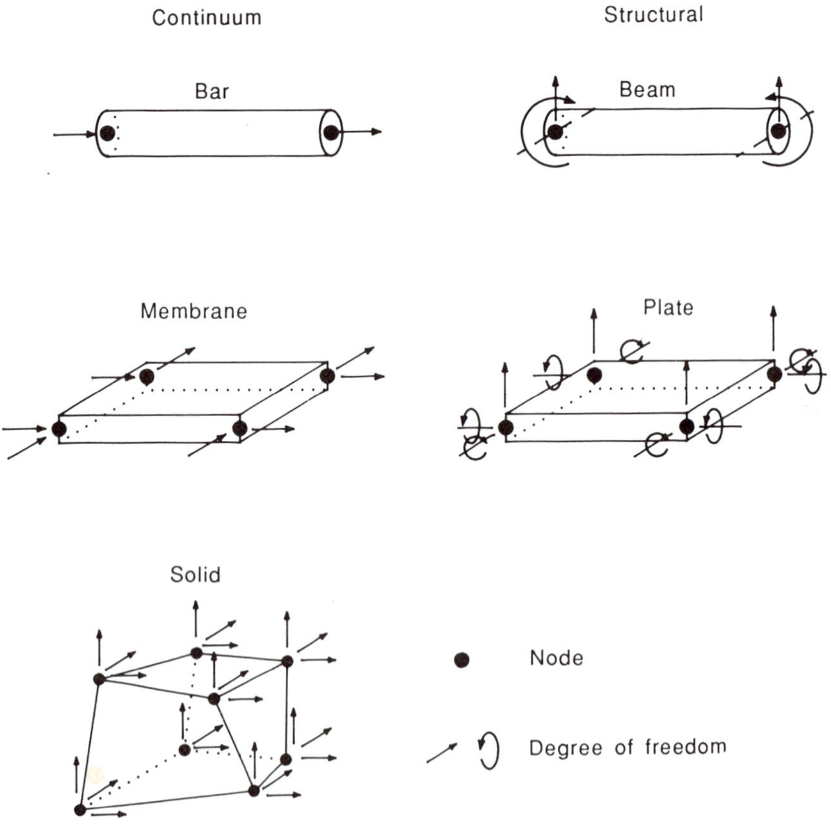

Fig. 1.2 *Types of element.*

surfaces, lie on the next level, and three-dimensional 'lumps' of material, which cannot be further idealized, appear on the lowest level.

Within the upper two levels, that is the one and two-dimensional groupings, there is a second (vertical) division based on function. It distinguishes between those elements on the left which carry in-plane loads by 'stretching' (bars and membranes) and those on the right which carry lateral loads by bending (beams and plates). This can be somewhat confusing since seemingly identical elements (a beam and a bar, for example) have the same number of nodes and the same material properties but quite different functions. It is a distinction which must be clearly drawn, however, if errors of application are to be avoided. In the remainder of this text those elements representing bars, membranes and solid lumps of material will be referred to as **elastic continuum elements** or **continuum elements** for short. Their formulation depends only upon the most primitive relationships between stress and strain in an elastic continuum. Elements which carry loads in bending (beams and plates) are termed **structural elements**. They are more complex since their formulation incorporates aspects of beam or plate theory ('plane sections remain plane...' and so on) and involves curvatures and moments rather than simple stresses and strains.

It is a general characteristic of continuum elements that their displaced states are completely defined by translational displacements at the nodes. These are indicated by straight arrows in Fig. 1.2. A structural element, on the other hand, usually requires that both translations *and* rotations are specified at each node. Such rotations are indicated by curved arrows in Fig. 1.2. Whether translational or rotational, the displacement parameters at each node are termed the **degrees of freedom** of the element and become the degrees of freedom of any finite element model of which it is a part.

It must be emphasized that the individual elements shown in Fig. 1.2 are 'typical' members of much larger groups. In the element library of a large finite element code, for example, each of the elements shown in Fig. 1.2 would be accompanied by others of the same general type but with different topologies or nodal displacements. There are also combined elements which are formed by adding 'continuum' and 'structural' elements together. By superimposing a bar element and a beam element, for example, we obtain a **frame element** which is capable of sustaining loads both in bending and in tension. Similarly, by superimposing a membrane element and a plate element we obtain a **shell element** capable of carrying both lateral and in-plane loads.

1.3 WHEN TO USE FINITE ELEMENTS IN SOLID AND STRUCTURAL CALCULATIONS

Although the finite element method is now available to general engineers for routine design calculations, certainly those involving linear stresses and deflections, its use is inappropriate in situations where acceptable results can be obtained by using traditional methods of analysis. In an engineering degree programme, such methods form the basis of courses with titles such as 'strength of materials', 'mechanics of materials', 'solid mechanics' and 'machine design'. To simplify subsequent discussion, they will be referred to here simply as 'strength of materials methods'. Let us assume that the reader has already benefitted from instruction of this type and is able to predict with reasonable accuracy the stresses and deflections for a catalogue of simple shapes and loadings. Prismatic bars subject to extension, bending, torsion and shear, for example, should present no difficulty, nor should thin-walled axisymmetric containers subject to external or internal pressures, thick walled cylinders or rotating discs. Simple frames with pinned or fixed joints and perhaps no more than half a dozen redundant forces should also be amenable to systematic, if somewhat tedious, hand calculation. Some readers may also have encountered traditional plate theory which, for practical purposes, yields useful solutions only for rectangular or circular plates subject to a limited set of lateral loads and edge conditions. The more adventurous (and more mathematically inclined) may also have touched upon the useful solutions which emerge from elasticity theory. These include plane stress/strain solutions for deep beams and curved bars, stress concentrations in the vicinity of holes and notches, Hertzian contact stresses, and so on (further discussion of most of the topics listed above is contained in [10]).

Armed with an ability to calculate stresses and deflections for these standard cases and assisted by stress concentration factors for local geometrical irregularities, design codes for particular classes of object and substantial overall safety factors, an experienced engineer may safely stress many components without recourse to more sophisticated methods of analysis.

There are, however, important situations in which this traditional approach is less than satisfactory. The most significant constraint is the limited nature of the geometries to which it can be applied. Many engineering components are designed (for reasons that have little to do with design for strength) to be of irregular shape and to contain inconvenient holes, cutouts, fillets, changes of thickness, etc. Even nominally prismatic components, such as shafts or beams are often afflicted with troublesome geometric irregularities of this type. The standard approach for coping with such problems is to use a crude geometrical approximation, which *does* correspond to one of the standard cases, and to acknowledge its imperfect correspondence to the real object by incorporating approximate 'stress concentration' factors, or simply by including a large safety factor in the ensuing calculations. This is quite acceptable in many instances but generally produces a conservative design. In situations where irregular objects must be designed to a relatively small margin of safety, such calculations are therefore of doubtful utility. It is in such cases that the finite element method provides a valuable tool for analysis. Virtually any geometrical shape can be modelled satisfactorily using an appropriate finite element mesh. The shape of an object is in fact quite immaterial as far as the analysis itself is concerned, although it may take more effort on the par the user to generate data for an irregular object than for a simple one. example of a component whose shape precludes simple and accurate strength of materials analysis — a doubly curved impeller blade — but which is quite simple to mesh and analyse using finite elements, is shown in Fig. 1.3. The mesh which was used in the analysis is formed from three dimensional 'continuum' elements similar to the block element of Fig. 1.2. A second example of an object which defies accurate analysis by traditional means is the fan runner shown in Fig. 1.4. It is fabricated from steel sheet but is of an overall shape which makes any form of traditional analysis using membrane or plate theory virtually impossible. A finite element analysis is not particularly difficult however and was performed in the current instance by dividing the runner into two-dimensional shell elements formed by superposing membrane and plate elements of the type shown of Fig. 1.2 (only one sixteenth of the runner was modelled, due to its cyclical symmetry).

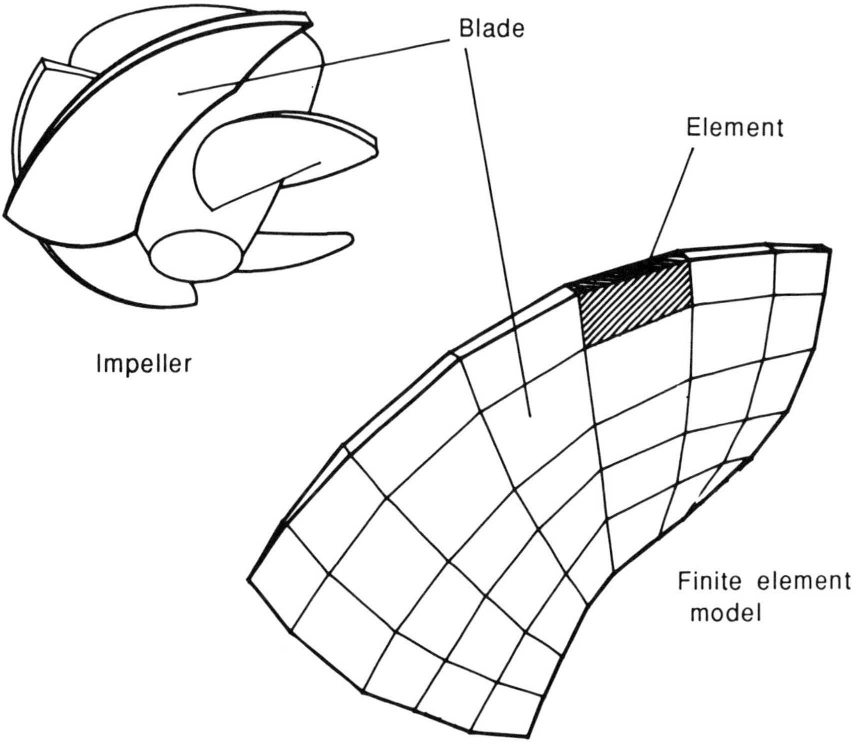

Fig. 1.3 *Finite element model of an impeller blade.*

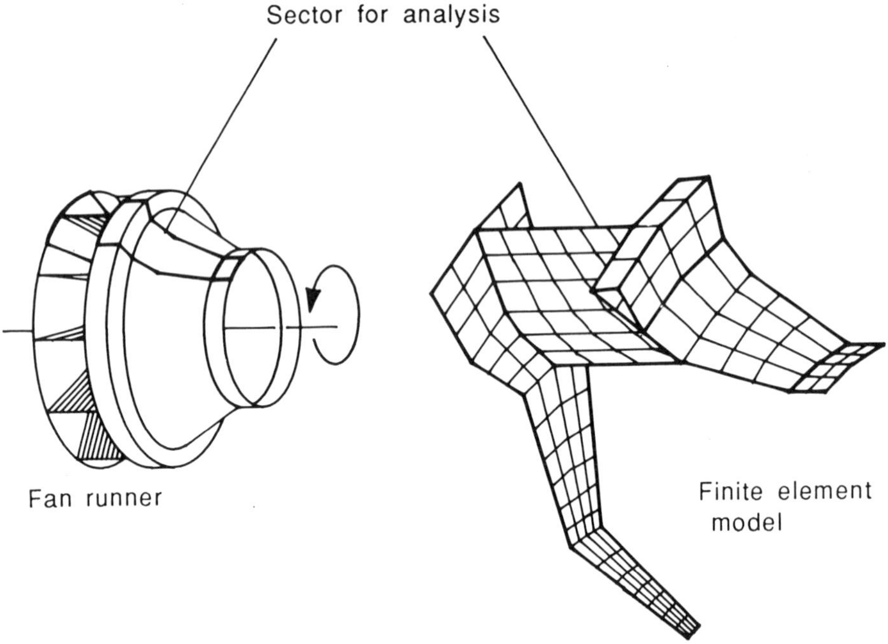

Fig. 1.4 *Finite element model of a fan runner.*

A quite different situation in which the 'strength of materials' approach offers little hope of a practical solution is that in which a structure is formed from a large number of individual components. Each may be simple to analyze in isolation, but when coupled together they form a system of great complexity. The welded tubular chassis shown in Fig. 1.5 is a case in point. Although constructed entirely of prismatic tubes, each of which is simple to analyze as an isolated component, it possesses a high degree of static indeterminacy and is quite impossible to analyse as a whole unless computational methods are used. In this case a finite element model was constructed in which each tubular member was modelled by a single composite element formed by superimposing a bar and a beam.

1.4 FORM OF THE FINITE ELEMENT EQUATIONS

The finite element method generates discrete models for continuous systems. In some instances the discretization is very obvious (in a framed structure for example, such as that of Fig. 1.5, where each member becomes a separate element) in others it requires considerable engineering insight. Some charac-

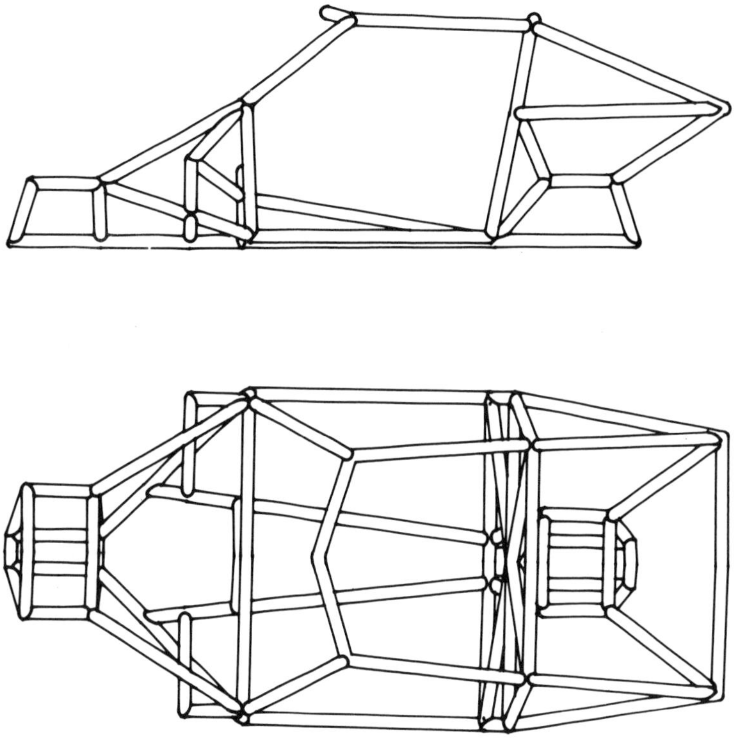

Fig. 1.5 *Finite element model of a tubular chassis.*

teristics of the discrete algebraic equations which result from any finite element representation can be deduced from the behaviour of a simple, two-dimensional, framed structure, such as that shown in Fig. 1.6. Suppose that the framework is pin-jointed and that all loads are applied at the joints so that no bending occurs in the individual members. The deflection of the entire structure is then defined, without approximation, by the displacements of the joints. No attempt is made at this stage to develop a quantitative model for such a system, but merely to get some feel for the algebraic representation which must inevitably result.

Towards this end let us reintroduce some terminology from the previous section. The bars are regarded as 'elements', the points at which they are connected as 'nodes' and the displacements at each node in the horizontal and vertical directions (if this were a three-dimensional model there would be a further nodal displacement perpendicular to the plane of the paper) as 'degrees of freedom'. The latter are denoted by the symbols $\delta_1, \delta_2, \ldots, \delta_n$. A node which is fixed so that it cannot move in a particular direction (for example node A which cannot move in the vertical direction) has no degree of freedom

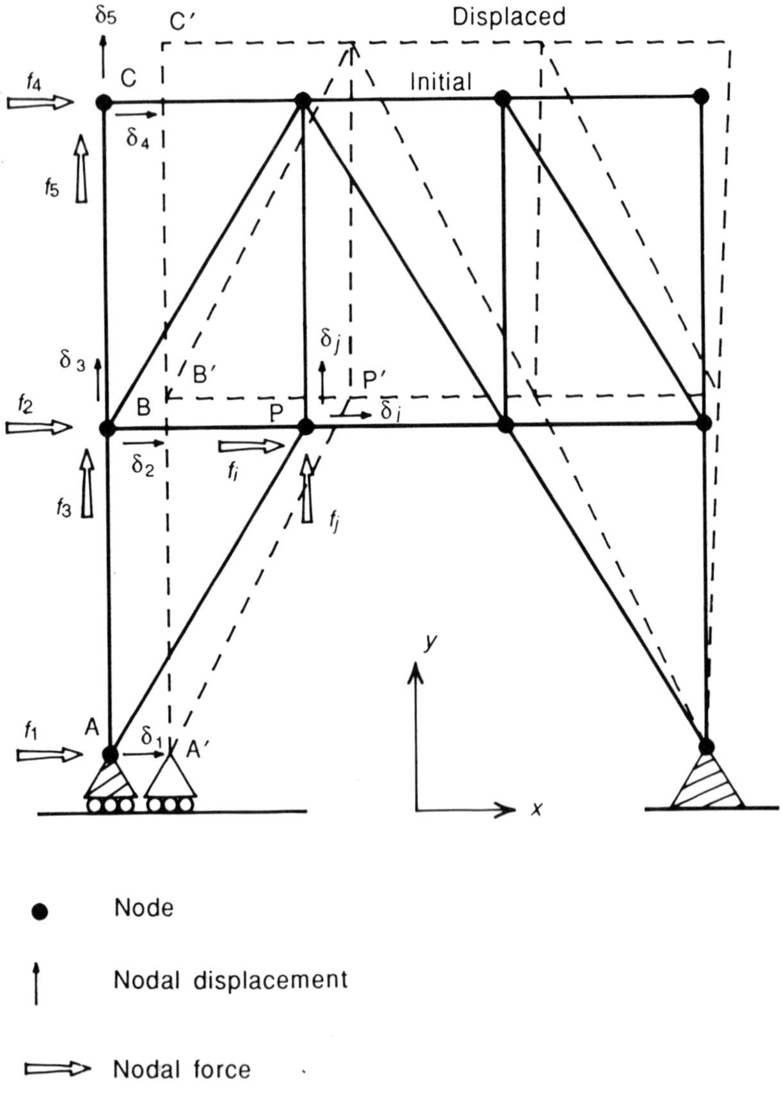

Fig. 1.6 *Discrete model of a plane framework.*

assigned to that displacement. The subscript attached to each degree of freedom, the i in δ_i, is termed the **degree of freedom number**. In Fig. 1.6, for example, the degree of freedom numbers assigned to nodes A, B, and C on the left hand vertical boundary of the structure are 1, 2, . . . , 5. At some general joint P within the structure, the degree of freedom numbers are i and j. Every unconstrained degree of freedom must be numbered and no number may be used twice.

The displaced locations of the nodal points of the structure are denoted by a prime (the superscript $'$). For example, node A displaces a distance δ_1 in the x direction, and moves to A'; node B displaces distances δ_2 and δ_3 in the x and y directions and moves to B', and so on. The external loads which produce this deformation are denoted by $f_1, f_2, \ldots, f_n$ and are ordered in the same way as the degrees of freedom, that is, f_1 is the load applied at node A in the direction of the displacement δ_1, f_2 is applied at node B in the direction of displacement δ_2 and so on. In practice, the value of many of these nodal forces is zero since external loads are applied only at selected points.

Without performing any quantitative analysis, the nature of the relationship which exists between the nodal forces and the nodal displacements can be deduced, given only that the system is linear. We can then take it as axiomatic:

(i) that the displacement at any point in the structure due to the combined application of all of the external loads is equivalent to the sum of the displacements at the same point due to the application of each load separately; and
(ii) that the displacement at any point due to the application of each load varies linearly with the magnitude of that load.

These statements are given mathematical expression by defining a quantity δ_{ij} as the contribution to the displacement δ_i which results from the application of a nodal force f_j in the absence of other loads. Statement (i) may then be rewritten

$$\begin{aligned} \delta_1 &= \delta_{11} + \delta_{12} + \delta_{13} + \ldots + \delta_{1n}, \\ \delta_2 &= \delta_{21} + \delta_{22} + \delta_{23} + \ldots + \delta_{2n}, \\ &\vdots \\ \delta_n &= \delta_{n1} + \delta_{n2} + \delta_{n3} + \ldots + \delta_{nn}. \end{aligned} \tag{1.1}$$

Statement (ii) implies the existence of a constant of proportionality which determines the contribution made by the load f_j to the displacement δ_i. This constant, c_{ij} say, is termed the 'flexibility influence coefficient' and defines a relationship

$$\delta_{ij} = c_{ij} f_j. \tag{1.2}$$

In the case of a system with n degrees of freedom there are n^2 such coefficients. Their evaluation is clearly a formidable task given that they depend upon the structural properties of the entire system (each coefficient may be interpreted physically as the displacement at one node due to a unit load applied at another). Leaving aside, for the moment, the problem of actually calculating these quantities, we can substitute equation 1.2 into equation 1.1 to produce a system of linear equations which relate the nodal displacements to the nodal forces. These are

$$\begin{aligned}\delta_1 &= c_{11}f_1 + c_{12}f_2 + \ldots + c_{1n}f_n,\\ \delta_2 &= c_{21}f_1 + c_{22}f_2 + \ldots + c_{2n}f_n,\\ &\vdots \qquad\qquad\qquad\qquad\quad ,\\ \delta_n &= c_{n1}f_1 + c_{n2}f_2 + \ldots + c_{nn}f_n.\end{aligned} \tag{1.3}$$

They may be expressed more concisely in matrix form as

$$\boldsymbol{d} = \boldsymbol{C}\boldsymbol{f} \tag{1.4}$$

where

$$\boldsymbol{d} = \begin{bmatrix}\delta_1\\ \delta_2\\ \vdots\\ \delta_n\end{bmatrix}, \quad \boldsymbol{C} = \begin{bmatrix}c_{11} & c_{12} & \ldots & c_{1n}\\ c_{21} & c_{22} & \ldots & \\ \vdots & & & \\ c_{n1} & & \ldots & c_{nn}\end{bmatrix}, \text{ and } \boldsymbol{f} = \begin{bmatrix}f_1\\ f_2\\ \vdots\\ f_n\end{bmatrix}. \tag{1.5}$$

The matrix $\boldsymbol{C}$ has n rows and n columns and is termed the **flexibility matrix**. Matrices $\boldsymbol{f}$ and $\boldsymbol{d}$ are termed the **nodal force vector** and the **nodal displacement vector**, respectively. If the influence coefficients c_{ij} are known, and 'force' methods of structural analysis set out to calculate these coefficients directly, the analysis is effectively complete since the right hand side of equation 1.4 can be evaluated explicitly for a given set of external loads to give the deflections of the structure at all points. This is not the approach adopted in a finite element model.

To understand the structure of the equivalent finite element equations, it is helpful to consider the equations which would result if equation 1.4 were inverted. Provided that the matrix $\boldsymbol{C}$ is nonsingular, and in practice this will always be the case for a properly supported structure, both sides of equation 1.4 can be premultiplied by the inverse of $\boldsymbol{C}$ to give

$$\boldsymbol{C}^{-1}\boldsymbol{d} = \boldsymbol{f},$$

The matrix $\boldsymbol{C}^{-1}$ is the 'stiffness matrix' of the system. It is usually denoted by the symbol $\boldsymbol{K}$, and the above equation is then written:

$$\boldsymbol{K}\boldsymbol{d} = \boldsymbol{f}, \tag{1.6}$$

If the components of $\boldsymbol{K}$, rather than $\boldsymbol{C}$, were known, the n simultaneous equations represented by equation 1.6 could then be solved to give the unknown nodal displacements (the components of $\boldsymbol{d}$) for a given set of external forces (the components of $\boldsymbol{f}$). Large sets of linear equations like this, although immensely time consuming to solve by hand, are quickly and easily solved using modern computers. There is, of course, little to be gained by doing so if the stiffness matrix $\boldsymbol{K}$ has itself been obtained by inverting the flexibility matrix $\boldsymbol{C}$, since the whole process could be avoided simply by returning to equation 1.4 and evaluating the nodal displacements directly. In a finite element formulation, however, the coefficients of the stiffness matrix $\boldsymbol{K}$ are obtained directly from the structure without reference to the flexibility coefficients. This is done using an 'assembly' process which methodically evaluates the contributions from each element. The solution of equation 1.6 then gives the nodal displacements of the structure. Once these have been found, the strains within each element (the elongation of each bar in the present case) may be determined explicitly and from these the stresses. The finite element model represents, in this sense, a 'displacement' method of analysis, in that the displacements of the structure are the primary unknown quantities and are calculated first, the stresses and strains being determined at a later stage in the solution.

The method outlined so far is relatively simple to comprehend when the various elements of the structure (bars in this case) are connected at a well defined set of nodes. This is clearly the case for framed structures but not so obviously true for two or three-dimensional continua. As we will see subsequently, however, the same general approach can be applied in these cases, provided that an element of approximation is introduced into the model.

1.5 SCOPE AND STRUCTURE OF THE CURRENT TEXT

The breadth of finite element applications (section 1.1) defies comprehensive treatment in an introductory text. Coverage here, therefore, is limited to solid and structural problems. More extensive texts which do cover the whole range of finite element activity are readily available for those seeking a broader treatment. Notable among them is the definitive work by Zienkiewicz and Taylor [11] which runs now to two substantial volumes in its fourth edition. The 'handbook' of Norrie and Kardestuncer [8], is also a good starting point for excursions into the nonstructural area.

Even given the restriction to solids and structures, a comprehensive treatment of all finite element formulations in this area is beyond the scope of the current text. It is in no sense a 'state of the art' treatment. A number of (more substantial) texts which do qualify for this description are available for those who wish to proceed further [11–13].

The objective of the current treatment is simply to prepare the uninitiated reader to use finite elements intelligently for relatively straightforward analyses. Coverage is limited primarily to static linear problems with a treatment of dynamic analysis only at its most elementary level and no treatment at all of nonlinear problems. The finite element method is regarded very much as a tool for practical analysis rather than as a theoretical entity in its own right, a contemporary alternative to 'strength of materials' calculations rather than a general mathematical approach for the solution of field problems. This reflects the manner in which it is applied in routine engineering analysis. That is to say, programs are used rather than written and an understanding of the capabilities and limitations of the method and of individual elements and element types is generally more important than an ability to write or modify finite element code. The emphasis on programming which has been a major preoccupation in many previous books is therefore absent in this one. Once again, the reader is well provided elsewhere with comprehensive assistance in programming the method should this be required [14, 15].

Many engineering graduates who now have access to the finite element method for design calculations, do not possess a particularly strong background in advanced engineering mathematics or indeed in continuum mechanics at an advanced level, prerequisites which one might reasonably expect, for example, in graduate students or specialist researchers. This presents some difficulty in preparing an introductory text addressed to this broader category of readership. In the present instance, it has been assumed that the mathematical background of the average reader extends no further than standard undergraduate coursework in linear algebra, vector calculus, and strength of materials at a fairly basic level. The finite element concept has accordingly been presented in a largely self-contained manner with coverage of the theoretical aspects confined to those specifically applicable to linear elastic analysis. These are fully developed from first principles within the text. Practicing engineers and non-specialist undergraduates should therefore encounter little that they are not already familiar with or that cannot be picked up easily during a first reading. Those who are not conversant with the general theory of elasticity, for example, or with the structural relationships which govern the behaviour of beams and plates are provided with a simple review of these topics in Chapter 2. This is not complete but provides an adequate basis for the Rayleigh–Ritz approach which is used as an intuitive theoretical framework for all applications of the finite element method throughout the book.

In the main body of the text, different categories of element are discussed in a progression which follows roughly the sequence of Fig. 1.2 working from top to bottom and left to right. Continuum elements are introduced first (Chapters 4 to 8) starting with the simplest linear formulations and working up to the more complex, higher-order elements. This is followed by a treatment of beam and plate formulations (Chapters 9 and 10) and the book concludes with a brief introduction to linear dynamic analysis (Chapter 11).

REFERENCES

[1] Clough, R. W. (1960) The finite element in plane stress analysis, Proceedings second conference on electronic computation, *American Society of Civil Engineers*, pp. 345–77.

[2] Courant, R. (1943) Variational methods for the solution of problems of equilibrium and vibration. *Bulletin of the American Mathematical Society*, **49**, 1–23.

[3] Turner, M. J., Clough, R. W., Martin, H. C. and Topp, L. J. (1956) Stiffness and deflection analysis of complex structures. *Journal of Aeronautical Science*, **23**, 805–23.

[4] Clough, R. W. (1980) The finite element method after twenty-five years. A personal view. *Computers and Structures*, **12**, 361–70.

[5] Norrie, D. and De Vries, G. (1976) *A Finite Element Bibliography*, IFI/Plenum, New York.

[6] Zienkiewicz, O. C. and Cheung, Y. K. (1965) Finite elements in the solution of field problems. *The Engineer*, **220**, 507–10.

[7] Szabo, B. E. and Lee, G. C. (1961) Derivation of stiffness matrices for problems in plane elasticity by Galerkin's method. *International Journal for numerical methods in engineering* **1**, 301–10.

[8] Norrie, D and Kardestuncer, H. (eds) (1987) *The Finite Element Handbook*, McGraw Hill, New York.

[9] Baran, N. M. (1988) *Finite Element Analysis on Microcomputers*, McGraw Hill, New York.

[10] Ugural, A. C. and Fenster S. K. (1981) *Advanced Strength and Applied Elasticity*, Edward Arnold, London.

[11] Zienkiewicz, O. C. and Taylor, R. L. (1990/1991) *The Finite Element Method* (4th edn), Vol. 1, *Basic Formulation and Linear Problems*, Vol. 2, *Solid and Fluid Mechanics Dynamics and Nonlinearity*, McGraw Hill, London.

[12] Hughes, T. J. R. (1987) *The Finite Element Method, Linear Static and Dynamic Finite Element Analysis*, Prentice-Hall, Englewood Cliffs NJ.

[13] Cook, R. D., Malkus, D. S. and Plesha, M. E. (1989) *Concepts and Applications of Finite Element Analysis*, John Wiley & Sons, New York.

[14] Hinton, E. and Owen, D. R. J. (1980) *Finite Element Programming*, Academic Press, London.

[15] Smith, I. M. and Griffiths, D. V. (1988) *Programming the Finite Element Method*, John Wiley & Sons, Chichester.

2

Preliminaries

The finite element method will be regarded in all that follows as a technique for the solution of problems in linear elasticity. Before embarking upon a development of the method along these lines, it is helpful to look at the nature of the exact problem for which solutions will be sought. A brief review of the theory of linear elasticity is presented in the remainder of this chapter. It is covered more or less from first principles and concludes with an elementary treatment of beam and plate theory.

2.1 THE CONTINUUM PROBLEM DEFINED

In determining the stresses and deformations of a solid body, it is impracticable to descend to the fine particulate structure of the material. In most engineering calculations, it is assumed that a sufficiently large number of particles exists within any volume of interest for the behaviour of the material to be independent of its fine structure. This is the 'continuum' assumption. The elastic behaviour of any portion of an object can therefore be modelled by a test specimen of the same material subject to the same forces. Empirical data obtained from simple test procedures may then be applied to the analysis of complex objects.

The behaviour of a deformable solid is governed by relationships between parameters which define its force and displacement states. These are usually taken to be 'stress' and 'strain'. A formal definition of these quantities follows in sections 2.2 and 2.3. Three conditions must then hold at all points within the body. These are as follows:

(i) Equilibrium
(ii) Compatibility
(iii) Constitutive

The first requires that a state of mechanical equilibrium exists within the body and on its boundary. For static loadings, this means that the forces and moments acting on any portion of the body must satisfy static equilibrium. The effects of 'internal' forces (forces exerted by neighbouring particles of the body on each other) and 'external' loads (applied to the body by some external agent) must both be taken into account. In simple problems, the equilibrium condition can often represented by a 'free body' diagram, such as that of Fig.

2.1(a), in which a subportion of a body is held in equilibrium by internal and external forces. Provided that the distribution of internal forces is sufficiently straightforward, they may often be completely determined by such an approach. This forms the basis of much elementary analysis in strength of materials. In more complex problems, the conditions of equilibrium are seldom amenable to such treatment and must be expressed in terms of partial differential equations. These are discussed in section 2.2.

The second condition that must hold within any deformable body is that of compatibility. This requires that the deformation is physically permissible. It is synonymous, in most instances, with a requirement that displacements are

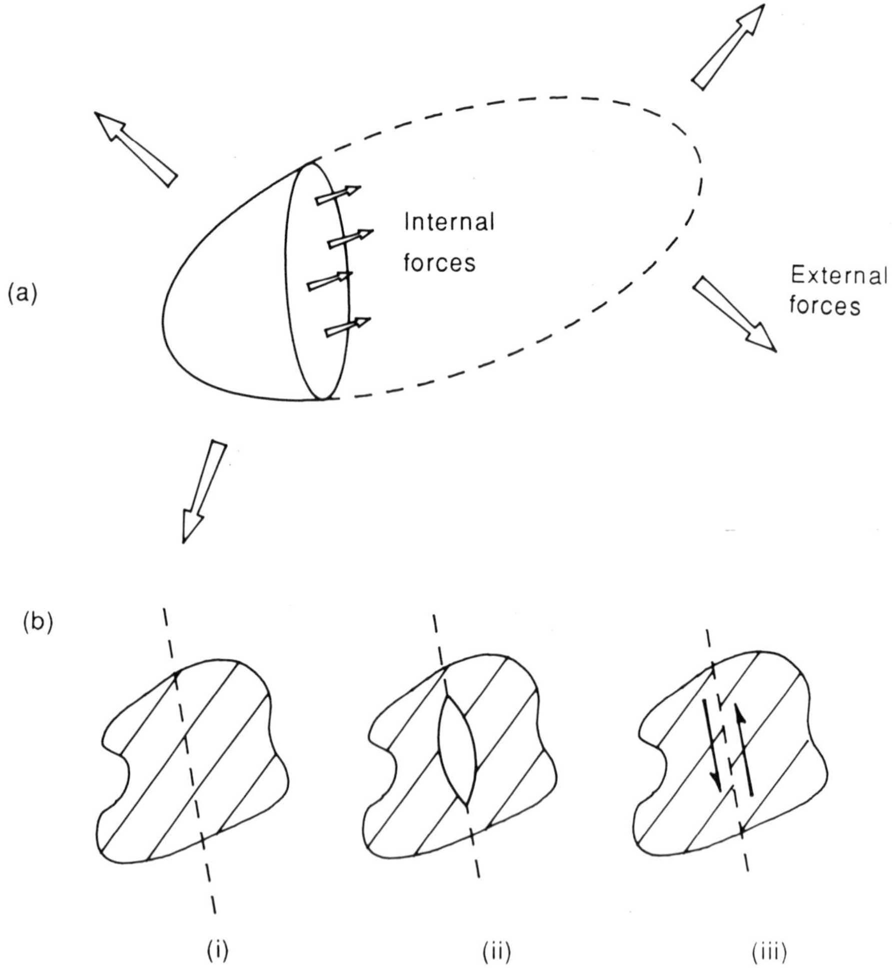

Fig. 2.1 *(a) Static equilibrium of a 'free body' and (b) Compatibility of displacement: (i) undeformed, (ii) incompatible normal displacement, (iii) incompatible tangential displacement.*

continuous. If this were not the case, adjacent particles could displace by different amounts to form cavities or slip planes as indicated in Fig. 2.1(b), violating the usual notion of a solid body.

A practical problem which arises in forming a mathematical statement of the compatibility requirement is that the deformation of the body is usually characterized not by the displacements themselves, but by nondimensional quantities termed 'strains' (defined in section 2.3). The compatibility requirement must therefore be translated into an equivalent condition in terms of strain. The relationship between strain and displacement, and the consequent strain-compatibility equations are derived in section 2.3.2.

The continuum problem is 'closed' (made complete) by a suitable set of 'constitutive' equations which relate the internal forces (characterized by stress) to the local deformation (characterized by strain). In the case of a linearly elastic material, the relationship is linear with constants of proportionality derived from empirical data. The constitutive relationship between stress and strain for a homogeneous, isotropic, elastic material is derived in section 2.4.

2.2 ANALYSIS OF STRESS

2.2.1. Definition and notation

The intensity of the internal forces within a body is termed 'stress'. Stress can be defined using the following construction. Consider a point P within a body in equilibrium under the action of external loads $\boldsymbol{F}_1, \boldsymbol{F}_2, \ldots$ (Fig. 2.2(a)). A plane, A, passing through P divides the body into two portions which are held together by internal forces. The force resultant which acts on the left hand portion of the body over a small subarea ΔA in the vicinity of P is denoted by the vector $\Delta \boldsymbol{F}$ (Fig. 2.2(b)). By representing $\Delta \boldsymbol{F}$ as the sum of orthogonal components, ΔF_1, ΔF_2 and ΔF_3 in directions 1, 2 and 3, where the '1' direction is taken to be an outward normal to ΔA, we can define stress components σ_1, τ_{12} and τ_{13} as

$$\sigma_1 = \lim_{\Delta A \to 0} \left(\frac{\Delta F_1}{\Delta A} \right), \quad \tau_{12} = \lim_{\Delta A \to 0} \left(\frac{\Delta F_2}{\Delta A} \right) \text{ and } \tau_{13} = \lim_{\Delta A \to 0} \left(\frac{\Delta F_3}{\Delta A} \right) \qquad (2.1a)$$

In this notation, the symbols σ and τ represent normal and shear components, respectively. The single subscript on a normal component denotes both the direction of the plane on which it is defined and the direction in which it acts. The two subscripts on the shear components denote the direction of the plane (the first subscript) and the direction in which it acts (the second subscript). The normal stress σ_1, for example, is the force per unit area in the 1 direction acting on a plane facing in the 1 direction. The shear stress τ_{12} is the force

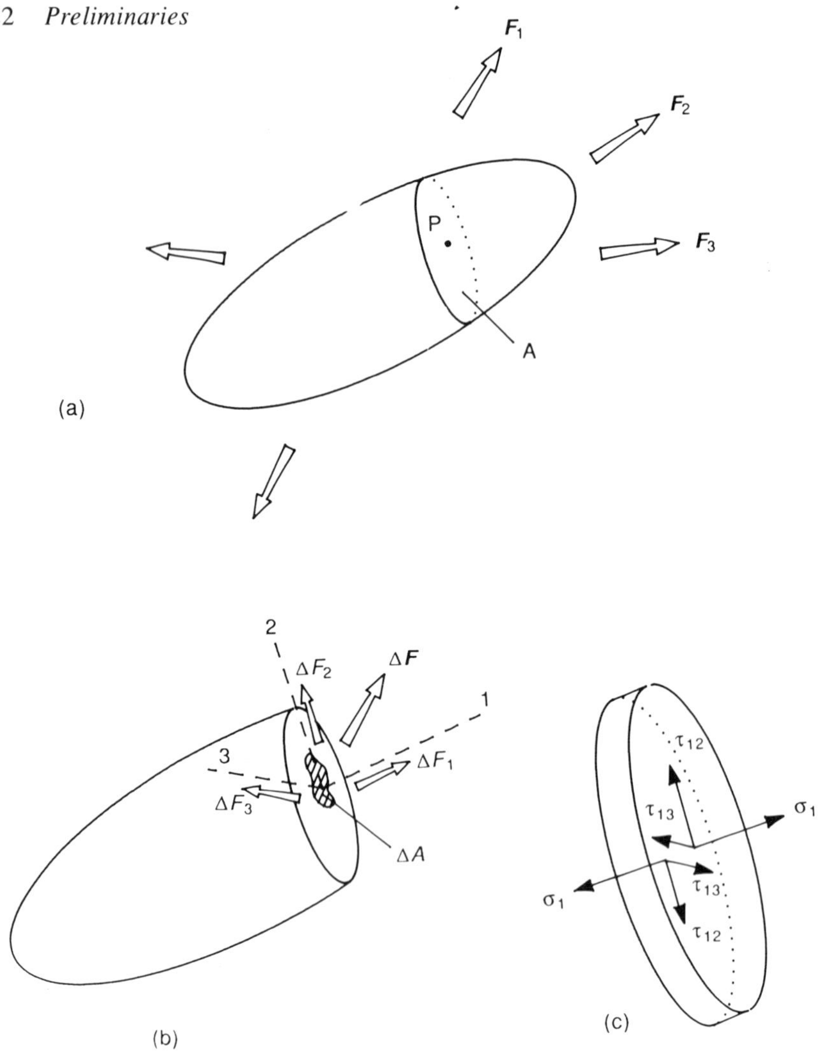

Fig. 2.2 *Stress at a point. (a) The plane A, (b) internal forces acting on ΔA, and (c) stress components on a thin slice in the plane of A.*

intensity in the 2 direction acting on a plane facing in the 1 direction and so on. In the particular case when the internal forces are distributed uniformly over the entire plane, the limiting procedure is no longer necessary and the stresses are given by averaged values,

$$\sigma_1 = \frac{F_1}{A}, \quad \tau_{12} = \frac{F_2}{A} \quad \text{and} \quad \tau_{13} = \frac{F_3}{A}, \tag{2.1b}$$

where F_1, F_2 and F_3 are the resultant force components on A. This more straightforward definition forms the basis for much simple analysis.

The right hand portion of the body might just as easily have been used in the above construction as the left hand portion. An equal and opposite force ($-\Delta \boldsymbol{F}$) would then have acted on the same subarea facing in the negative 1 direction, and stress components with the same numerical values would be defined at the same point but acting in the negative 1, 2 and 3 directions. This is illustrated in Fig. 2.2(c) which shows equal and opposite stresses acting on either side of a thin slice of material lying in the plane of A. Note that a positive value of σ_1 implies a tensile stress in the usual sense whereas a negative value implies a compressive stress.

2.2.2 Orthogonal components of stress

A local state of stress is conveniently visualized in terms of the stresses acting on the faces of a small orthogonal block enclosing the point of interest. If the

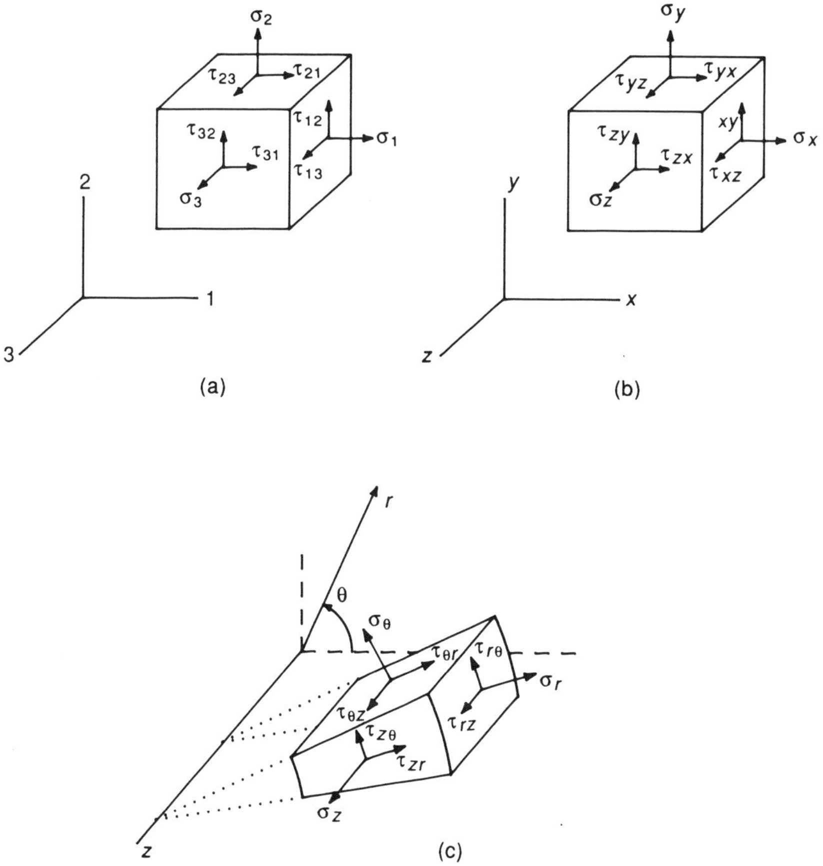

Fig. 2.3 *Stress at a point. (a) General orthogonal components, (b) cartesian components, and (c) cylindrical polar components.*

faces are orientated perpendicular to directions, 1, 2 and 3, the state of stress is characterized by nine components $\sigma_1, \sigma_2, \sigma_3, \tau_{12}, \tau_{21}, \tau_{13}, \tau_{31}, \tau_{23}$, and τ_{32}, as shown in Fig. 2.3(a). In the case of a cartesian coordinate system, in which a 'natural' set of orthogonal directions is defined by the x, y and z axes, the subscripts 1, 2 and 3 are replaced by x, y and z and the stress components become $\sigma_x, \sigma_y, \sigma_z, \tau_{xy}, \tau_{yx}, \tau_{xz}, \tau_{zx}, \tau_{zy}$, and τ_{yz} (Fig. 2.3(b)).

Curvilinear systems may also be used to define local, orthogonal directions and hence orthogonal components of stress. An example which will prove useful later on is that of a cylindrical polar system (r, θ, z) for which the appropriate orthogonal stress components $\sigma_r, \sigma_\theta, \sigma_z, \tau_{r\theta}, \tau_{\theta r}, \tau_{rz}, \tau_{zr}, \tau_{z\theta}$, and $\tau_{\theta z}$, are shown in Fig. 2.3(c) (in such a coordinate system, the r, θ and z directions are defined by increasing r, θ or z, respectively, while the other coordinates are held constant).

Whatever orthogonal axes are used, it appears at this stage that nine stress components are required to define the state of stress at a point. As we shall see shortly, only six of these are, in fact, independent. Moreover, the number is often reduced further by purely physical considerations. In the case of a 'thin' body, for example, loaded with in-plane forces in the x–y plane (Fig. 2.4) the only significant stress components are τ_{xy}, τ_{yx}, σ_x and σ_y. Their variation through the thickness of the body is generally negligible and a state of 'plane stress' is then said to exist.

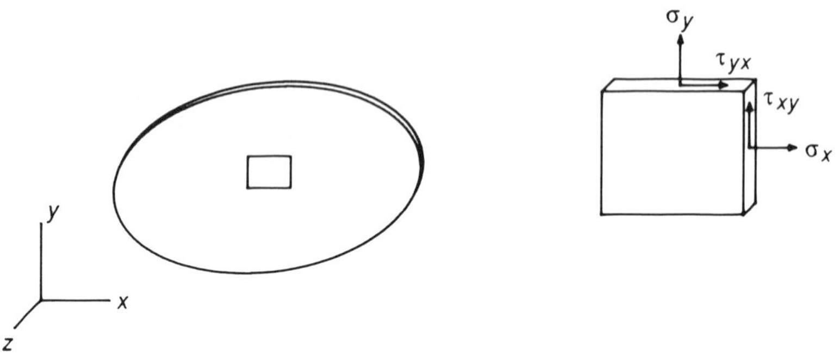

Fig. 2.4 *Cartesian stress components, plane stress.*

Another two-dimensional stress state is that which occurs in a body which is long in the out-of-plane (z) direction (Fig. 2.5). This is distinguished from a state of plane stress by the existence of a nonzero normal stress σ_z in the out-of-plane direction, and is termed a state of 'plane strain'.

A somewhat similar state of stress, in terms of the number and orientation of the nonzero stress components, is that within an axisymmetric body subject to axisymmetric loads. Considerations of symmetry then require that the stresses on any slice through the axis of symmetry are invariant with θ (Fig. 2.6). The only nonzero stress components are then $\sigma_r, \sigma_z, \sigma_\theta, \tau_{rz}$ and τ_{zr}. This is referred to as an 'axisymmetric' state of stress.

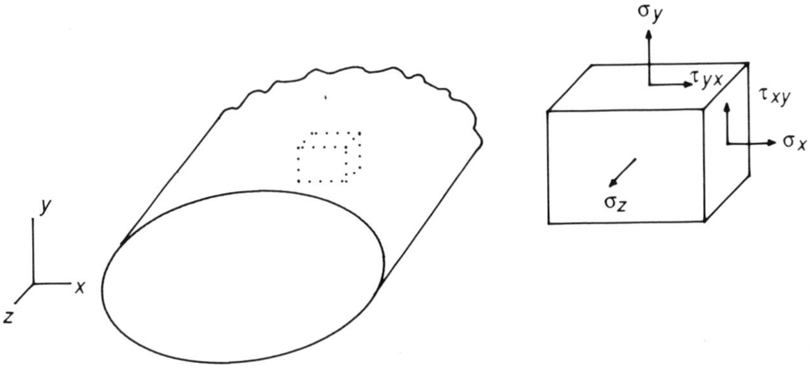

Fig. 2.5 *Cartesian stress components, plane strain.*

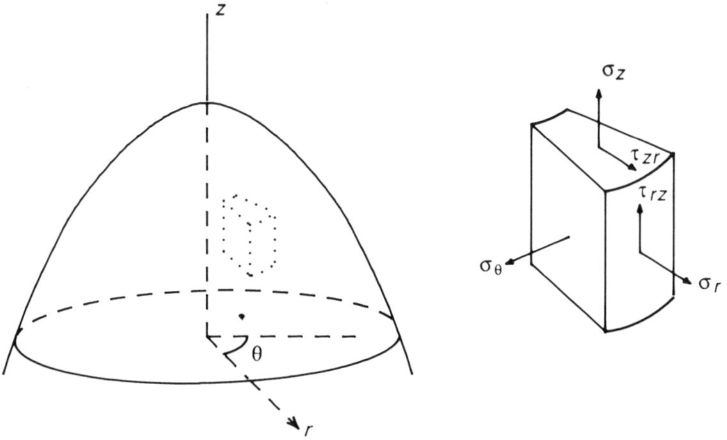

Fig. 2.6 *Cylindrical stress components, axisymmetric stress.*

2.2.3 Variation of stress within a body, the stress equations of equilibrium

The values of stress within a body generally vary from point to point. They do so, however, in a manner which ensures that equilibrium is maintained for the body as a whole and for all subportions of it. The relationships which must be satisfied by the stress components in order that equilibrium is satisfied are expressed as partial differential equations, obtained by applying the conditions of static equilibrium to a small differential element. The resulting equations depend upon the coordinate system used and are derived here for a two-dimensional cartesian system. They are subsequently generalized to three dimensions without further proof.

Consider a small rectangular block of sides δx and δy and of unit thickness in the z direction (Fig. 2.7). It is subject to stresses σ_x, σ_y, τ_{xy} and τ_{yx}, which vary with x and y but are independent of z (the element is in a state of plane stress or plane strain). An external 'body force' (gravity, for example) acts at all points within the body and has components f_x and f_y per unit volume in the x and y directions. The variation in stress within the element is approximated, to the first order in δx or δy, by two terms of a truncated Taylor expansion. For example, the values of σ_x at two points which are separated spatially by a distance δx are defined as σ_x and $\sigma_x + (\partial\sigma_x/\partial x)\,\delta x$. Similar expressions for the variations in σ_y, τ_{xy} and τ_{yx} are indicated in Fig. 2.7. To the same degree of approximation, the body force can be replaced by a concentrated load with components $F_x\,(=f_x\,\delta x\delta y)$ and $F_y\,(=f_y\,\delta x\delta y)$ acting at the centroid of the element.

Moment equilibrium is satisfied if the resultant moment acting on the element about any axis perpendicular to the x–y plane equates to zero. Taking an axis through the lower left hand corner gives

$$(\tau_{yx}\,\delta x)\,\delta y - (\tau_{xy}\,\delta y)\,\delta x + \text{ higher order terms in } \delta x \text{ and } \delta y = 0,$$

where the higher order terms contain factors, δx^3, δy^3, $\delta x\delta y^2$ and so on. Dividing through by $\delta x\delta y$ and taking the limit as δx and δy tend to zero, we obtain

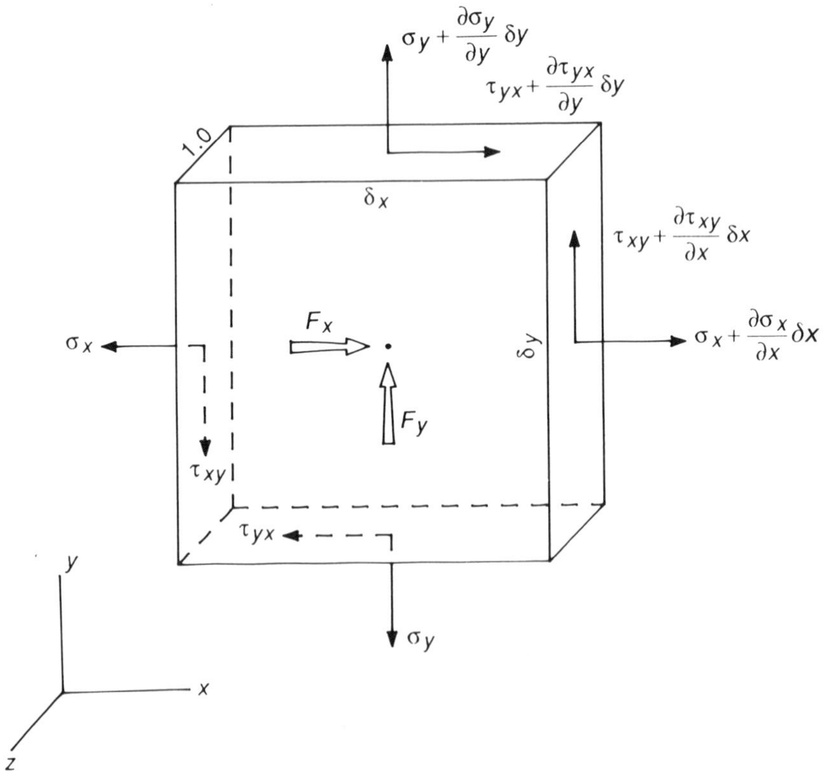

Fig. 2.7 *Stress equilibrium of a cartesian element.*

$$\tau_{yx} = \tau_{xy} \tag{2.2}$$

A similar relationship holds for all orthogonal stress components and may be stated more generally as

$$\tau_{ij} = \tau_{ji} \tag{2.3}$$

This property is termed the 'complementarity of shear stress' and it immediately reduces the number of unknown components of stress at a point from nine to six, there being only three independent shear components.

Force equilibrium for the element of Fig. 2.7 is satisfied if the resultant force acting on it is zero. Resolving forces in the x direction, we obtain

$$\left\{(\sigma_x + \frac{\partial \sigma_x}{\partial x}\,\delta x) - \sigma_x\right\}\delta y + \left\{(\tau_{yx} + \frac{\partial \tau_{yx}}{\partial y}\,\delta y) - \tau_{yx}\right\}\delta x + f_x\,\delta x \delta y = 0.$$

or, dividing by $(\delta x \delta y)$ and cancelling terms

$$\frac{\partial \sigma_x}{\partial x} + \frac{\partial \tau_{yx}}{\partial y} + f_x = 0. \tag{2.4}$$

The analogous equation in the y direction is

$$\frac{\partial \tau_{xy}}{\partial x} + \frac{\partial \sigma_y}{\partial y} + f_y = 0 \tag{2.5}$$

When a full three-dimensional stress state is considered, the corresponding equations are

$$\begin{aligned}
&\frac{\partial \sigma_x}{\partial x} + \frac{\partial \tau_{yx}}{\partial y} + \frac{\partial \tau_{zx}}{\partial z} + f_x = 0,\\
&\frac{\partial \tau_{xy}}{\partial x} + \frac{\partial \sigma_y}{\partial y} + \frac{\partial \tau_{zy}}{\partial z} + f_y = 0,\\
&\frac{\partial \tau_{xz}}{\partial x} + \frac{\partial \tau_{yz}}{\partial y} + \frac{\partial \sigma_z}{\partial z} + f_z = 0.
\end{aligned} \tag{2.6}$$

These are the stress equations of equilibrium in cartesian coordinates and hold for any stress field that satisfies the conditions of static equilibrium.

Similar equations can be derived for curvilinear coordinate systems. We state here, without proof, the equations obtained in this way for a cylindrical polar system (r, θ, z). They are

$$\begin{aligned}
&\frac{\partial \sigma_r}{\partial r} + \frac{1}{r}\frac{\partial \tau_{\theta r}}{\partial \theta} + \frac{\partial \tau_{zr}}{\partial z} - \frac{\sigma_\theta - \sigma_r}{r} + f_r = 0,\\
&\frac{\partial \tau_{r\theta}}{\partial r} + \frac{1}{r}\frac{\partial \sigma_\theta}{\partial \theta} + \frac{\partial \tau_{z\theta}}{\partial z} + \frac{2\tau_{\theta r}}{r} + f_\theta = 0,
\end{aligned} \tag{2.7}$$

$$\frac{\partial \tau_{rz}}{\partial r}+\frac{1}{r}\frac{\partial \tau_{\theta z}}{\partial \theta}+\frac{\partial \sigma_z}{\partial z}+\frac{\tau_{rz}}{r}+f_z=0,$$

where the *body* force components f_r, f_θ and f_z are forces per unit volume acting in the r, θ and z directions. A full derivation of these equations and a more complete treatment of stress equilibrium in other coordinate systems can be found in [1].

2.2.4 Stresses on oblique planes, boundary equilibrium

Stress equilibrium on the boundary of a body requires that the internal state of stress exactly balances the external pressure or traction applied to the surface. If the bounding surface does not correspond to one of the orthogonal planes used to specify the stress field, this requires that the normal and tangential stress components on the surface are written first in terms of the known orthogonal components. Expressions which enable us to do this are now derived. In the process, we will also establish an important associated result; that the stress components on *any* plane are entirely determined by a complete set of orthogonal components.

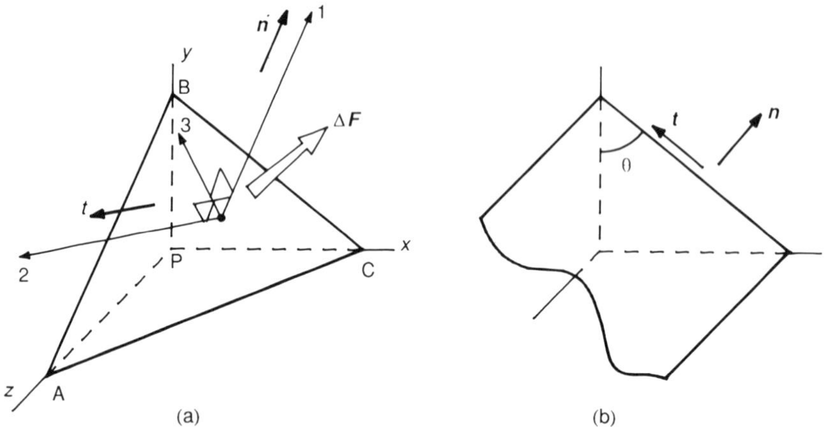

Fig. 2.8 *Stress on an oblique plane. (a) Three-dimensional, and (b) two-dimensional.*

Consider the stresses acting on a small tetrahedron with three orthogonal faces perpendicular to the x, y and z axes, and an oblique face which is defined by an outward unit normal $\boldsymbol{n} = (l_1, m_1, n_1)$ (Fig. 2.8(a)). Let us assume that the cartesian stress components are known, and that we wish to calculate the normal and shear components σ_1, τ_{12} and τ_{13} on the oblique face ABC. Overall equilibrium of the tetrahedral element dictates that the forces acting on the faces APB, APC and BPC equilibrate with the force $\Delta \boldsymbol{F}$ acting on ABC. Given that the normal and tangential stresses on APB, APC, and BPC are simply the cartesian stress components; $-\sigma_x$, $-\tau_{xz}$ and $-\tau_{xy}$ on APB and so on, and that the areas of these faces are $l_1 \Delta A$, $m_1 \Delta A$ and $n_1 \Delta A$ respectively (being the projections of ΔA, the area of the oblique face, on the *zy*, *zx* and *xy*-planes), the condition of force equilibrium can be written in cartesian vector form as

$$\Delta \boldsymbol{F} + l_1 \Delta A(-\sigma_x, -\tau_{xy}, -\tau_{xz}) + m_1 \Delta A(-\tau_{yx}, -\sigma_y, -\tau_{yz}) + n_1 \Delta A(-\tau_{zx}, -\tau_{zy}, -\sigma_z) = 0.$$

Division by ΔA and the definition of a vector stress, $\boldsymbol{s}$, as the limit of $\Delta \boldsymbol{F}/\Delta A$ as $\Delta A \to 0$ gives

$$\boldsymbol{s} = (l_1 \sigma_x + m_1 \tau_{yx} + n_1 \tau_{zx},\ l_1 \tau_{xy} + m_1 \sigma_y + n_1 \tau_{zy},\ l_1 \tau_{xz} + m_1 \tau_{yz} + n_1 \sigma_z). \quad (2.8)$$

The normal and tangential components of $\boldsymbol{s}$ then correspond to direct and shear stresses on the oblique plane. In other words, the normal stress in the 1 direction (Fig. 2.8(a)) is obtained by resolving $\boldsymbol{s}$ in the direction of $\boldsymbol{n}$. This gives

$$\sigma_1 = \boldsymbol{s} \cdot \boldsymbol{n} = l_1^2 \sigma_x + m_1^2 \sigma_y + n_1^2 \sigma_z + 2 l_1 m_1 \tau_{xy} + 2 l_1 n_1 \tau_{xz} + 2 n_1 m_1 \tau_{yz}. \quad (2.9)$$

Similarly, the shear stress τ_{12} is obtained by resolving $\boldsymbol{s}$ in the direction of the unit vector $\boldsymbol{t}$ lying along the 2 axis. If $\boldsymbol{t}$ has direction cosines (l_2, m_2, n_2), this gives

$$\tau_{12} = \boldsymbol{s} \cdot \boldsymbol{t} = l_1 l_2 \sigma_x + m_1 m_2 \sigma_y + n_1 n_2 \sigma_z + \tau_{xy}(l_1 m_2 + l_2 m_1)$$
$$+ \tau_{xz}(l_1 n_2 + l_2 n_1) + \tau_{yz}(n_1 m_2 + n_2 m_1). \quad (2.10)$$

A similar expression can clearly be obtained for τ_{13}. For the particular case when $\boldsymbol{n}$ and $\boldsymbol{t}$ lie in the *x*–*y* plane (Fig. 2.8(b)) so that $\boldsymbol{n} = (\cos\theta, \sin\theta, 0)$ and $\boldsymbol{t} = (-\sin\theta, \cos\theta, 0)$, expressions 2.9 and 2.10 reduce to

$$\sigma_1 = \sigma_x \cos^2\theta + \sigma_y \sin^2\theta + 2\tau_{xy} \sin\theta \cos\theta,$$
$$(2.11)$$
$$\tau_{12} = (\sigma_y - \sigma_x) \sin\theta \cos\theta + \tau_{xy}(\cos^2\theta - \sin^2\theta).$$

These should already be familiar to the reader as the basis for **Mohr's circle** of stress in two dimensions. They are also instrumental in establishing the existence of principal planes of stress in two dimensions, that is, orthogonal planes free of shear stress on which the direct stress takes a maximum or minimum value. These have a parallel in three dimensions. The full stress transformation equations (2.9 and 2.10) can then be used to establish the existence of *three* orthogonal planes with similar properties. The principal stresses are the roots of a cubic equation and are somewhat harder to evaluate than their two-dimensional equivalents (Appendix B of [2]). Proof of their existence is left to the reader as an exercise (see Problem 5 at the end of this chapter).

2.3 ANALYSIS OF DEFORMATION

2.3.1 Strain, definitions and notation

It is the relative, rather than the absolute, displacements within an elastic body which generate internal forces. The displacement of each point relative to its neighbours is therefore the quantity which is of greatest interest to us when choosing parameters with which to describe the state of deformation within such a body. Parameters, termed strains, which quantify relative displacements of this type are defined in the following way.

Consider the deformation of orthogonal fibres PA and PB in the vicinity of a point P. Initially, they lie parallel to directions 1 and 2 but deform to positions P′A′ and P′B′ under load, as shown in Fig. 2.9. The direct strain, e_1, in the direction 1, at the point P, is defined to be the elongation per unit length of the fibre PA in the limit as its initial length becomes vanishingly small. In other words, if PA is of initial length l_0 and undergoes an extension Δl, the direct strain e_1 is defined as

$$e_1 = \lim_{l_0 \to 0} (\Delta l / l_0). \tag{2.12}$$

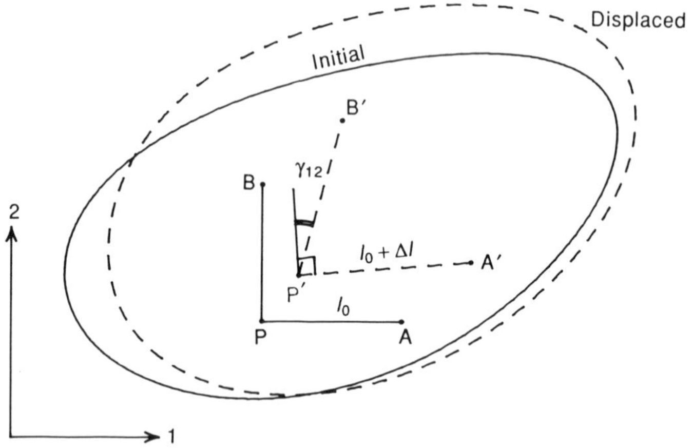

Fig. 2.9 *Local deformation of orthogonal fibres.*

Technically this is the nominal direct strain. When large deformations occur, that is when Δl is significant compared to l_0, a 'true strain', ε_1, can be defined by integrating the instantaneous 'nominal strain' to give

$$\varepsilon_1 = \int_{l_0}^{l_0+\Delta l} \frac{dl}{l} = \log_e\left(1 + \frac{\Delta l}{l_0}\right). \tag{2.12b}$$

If $\Delta l/l_0$ is small, the true and nominal strains differ only by terms of order $(\Delta l/l_0)^2$. These will be assumed to be negligible in all that follows so that strains can be taken to be 'nominal' rather than 'true' unless otherwise stated.

The direct strain, as defined above, quantifies the tensile or compressive local deformation in a particular direction. A different type of deformation is characterized by the rotation of orthogonal fibres relative to one another. This is represented in Fig. 2.9 and is quantified by the shear strain component, γ_{12}, defined to be the *decrease* in radian angle between fibres PA and PB initially lying parallel to orthogonal 1 and 2 directions.

Extrapolating these notions of direct and shear strain to three orthogonal directions, 1, 2 and 3, we can define nine components of strain at a point. These are e_1, e_2, e_3 ,γ_{12}, γ_{21}, γ_{13}, γ_{31}, γ_{23} and γ_{32}. Shear strains, however, are complementary ($\gamma_{ij} = \gamma_{ji}$) by the very nature of their definition, so that the number of *independent* strain components reduces to six. As with stress components, the subscripts 1, 2 and 3 are often replaced by coordinate descriptors. A full set of cartesian strain components, for example, is e_x, e_y, e_z, γ_{xy} $(=\gamma_{yx})$, γ_{xz} $(=\gamma_{zx})$ and γ_{yz} $(=\gamma_{zy})$. In a cylindrical polar system, the analogous components are e_r, e_θ, e_z, $\gamma_{r\theta}$ $(=\gamma_{\theta r})$, $\gamma_{\theta z}$ $(=\gamma_{z\theta})$ and γ_{rz} $(=\gamma_{zr})$.

2.3.2 Strain–displacement relationship, compatibility of strains

In the preceding section, strain components were defined as physical quantities associated with relative displacements of neighbouring points. They can also be expressed directly in terms of the displacement field and its derivatives.

Consider a deformable solid with a cartesian displacement field (u, v, w). That is to say, the cartesian components of displacement at the point (x, y, z) are $u(x, y, z)$ in the x direction, $v(x, y, z)$ in the y direction, and $w(x, y, z)$ in the z direction. The displacements of two neighbouring points P and Q separated by a distance δ_x are denoted by (u, v, w) and ($u + \delta u$, $v + \delta v$, $w + \delta w$) as indicated in Fig. 2.10(a). Since PQ is a 'fibre initially lying in the x direction', the strain component e_x is, by definition, the limit of the quantity $((P'Q' - PQ)/PQ)$ as the length of PQ tends to zero (the primes, superscript$'$, denote a displaced location). Provided that the strains are small, that is provided that δu, δv and δw are much smaller than δx, $P'Q'$ is given, to the first order in δu, δv and δw, by $P'Q' = \delta x + \delta u$. The expression for e_x then becomes

$$e_x = \lim_{\delta x \to 0} \left(\frac{\delta u}{\delta x} \right) \quad \text{or} \quad e_x = \frac{\partial u}{\partial x}, \tag{2.13a}$$

where $\partial/\partial x$ denotes partial differentiation with respect to x in the usual sense. Similarly, e_y and e_z are given by

$$e_y = \frac{\partial v}{\partial y} \quad \text{and} \quad e_z = \frac{\partial w}{\partial z}. \tag{2.13b}$$

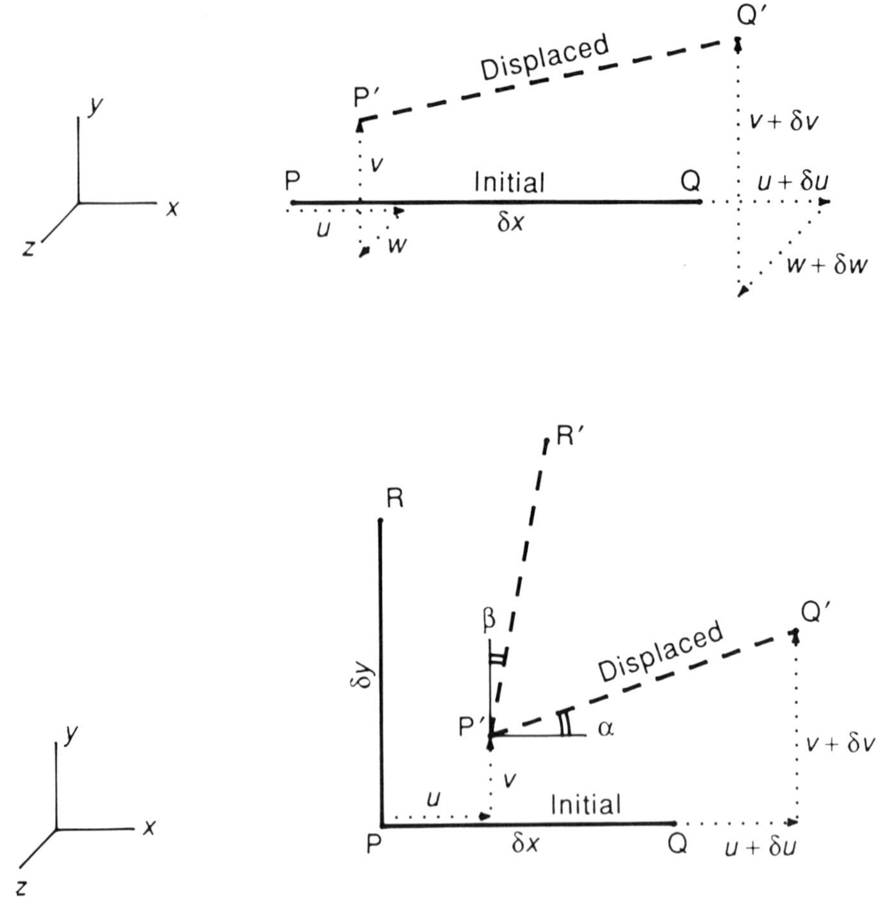

Fig. 2.10 *Geometry of the strain–displacement relationship. (a) Direct strain, and (b) shear strain.*

These are the **strain–displacement equations** for direct strain in a cartesian coordinate system.

The components of shearing strain may also be expressed in terms of displacement if we consider the relative deformation of orthogonal fibres PQ and PR which lie initially parallel to the x and y-axes. After deformation they displace to $P'Q'$ and $P'R'$ as shown in Fig. 2.10(b). By definition, the shearing strain γ_{xy} is the decrease in radian angle between them, and is equal, in the notation of Fig. 2.10(b), to the sum of the angles α and β. If we denote the displacements of P by u and v, and those of Q by $u + \delta u$ and $v + \delta v$, elementary trigonometry tells us that

$$\tan \alpha = \frac{\partial v}{\delta x + \delta u} \, .$$

Assuming that δu is small compared with δx, and retaining only the leading term in the denominator, we obtain

$$\alpha \simeq \tan\alpha \simeq \frac{\partial v}{\partial x},$$

or, taking the limit as $\delta x \rightarrow 0$

$$\alpha = \frac{\partial v}{\partial x}.$$

Similarly, β is equal to $\partial v/\partial y$ and the shear strain γ_{xy} is given by

$$\gamma_{xy} = \alpha + \beta = \frac{\partial u}{\partial y} + \frac{\partial v}{\partial x}. \tag{2.14a}$$

A similar construction for γ_{xz} and γ_{yz} yields

$$\gamma_{xz} = \frac{\partial u}{\partial z} + \frac{\partial w}{\partial x} \text{ and } \gamma_{yz} = \frac{\partial v}{\partial z} + \frac{\partial w}{\partial y}. \tag{2.14b, c}$$

Expressions 2.13 and 2.14 define the six cartesian strain components in terms of three independent variables, u, v and w. This implies that the strains are not themselves independent and cannot be chosen as arbitrary functions of x, y and z. Consider for example a two-dimensional displacement field with in-plane cartesian strains

$$e_x = \frac{\partial u}{\partial x},\ e_y = \frac{\partial v}{\partial y} \text{ and } \gamma_{xy} = \frac{\partial u}{\partial y} + \frac{\partial v}{\partial x},$$

Double differentiation of these expressions with respect to x and/or y followed by the elimination of u and v yields the identity

$$\frac{\partial^2 e_x}{\partial y^2} + \frac{\partial^2 e_y}{\partial x^2} - \frac{\partial^2 \gamma_{xy}}{\partial x \partial y} = 0. \tag{2.15a}$$

This is the two-dimensional 'compatibility' condition. Implicit in its derivation is the assumption that the displacement field is differentiable and hence continuous. It can be regarded, in effect, as an assurance that the strain field derives from a physically acceptable ('compatible' in the sense of Fig. 2.1) displacement field. In the fully three-dimensional case, six compatibility conditions must be satisfied. These are

$$\frac{\partial^2 e_x}{\partial y^2} + \frac{\partial^2 e_y}{\partial x^2} - \frac{\partial^2 \gamma_{xy}}{\partial x \partial y} = 0,$$

$$\frac{\partial^2 e_y}{\partial z^2} + \frac{\partial^2 e_z}{\partial y^2} - \frac{\partial^2 \gamma_{yz}}{\partial y \partial z} = 0,$$

$$\frac{\partial^2 e_x}{\partial z^2}+\frac{\partial^2 e_z}{\partial x^2}-\frac{\partial^2 \gamma_{xz}}{\partial x \partial z}=0,$$

$$\frac{\partial^2 \gamma_{yz}}{\partial x^2}-\frac{\partial^2 \gamma_{xz}}{\partial x \partial y}-\frac{\partial^2 \gamma_{xy}}{\partial x \partial z}+2\frac{\partial^2 e_x}{\partial y \partial z}=0, \tag{2.15b}$$

$$\frac{\partial^2 \gamma_{xz}}{\partial y^2}-\frac{\partial^2 \gamma_{yz}}{\partial y \partial x}-\frac{\partial^2 \gamma_{xy}}{\partial y \partial z}+2\frac{\partial^2 e_y}{\partial x \partial z}=0,$$

$$\frac{\partial^2 \gamma_{xy}}{\partial z^2}-\frac{\partial^2 \gamma_{yz}}{\partial z \partial x}-\frac{\partial^2 \gamma_{xz}}{\partial z \partial y}+2\frac{\partial^2 e_z}{\partial x \partial y}=0.$$

The derivation of analogous relationships for non-cartesian coordinate systems proceeds along similar lines. In a cylindrical polar system, for example, the strain components are given in terms of the displacement field by

$$e_r=\frac{\partial u_r}{\partial r}, \quad e_\theta=\frac{\partial u_\theta}{r\partial \theta}+\frac{u_r}{r}, \quad e_z=\frac{\partial u_z}{\partial z},$$

$$\gamma_{rz}=\frac{\partial u_r}{\partial z}+\frac{\partial u_z}{\partial r}, \qquad \gamma_{\theta z}=\frac{\partial u_\theta}{\partial z}+\frac{\partial u_z}{r\partial \theta}, \tag{2.16}$$

and

$$\gamma_{r\theta}=\frac{\partial u_\theta}{\partial r}-\frac{u_\theta}{r}+\frac{\partial u_r}{r\partial \theta},$$

where u_r, u_θ and u_z are displacements in the r, θ and z directions.

The derivation of the two-dimensional compatibility equation in cylindrical polar coordinates is left to the reader as an exercise (see Problem 8 at the end of this chapter). Comprehensive treatment of the strain–displacement relationship and the resulting compatibility equations in a variety of coordinate systems is to be found in [1].

2.4 STRESS–STRAIN RELATIONSHIP

2.4.1 Test behaviour, Hooke's law

The elastic behaviour of most engineering materials is derived from one dimensional test data. These are collected from measurements of the extension, Δl, of a prismatic test specimen of initial length l_0 and cross-section A_0, subject to a known axial load F. The arrangement is represented in Fig. 2.11(a). The axial stress σ $(=F/A_0)$ is plotted against the corresponding axial strain e $(=\Delta l/l_0)$ to give curves of the type shown in Fig. 2.11(b). These are substantially independent of specimen size provided that the measured section is sufficiently distant from the end attachments.

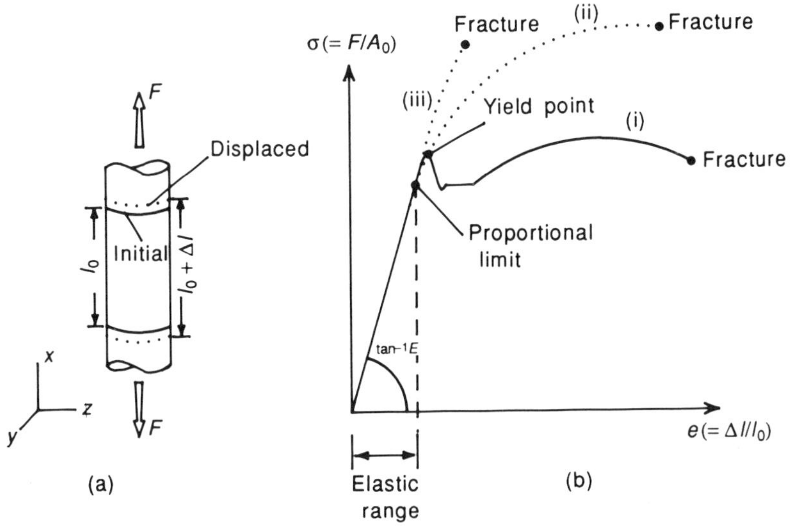

Fig. 2.11 *One-dimensional elastic behaviour. (a) The tensile test, and (b) the stress-strain curve.*

The stress–strain curves obtained in this way generally resolve themselves into two or more distinct regions. The first of these, and the only one of interest to us here, is the linearly elastic region close to the origin. Within it, the curve approximates a straight line and is reversible (when the load is removed the test piece unloads along the same path). This linear region is bounded above by the 'proportional limit', beyond which the behaviour of the material is nonlinear and (usually) irreversible. The stress–strain relationship in the non-linear region varies from material to material. In low and medium carbon steels, for example, the onset of nonlinearity is followed by a distinct 'yield point' beyond which large plastic deformations occur before ultimate fracture (curve (i)). In other ductile metals, the transition from linear to nonlinear behaviour is relatively smooth (curve (ii)), whereas in brittle materials (cast iron or ceramics, for example) nonlinear behaviour, if it exists at all, does so only for a small stress range close to ultimate fracture (curve (iii)). In all cases, however, we can isolate a portion of the stress–strain curve that is, for all intents and purposes, linear and reversible. The slope of this curve defines the Young's modulus, E, of the material, and the observed relationship between σ and e can therefore be written

$$e = \frac{\sigma}{E}. \tag{2.17}$$

This can be extended to two and three-dimensional situations provided that the material is 'isotropic'. In other words, provided that it has no preferred plane or direction such as occurs for example in timber of fibre reinforced composites. The process then is somewhat more complicated and the reader must look elsewhere for further details [3, 4]. In the isotropic case, however, the strain in any lateral direction is observed to be proportional to the axial strain but of the opposite sign. The constant of proportionality is termed Poisson's ratio and is denoted by the symbol ν. If the axis of the test specimen is taken as the x direction of a cartesian system (Fig. 2.11(a)), the observed lateral strains e_y and e_z are therefore given by

$$e_y = e_z \doteq -\nu e_x$$

or, incorporating equation 2.17 and putting $\sigma = \sigma_x$,

$$e_y = e_z = -\nu \frac{\sigma_x}{E}. \tag{2.18a}$$

Since the material is isotropic, the x, y and z axes can be interchanged without altering the algebraic form of these relationships. In other words, if a stress σ_y were applied to a sample of the same material, strains e_y, e_x and e_z would result, where

$$e_y = \frac{\sigma_y}{E}, \text{ and } e_x = e_z = -\nu \frac{\sigma_y}{E}. \tag{2.18b}$$

Similarly, a stress σ_z would cause strains

$$e_z = \frac{\sigma_z}{E} \text{ and } e_x = e_y = -\nu \frac{\sigma_z}{E}. \tag{2.18c}$$

Provided that the linearity of the material extends to the three-dimensional case, the strains due to the simultaneous application of σ_x, σ_y and σ_z are then obtained by summing the contributions from expressions 2.18. This yields the three-dimensional, elastic stress–strain equations for an isotropic material;

$$e_x = \frac{\sigma_x}{E} - \nu \frac{\sigma_y}{E} - \nu \frac{\sigma_z}{E},$$

$$e_y = -\nu \frac{\sigma_x}{E} + \frac{\sigma_y}{E} - \nu \frac{\sigma_z}{E}, \tag{2.19}$$

$$e_z = -\nu \frac{\sigma_x}{E} - \nu \frac{\sigma_y}{E} + \frac{\sigma_z}{E}.$$

Although cartesian components of stress and strain have been used in the above derivation, the resulting expressions are valid in any orthogonal system. A more general statement is obtained by replacing x, y and z by indices 1, 2 and 3 and rearranging somewhat the terms on the right hand side to give

$$e_1 = \frac{1}{E}[\sigma_1 - \nu\sigma_2 - \nu\sigma_3],$$

$$e_2 = \frac{1}{E}[\sigma_2 - \nu\sigma_3 - \nu\sigma_1], \tag{2.20}$$

$$e_3 = \frac{1}{E}[\sigma_3 - \nu\sigma_1 - \nu\sigma_2].$$

2.4.2 Elastic stress–strain relationship in shear

The stress–strain behaviour of an isotropic material in shear may readily be deduced from the direct stress–strain relationships already derived. This is done by considering the deformation of a square element ABCD which is subject to orthogonal tensile and compressive stresses, σ and $-\sigma$, as shown in Fig. 2.12(a). The stresses acting on a second square EFGH inscribed at 45 degrees inside ABCD are then a shear stress τ, of magnitude σ, and a normal stress of magnitude zero (these values are obtained using equations 2.11).

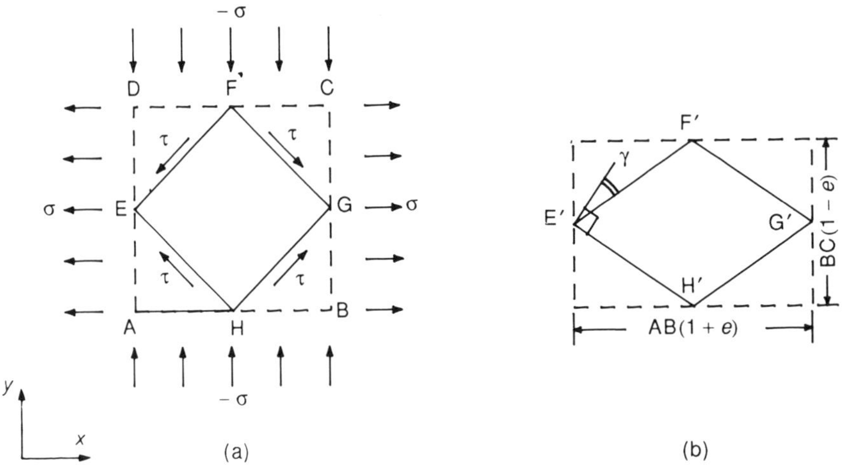

Fig. 2.12 *Elastic behaviour in pure shear. (a) Stress, and (b) deformation.*

The direct strains parallel to AB and BC are then given by expressions 2.19 with $\sigma_x = \sigma$, $\sigma_y = -\sigma$ and $\sigma_z = 0$. That is,

$$e_x = -e_y = e \tag{2.21}$$

where $e = (1 + \nu)\sigma/E$. The resulting deformed shape of the outer and inner squares is shown in Fig. 2.12(b). The deformation of the inner square EFGH, in particular, is one of pure shear, the associated shear strain being denoted by γ. Provided that all strains are small, elementary trigonometry yields the result that $\gamma = 2e$. The relationship between the shearing strain γ experienced by the inner square EFGH and the shearing stress τ ($= \sigma$) applied to it, can then be obtained directly by rewriting equation 2.21 as

$$\gamma = \frac{\tau}{G} \tag{2.22}$$

where $G = E/2\,(1+\nu)$. The square ABCD may be discarded entirely at this stage and equation 2.22 becomes a statement of the elastic stress–strain relationship in shear. G is the shear modulus of the material. In the general case defined by three orthogonal axes 1, 2 and 3, the same relationship holds for each orthogonal pair, that is

$$\gamma_{12} = \frac{\tau_{12}}{G}, \quad \gamma_{13} = \frac{\tau_{13}}{G}, \quad \gamma_{23} = \frac{\tau_{23}}{G} \tag{2.23}$$

The addition of these equations to the direct stress–strain relationship (equations 2.20) completes the statement of Hooke's law for an isotropic, linearly elastic material.

2.4.3 Thermal effects

The stress–strain relationships derived in the preceding sections must be modified when temperature variations are present. Their effect can be incorporated quite simply in the following way. When an unstressed element of an isotropic, elastic material is subject to a temperature rise T, it expands uniformly in all directions. A fibre initially of length l_0 then extends by an amount $\alpha l_0 T$ where α is the coefficient of thermal expansion. This is independent of the orientation of the fibre. From the definition of direct strain (equation 2.12) this is equivalent to a direct strain, e^{T}, where $e^{\mathrm{T}} = \alpha T$.

In the case of three orthogonal directions 1, 2, and 3, the effect of a temperature rise is therefore to cause 'thermal' strains

$$e_1^{\mathrm{T}} = e_2^{\mathrm{T}} = e_3^{\mathrm{T}} = \alpha T. \tag{2.24}$$

These must be added to any elastic strains which result from the action of stresses applied to the same element. The total strain experienced by the element is therefore obtained by combining expressions 2.24 and 2.20. This yields a modified form of equations 2.20 which include thermal effects. These are,

$$\begin{aligned} e_1 &= \frac{1}{E}\,[\sigma_1 - \nu\sigma_2 - \nu\sigma_3] + \alpha T, \\ e_2 &= \frac{1}{E}\,[\sigma_2 - \nu\sigma_1 - \nu\sigma_3] + \alpha T, \\ e_3 &= \frac{1}{E}\,[\sigma_3 - \nu\sigma_2 - \nu\sigma_1] + \alpha T. \end{aligned} \tag{2.25}$$

The thermal expansion or contraction of the material has no effect on its shear deformation. No modifications to the shear stress–strain relationship (equations (2.23)) are therefore required.

2.5 BEAMS AND PLATES

Beam theory and plate theory provide an effective means of reducing three-dimensional problems to more tractable calculations in one or two dimensions. They are applicable to bodies which are initially prismatic (in the case of a beam) or flat (in the case of a plate) and whose thickness is small compared with their overall dimensions. In such cases lateral loads are supported by 'bending', provided that the lateral deflection of the centroidal axis (of a beam) or mid-surface (of a plate) is small. The neglect of shear deformation through the depth of the beam or thickness of the plate is also implied. Provided that these assumptions are justified, the stresses at all points can be expressed directly in terms of the deformation and curvature of the centroidal axis or plane.

A brief review of these theories now follows. It is by no means comprehensive, nor does it touch upon the techniques used to obtain solutions to the resulting equations. The derivation and application of such solutions constitutes a substantial topic in its own right and the reader must look elsewhere for a more detailed treatment (references [5–7], for example). Our objective here is merely to introduce those expressions governing the behaviour of beams and plates which will be useful to us later in our treatment of structural finite elements.

2.5.1 Bernoulli–Euler beam theory, basic relationships

Classical (Bernoulli–Euler) beam theory applies to slender, prismatic beams of arbitrary cross-sectional shape. Certain properties of the cross-section are required in order to proceed. These include the cross-sectional area A, the second moment of area I, and the location of the centroid. It is assumed here that these quantities are already known, techniques for their evaluation being covered in most introductory strength of materials texts (see Appendix C of [2], for example). Consider then a beam of cross-sectional area A and second moment of area I (about the y-axis), whose centroidal axis defines the x-axis of a cartesian coordinate system (Fig. 2.13(a)). The beam bends about the y-axis under the action of a bending moment M and a shear force Q. An axial load P is also included for completeness. The resulting displacements of the centroidal axis are $u_0(x)$ and $w_0(x)$ in the x and z directions, respectively (Fig. 2.13(b)).

It has been assumed here that a bending moment and shear force acting in the x-z plane, will of necessity cause a centroidal deflection only in that plane.

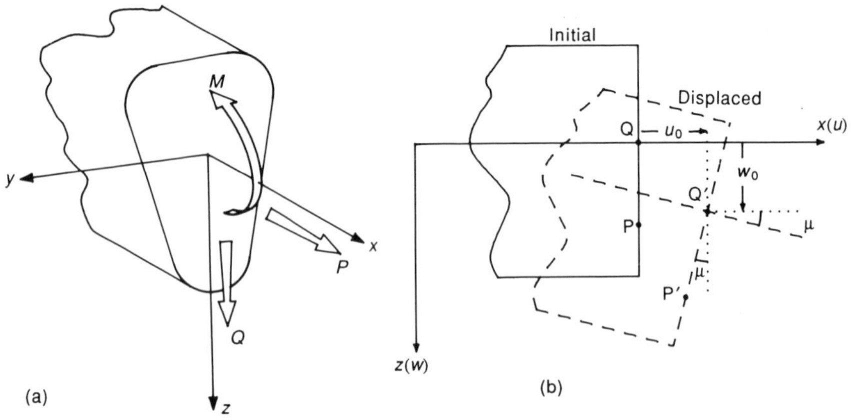

Fig. 2.13 *Bernoulli–Euler beam theory. (a) Forces and moments on a typical section, and (b) deformation in the plane of bending.*

This is true provided that y and z are 'principal' axes of the section. These are defined as axes for which

$$\int_A (yz)\,\mathrm{d}A = 0 \tag{2.26}$$

This is certainly the case when either y or z is an axis of symmetry but is not obviously so otherwise. Principal axes *do* however exist for all cross-sections, although they are somewhat harder to locate when no symmetry is present (see Appendix C of [2]). For the purposes of the present discussion, it is assumed that principal axes *have* been located and do indeed correspond to the y and z axes of Fig. 2.13(a).

Returning to the deformation of the beam (Fig. 2.13(b)), the fundamental assumption of beam theory requires that *plane sections which are initially perpendicular to the centroidal axis remain plane and perpendicular to that axis after deformation.* This fixes the displacement of all points within the beam once the deflection of the centroidal axis is specified. Consider, for example, the displacement of two points Q and P which initially lie on the x-axis and a distance z below it. Their displaced locations are denoted by Q' and P'. By definition, Q experiences a translation $u_0(x)$ in the x-direction, and $w_0(x)$ in the z-direction. The displacement of P, is a combination of the translation of Q plus the rotation of PQ through an angle μ. Provided that μ is small, the cartesian displacements u and v at the point P are then given by

$$u(x, y, z) = u_0(x) - z\mu \quad \text{and} \quad w(x, y, z) = w_0(x) \tag{2.27}$$

or, replacing μ by $\mathrm{d}w_0/\mathrm{d}x$

$$u(x, y, z) = u_0(x) - z(\mathrm{d}w_0/\mathrm{d}x)\ . \tag{2.28}$$

Equations 2.27 and 2.28 define a displacement field throughout the beam. They can therefore be substituted into the cartesian, strain–displacement equations (equations 2.13 and 2.14) to give strains

$$e_x = \frac{\partial u}{\partial x} = \frac{\mathrm{d}u_0}{\mathrm{d}x} - z\frac{\mathrm{d}^2 w_0}{\mathrm{d}x^2},$$

$$e_z = \frac{\partial w}{\partial z} = \frac{\partial w_0}{\partial z} = 0, \tag{2.29}$$

and
$$\gamma_{xz} = \frac{\partial u}{\partial z} + \frac{\partial w}{\partial x} = -\frac{\mathrm{d}w_0}{\mathrm{d}x} + \frac{\mathrm{d}w_0}{\mathrm{d}x} = 0\,.$$

The only nonzero component, e_x, can then be rewritten

$$e_x = e^0 + e^b, \tag{2.30}$$

where $e^0\,(=\mathrm{d}u_0/\mathrm{d}x)$ is a constant longitudinal strain and $e^b\,(=-z\,(\mathrm{d}^2w_0/\mathrm{d}x^2))$ is a bending strain which varies linearly with depth. The first of these terms (e^0) is zero when the centroidal axis is unstrained, and equation 2.30 then reduces to

$$e_x = \frac{z}{R}, \tag{2.31}$$

where

$$\frac{1}{R} = -\frac{\mathrm{d}^2 w_0}{\mathrm{d}z^2}. \tag{2.32}$$

In terms of the geometry of the deformation, R is the local radius of curvature of the centroidal axis, and equation 2.31 can be obtained directly using purely geometrical arguments. It is, in fact, often presented in this way in strength of materials derivations of the same relationship. In the current instance, it is more convenient to use the differential strain–displacement equations, partly because they give a more concise formulation and partly because they render the extension to plate theory (to be attempted shortly) somewhat more straightforward.

The elastic stress–strain relationship can now be used to determine the stresses in the beam. Provided that the lateral stresses σ_y and σ_z are negligible in comparison to the axial stress (this is an acceptable approximation provided that the length-to-depth ratio of the beam is sufficiently large) σ_x is given by the uniaxial form of Hooke's law;

$$\sigma_x = E\,(e^0 + e^b) = \sigma^0 + \sigma^b, \tag{2.33}$$

where $\sigma^0(=Ee^0)$ is the uniform stress due to straining of the centroidal axis and $\sigma^b(=Ee^b=Ez/R)$ is the bending stress resulting from its curvature. These equilibrate with the force and moment resultants acting on the section provided that

$$P=\int_A \sigma_x \mathrm{d}A, \text{ and } M=\int_A z\sigma_x \mathrm{d}A.$$

Substitution of equation 2.33 into the above expressions and their evaluation using the identities; $\int_A z\mathrm{d}A=0$ (since the x-axis passes through the centroid of area), and $\int_A z^2\mathrm{d}A=I$ (by definition), then gives

$$P=EAe^0, \tag{2.34a}$$

and

$$M=EI/R \tag{2.34b}$$

which can be substituted into equation 2.33 to give the familiar result,

$$\sigma_x=\frac{P}{A}+\frac{Mz}{I}. \tag{2.35}$$

This is the standard strength of materials formula for the longitudinal stress in a bar subject to an axial load P and a bending moment M. The first term, P/A is the uniform stress due to the axial load and the second, Mz/I, the flexural stress due to bending. A second equation relating bending moment and deflection is obtained by replacing $1/R$ in equation 2.34b by $-\mathrm{d}^2w_0/\mathrm{d}x^2$ as permitted by equation 2.32, to give

$$\frac{\mathrm{d}^2w_0}{\mathrm{d}x^2}=-\frac{M}{EI}. \tag{2.36}$$

This is the 'deflection equation' for a beam and may be integrated for a given moment distribution $M(x)$, to yield the lateral deflection $w_0(x)$. Numerous exercises which involve the use of equations 2.35 and 2.36 to calculate the stresses and deflections in beams subject to a variety of loadings and support conditions are to be found in most strength of materials texts. We proceed no further in this direction here.

There is an inconsistency, however, in the expressions so far which deserves further comment. The analysis presented here implies (equations 2.29) that the shear strain, γ_{xz}, is identically zero. This is consistent with the assumption that a right angle is preserved between the line PQ (Fig. 2.13(b)) and the centroidal axis as the deformation takes place. That is to say, no change takes place in the angle between fibres lying initially in the x and z directions and consequently no shear strain γ_{xz} is generated. This appears to be an acceptable state of affairs until one recalls the presence of the shear force Q in Fig.

2.13(a). If γ_{xz} is zero, then τ_{xz} must also be zero (from Hooke's law) and the resultant shear force Q, which is statically equivalent to the shear stress distribution, must vanish identically. From elementary mechanics, however, Q is known to be proportional to the axial derivative of the bending moment and is therefore zero only if the bending moment is constant along the beam. We conclude that the current analysis is strictly correct only in the case of 'pure bending' when M is constant. In practice, however, it is acceptable for other bending moment distributions provided that the beam is slender, in which case, although there are indeed non-zero shear strains, their effect on axial stress and lateral deflection is small. The neglected shear stresses can themselves be estimated without much difficulty provided that the current pure bending expression for axial stress is assumed to be substantially correct, an assumption which gives unsatisfactory results only when the depth of the beam is significant in comparison to its length. A somewhat more complex approximate analysis, attributed to Timoshenko [8], may then be used. This permits the inclusion of the effects of shear deformation within a modified set of equations. The Timoshenko theory will be discussed in greater detail in Chapter 9 when finite elements for 'deep' beams including shear deformation are developed.

2.5.2 Kirchhoff's thin plate theory, basic relationships

A plate is an initially flat sheet whose thickness, t, is small compared with its other dimensions. It is assumed in all that follows that the initial position of the mid-surface of the plate forms the x–y plane of a cartesian coordinate system (Fig. 2.14). The subscript zero is again used to denote displacements of this plane, the lateral displacement (in the z direction) being denoted by $w_0(x, y)$, and the in-plane displacements by $u_0(x, y)$ and $v_0(x, y)$.

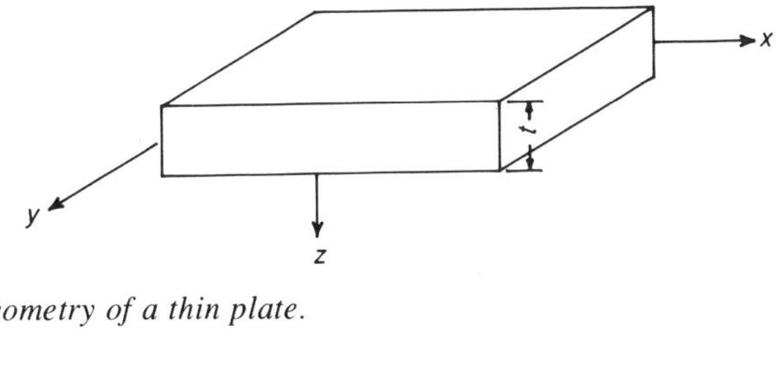

Fig. 2.14 *Geometry of a thin plate.*

The assumptions now made regarding the deflection of the plate are similar to those of beam theory. That is to say, straight lines initially normal to the mid–surface are assumed to remain straight and normal to that surface after deformation. A geometrical construction similar to that of Fig. 2.13(b) then yields the following expressions for $u(x, y, z)$, $v(x, y, z)$ and $w(x, y, z)$, the cartesian displacements at any point within the plate:

$$u(x, y, z) = u_0(x, y) - z\frac{\partial w_0}{\partial x},$$

$$v(x, y, z) = v_0(x, y) - z\frac{\partial w_0}{\partial y}, \quad (2.37)$$

$$w(x, y, z) = w_0(x, y).$$

These are clearly analogous to the expressions in equation 2.28 but contain partial, rather than ordinary, derivatives of $w_0(x, y)$. The strain–displacement equations can be applied to the above displacement field to give cartesian strain components:

$$e_x = \frac{\partial u}{\partial x} = \frac{\partial u_0}{\partial x} - z\frac{\partial^2 w_0}{\partial x^2},$$

$$e_y = \frac{\partial v}{\partial y} = \frac{\partial v_0}{\partial y} - z\frac{\partial^2 w_0}{\partial y^2}, \quad (2.38a)$$

and

$$\gamma_{xy} = \frac{\partial u}{\partial y} + \frac{\partial v}{\partial x} = \frac{\partial u_0}{\partial y} + \frac{\partial v_0}{\partial x} - 2z\frac{\partial^2 w_0}{\partial x \partial y}. \quad (2.38b)$$

Each of these comprises two distinct contributions. The first involves in-plane derivatives of u_0 and v_0 and arises from stretching and shearing of the centroidal plane. The second is a consequence of lateral (bending) deformation and varies linearly with depth, being proportional to the curvatures of the centroidal plane represented by the second derivatives $\partial^2 w_0/\partial x^2, \partial^2 w_0/\partial y^2$ and $\partial^2 w_0/\partial x \partial y$. These 'membrane' and 'bending' contributions are denoted by superscripts '^{0}' and '^{b}', respectively. Using this notation, equations 2.38 may be rewritten

$$e_x = e_x^0 + e_x^b,$$

$$e_y = e_y^0 + e_y^b, \quad (2.39a)$$

and

$$\gamma_{xy} = \gamma_{xy}^0 + \gamma_{xy}^b, \quad (2.39b)$$

where $e_x^0 = \partial u_0/\partial x$, $e_x^b = -z\partial^2 w_0/\partial x^2$ et cetera. Provided that the *through-thickness* direct stress, σ_z, is small in comparison to the in-plane stresses, and this is generally the case for thin plates as it is for slender beams, the cartesian stress components within the plate are then given in terms of the strains by Hooke's law (equations 2.22 and 2.23) with σ_z set equal to zero. This yields

$$e_x = \frac{1}{E}\left[\sigma_x - \nu\sigma_y\right],$$

$$e_y = \frac{1}{E}\left[\sigma_y - \nu\sigma_x\right], \qquad (2.40a)$$

and

$$\gamma_{xy} = \frac{1}{G}\,\tau_{xy} \qquad (2.40b)$$

which can be inverted to give

$$\sigma_x = \frac{E}{1-\nu^2}\left[e_x + \nu e_y\right],$$

$$\sigma_y = \frac{E}{1-\nu^2}\left[e_y + \nu e_x\right], \qquad (2.41a)$$

and

$$\tau_{xy} = \frac{E}{2(1+\nu)}\,\gamma_{xy}. \qquad (2.41b)$$

Once again, these expressions can be subdivided into contributions of the form:

(i) for membrane stresses

$$\sigma_x^0 = \frac{E}{1-\nu^2}\left[e_x^0 + \nu e_y^0\right],$$

$$\sigma_y^0 = \frac{E}{1-\nu^2}\left[e_y^0 + \nu e_x^0\right], \qquad (2.42a)$$

and

$$\tau_{xy}^0 = \frac{E}{2(1+\nu)}\,\gamma_{xy}^0, \qquad (2.42b)$$

(ii) for bending stresses

$$\sigma_x^{\mathrm{b}} = \frac{E}{1-\nu^2}\left[e_x^{\mathrm{b}} + \nu e_y^{\mathrm{b}}\right],$$

$$\sigma_y^{\mathrm{b}} = \frac{E}{1-\nu^2}\left[e_y^{\mathrm{b}} + \nu e_x^{\mathrm{b}}\right], \qquad (2.43a)$$

and

$$\tau_{xy}^{\mathrm{b}} = \frac{E}{2(1+\nu)}\,\gamma_{xy}^{\mathrm{b}}. \qquad (2.43b)$$

The membrane stresses (expressions 2.42) depend only upon the membrane strains and are constant with depth. They are two-dimensional equivalents of the uniform axial stress σ^0 encountered in beam theory. They also form an 'uncoupled' problem in the sense that they can be evaluated for a flat plate without reference to the bending stresses or deflections. They constitute, in effect, a state of plane stress.

The bending stresses (expressions 2.43) vary linearly with depth and are two-dimensional analogues of the flexural stress σ^b in a beam. When integrated over any normal plane through the plate, they give a zero force resultant (the stresses above and below the centroidal plane cancelling each other exactly) but are statically equivalent to a system of distributed moments. On planes perpendicular to the x and y-axes, for example, they are equivalent to distributed moments M_x, M_y and M_{xy} per unit length (Fig. 2.15(b)), where

$$M_x = \int_{-t/2}^{t/2} \sigma_x \, z\mathrm{d}z,$$

$$M_y = \int_{-t/2}^{t/2} \sigma_y \, z\mathrm{d}z, \tag{2.44a}$$

and

$$M_{xy} = \int_{-t/2}^{t/2} \tau_{xy} \, z\mathrm{d}z. \tag{2.44b}$$

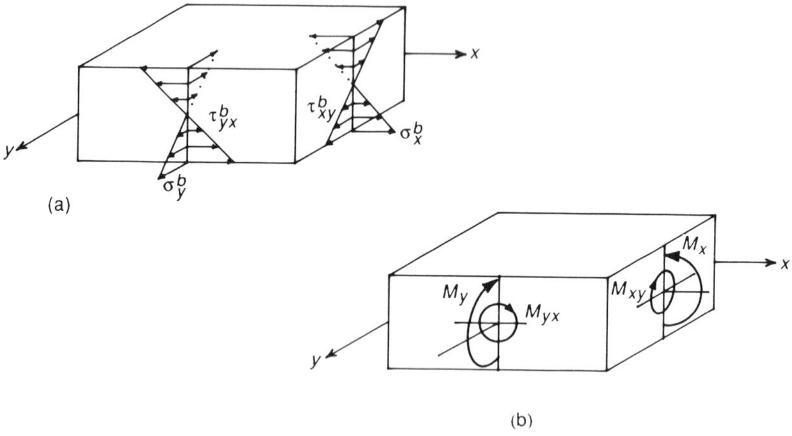

Fig. 2.15 *Stress and moment distribution in a plate. (a) Stresses, and (b) distributed moments.*

Substitution of expressions 2.42 and 2.43 into the above integrals followed by their integration with respect to z gives

$$M_x = -D\left[\frac{\partial^2 w_0}{\partial x^2} + \nu \frac{\partial^2 w_0}{\partial y^2}\right],$$

$$M_y = -D\left[\frac{\partial^2 w_0}{\partial y^2} + \nu\frac{\partial^2 w_0}{\partial x^2}\right], \quad (2.45a)$$

and

$$M_{xy} = -D(1-\nu)\frac{\partial^2 w_0}{\partial x \partial y}, \quad (2.45b)$$

where D, the 'flexural rigidity' of the plate, is given by

$$D = \frac{Et^3}{12(1-\nu^2)} \quad (2.46)$$

Equations 2.45 are the 'moment-curvature relations' for an isotropic elastic plate. If the distributed moments M_x, M_y and M_{xy} were known, we could, in theory, invert these equations to obtain the curvatures in terms of the applied loads and substitute these values into equation 2.43 to give the bending stresses within the plate. Unfortunately, the moments M_x, M_y and M_{xy} cannot be obtained directly using straightforward equilibrium arguments (a free body diagram, for example) but only as the solution of coupled partial differential equations involving the distributed moments, their associated distributed shear forces and the distributed load on the surface of the plate. These equations are not derived here, but when combined with equations 2.45, they lead to the classical 'plate equation'

$$\nabla^4 w_0 = p(x, y)/D \quad (2.47)$$

where ∇^4 is the biharmonic operator $(\partial^2/\partial x^2 + \partial^2/\partial y^2)^2$ and $p(x, y)$ is the distributed pressure on the upper surface of the plate. This fourth order equation can be solved subject to appropriate boundary conditions to give the lateral deflection w_0 and hence, using equations 2.45 and 2.43, the curvatures, distributed moments and bending stresses throughout the plate.

Fortunately, the derivation and solution of equation 2.47 is not essential to the present discussion. The finite element approach treats equilibrium requirements in quite a different way, and all the information that we need to formulate finite plate elements is contained already in equations 2.36–2.45. The reader is directed elsewhere for a more complete treatment of the traditional methods of solution. (The classic text by Timoshenko and Woinowsky-Krieger [5] is as good a place as any to start as are more recent texts by Donnell [6] and Szilard [7]).

REFERENCES

[1] Ford, H. (1963), *Advanced Mechanics of Materials*, 2nd edn, Longman, London.

[2] Ugural, A. C. and Fenster, S. K. (1987), *Advanced Strength and Applied Elasticity*, 3rd edn, Edward Arnold, London, chapter 1.

[3] Halpin, J. C., (1984) *Revised Primer on Composite Materials: Analysis*, Technomic, Lancaster PA.

[4] Hull, D., (1981). *An introduction to Composite Materials*, CUP, Cambridge.
[5] Timoshenko, S. and Woinowsky-Krieger, S. (1959) *Theory of Plates and Shells*, 2nd edn, McGraw-Hill, New York.
[6] Donnell, L. H., (1976) *Beams Plates and Shells*, McGraw-Hill, New York.
[7] Szilard, R., (1974) *Theory and Analysis of Plates, Classical and Numerical Methods*, Prentice-Hall, Englewood Cliffs, NJ.
[8] Timoshenko, S., (1956) *Strength of Materials, Part 1*, 3rd edn, Van Nostrand, New York.

PROBLEMS

1. The cartesian components of stress in an elastic body are given by:

   (a) $\sigma_x = y + 3z^2, \; \sigma_y = x + 2z, \; \sigma_z = 2x + y,$
   $\tau_{xy} = z^3, \; \tau_{zx} = y^2, \; \tau_{xy} = x^2.$

   (b) $\sigma_x = 20x^3 + y^3, \; \sigma_y = 30x^3 + 100, \; \sigma_z = 30y^2 + 30z^2,$
   $\tau_{xy} = 100 + 80y^2, \; \tau_{yz} = 0, \; \tau_{zx} = xz^3 + 30x^2y.$

   Do the above stress fields satisfy the conditions of static equilibrium in the absence of body forces? If not, what body force components are required in order that equilibrium *is* satisfied?

2. A stress field has the following cylindrical polar components:

   $$\sigma_r = A\,(1 + 2\ln r) + B + C/r^2,$$

   $$\sigma_\theta = A\,(3 + 2\ln r) + B - C/r^2,$$

   $$\sigma_z = D,$$

   $$\tau_{r\theta} = \tau_{z\theta} = \tau_{rz} = 0.$$

   Show that it satisfies the stress equations of equilibrium in the absence of body forces for all values of the constants A, B, C and D.

3. The stress field within an axisymmetric body is defined by the following cylindrical polar stress components:

   $$\sigma_r = A + \frac{B}{r^2} - (3 + \nu)\frac{\rho\omega^2 r^2}{8},$$

   $$\sigma_\theta = A - \frac{B}{r^2} - (1 + 3\nu)\frac{\rho\omega^2 r^2}{8},$$

   $$\tau_{r\theta} = \tau_{rz} = \tau_{z\theta} = \sigma_{zz} = 0,$$

   (A, B, ν, ρ and ω are constants).

Confirm that the above stress field satisfies equilibrium within a body of density ρ which is rotating about the z-axis with an angular velocity ω rad/s. (Suggestion: show that stress equilibrium is satisfied provided that a body force of magnitude $r\rho\omega^2$ per unit volume acts in the radial direction.)

4. The 'strength of materials' solution for the stress field within a cantilever beam of rectangular cross-section subject to a distributed load w per unit length (see figure) is given by:

$$\sigma_x = \frac{My}{I},$$

$$\tau_{xy} = \frac{Q}{2I}\left(y^2 - \frac{h^2}{4}\right),$$

$$\tau_{xz} = \tau_{yz} = \sigma_z = \sigma_y = 0,$$

where $M = -wx^2/2$, $Q = wx$ and $I = th^3/12$.

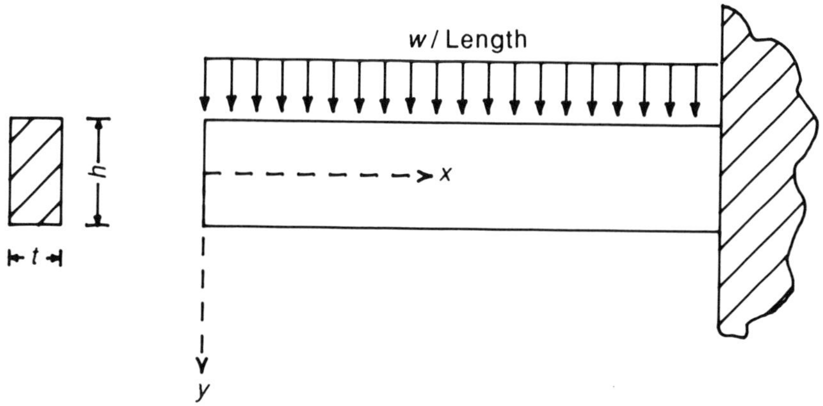

Problem 4.

Show that these stresses do *not* satisfy equilibrium in the absence of body forces. What body force is required to ensure that equilibrium *is* satisfied?

5. By writing the stress vector s of expression 2.8 in component form

$$s = (s_x, s_y, s_z),$$

show that equation 2.8 may be rewritten in matrix notation as

$$\begin{bmatrix} s_x \\ s_y \\ s_z \end{bmatrix} = \begin{bmatrix} \sigma_x & \tau_{yx} & \tau_{zx} \\ \tau_{xy} & \sigma_y & \tau_{zy} \\ \tau_{xz} & \tau_{yz} & \sigma_z \end{bmatrix} \begin{bmatrix} l_1 \\ m_1 \\ n_1 \end{bmatrix}$$

A 'principal plane' is defined to be one on which no shear stress acts. The normal stress acting on such a plane is termed a 'principal stress'. Show that a principal stress σ which acts in the direction of the unit vector (l, m, n) must satisfy the equation

$$\begin{bmatrix} \sigma_x & \tau_{yx} & \tau_{zx} \\ \tau_{xy} & \sigma_y & \tau_{zy} \\ \tau_{xz} & \tau_{yz} & \sigma_z \end{bmatrix} \begin{bmatrix} l \\ m \\ n \end{bmatrix} = \sigma \begin{bmatrix} l \\ m \\ n \end{bmatrix}. \tag{i}$$

Hence show that the principal stresses at any point in a solid body are the roots of the characteristic equation

$$\det |\boldsymbol{A} - \sigma \boldsymbol{I}| = 0,$$

where $\boldsymbol{A}$ is the 3×3 matrix on the left-hand side of equation (i) and $\boldsymbol{I}$ is the 3×3 identity matrix.

Determine the principal stresses and the planes upon which they act for the cartesian stress state

$$\sigma_x = 15, \ \sigma_y = -5, \ \sigma_z = -10, \ \tau_{xy} = -3, \ \tau_{xz} = 1, \ \tau_{yz} = 0.$$

6. The displacement field within a deformable solid is given by

$$u = (x^2 + 10) \times 10^{-2}, \ v = 2yz \times 10^{-2}, \ w = (z^2 - xy) \times 10^{-2},$$

(u, v and w are cartesian components of displacement in the usual notation). Determine the cartesian strain components at the point (0, 2, 1).

7. The displacement field within an axisymmetric body is given by

$$u_r = (r + 2/r) \times 10^{-2}, \quad u_\theta = u_z = 0,$$

where r, θ, and z are cylindrical polar coordinates. Obtain expressions for the strain components e_r, e_θ and $\gamma_{r\theta}$ and evaluate them at $r = 2$.

8. A two-dimensional displacement field has cylindrical polar components

$$u_r = u_r(r, \theta), \ u_\theta = u_\theta(r, \theta), \ u_z = 0.$$

Obtain expressions for the strain components e_r, e_θ, and $\gamma_{r\theta}$. Eliminate u_r and u_θ from these expressions and hence show that the two-dimensional compatibility condition in cylindrical polar coordinates is given by

$$r\frac{\partial^2 (re_\theta)}{\partial r^2} - r\frac{\partial e_r}{\partial r} + \frac{\partial^2 e_r}{\partial \theta^2} - \frac{\partial^2 (r\gamma_{r\theta})}{\partial r \partial \theta} = 0.$$

9. Does the two-dimensional strain field (below) satisfy the compatibility requirement in a continuous material?

$$e_x = c\,(z^2 + y^2), \ e_y = cy^2, \ \gamma_{xy} = 2cxy.$$

If so, obtain general expressions (including constants of integration) for the cartesian displacements u and v, assuming that they are independent of z.

10. Does the strain field (below) satisfy the compatibility requirement in three dimensions?

$$e_x = Cz(x^2 + y^2), \quad Ce_y = y^2 z, \quad \gamma_{xy} = 2Czxy.$$

$$e_z = \gamma_{xz} = \gamma_{yz} = 0, \quad (C = \text{const.}).$$

11. Under what conditions are the following expressions for the displacements and shear strain of a point compatible?

$$u = ax^2y^2 + bxy^2 + cx^2y,$$

$$v = ax^2y + bxy,$$

$$\gamma_{xy} = \alpha x^2 y + \beta xy + cx^2 + by.$$

12. A rectangular block of elastic material (Young's modulus E, Poisson's ratio ν) is subject to a compressive stress $\sigma_x = -p$ on its upper surface as illustrated below. The 'effective' Young's modulus, E_{eff}, is defined as the ratio σ_x/e_x. Obtain an expression for E_{eff} in terms of E and ν, for the two cases:
(a) the block is constrained in the y direction only.
(b) the block is constrained in the y and z directions.
What is the physical significance of the case $\nu = 0.5$?

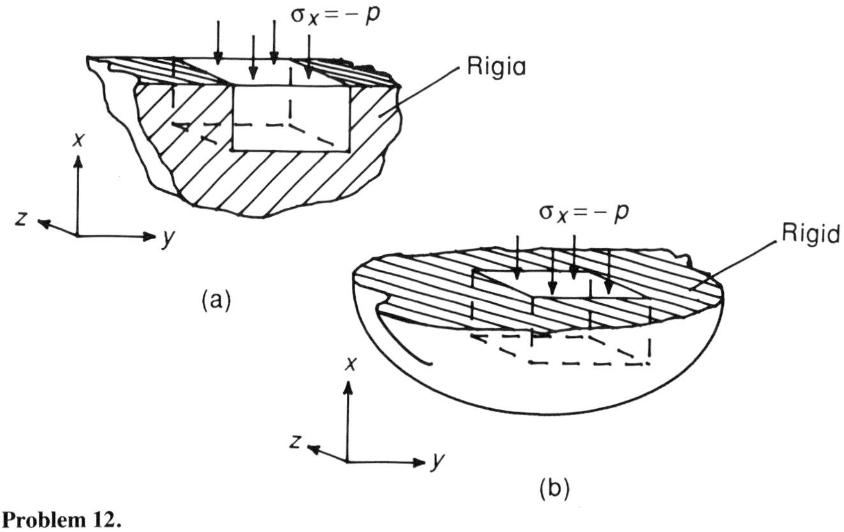

Problem 12.

13. A two-dimensional state of plane stress is defined by nonzero cartesian stress components σ_x, σ_x, and τ_{xy}. Show that the strain components are given by:

$$e_x = \frac{1}{E}\left[\sigma_x - \nu\sigma_y\right], \quad e_y = \frac{1}{E}\left[\sigma_y - \nu\sigma_x\right] \quad \text{and} \quad \gamma_{xy} = \frac{2\tau_{xy}}{E}(1+\nu)$$

Provided that the strains are compatible, show that the stress components must then satisfy the second order partial differential equation

$$\frac{\partial^2\sigma_x}{\partial y^2} + \frac{\partial^2\sigma_y}{\partial x^2} - \nu\left(\frac{\partial^2\sigma_x}{\partial x^2} + \frac{\partial^2\sigma_y}{\partial y^2}\right) - 2(1+\nu)\frac{\partial^2\tau_{xy}}{\partial x\partial y} = 0$$

[Suggestion: substitute the expressions for e_x, e_y and γ_{xy} into the two-dimensional compatibility equation (equation 2.15a).]

14. A state of plane-stress is defined by the cartesian stress components $\sigma_x = ky^2, \sigma_y = -kx^2, \tau_{xy} = 0$, ($k$ is a constant). Confirm that this is consistent with a compatible state of deformation (see problem 13) and obtain expressions for the strain components e_x, e_y and γ_{xy}. By integrating the strain–displacement equations, obtain also expressions (including constants of integration) for the cartesian displacement components u and v, assumed to be independent of z.

15. A thin prismatic bar of rectangular cross-section of Young's modulus E, Poisson's ratio ν and density ρ, is loaded by its own weight (a body force

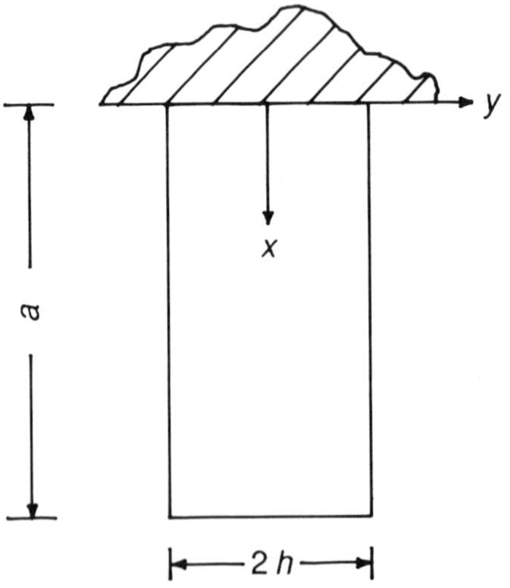

Problem 15.

ρg per unit volume acting downwards). A solution for the cartesian displacement field in the bar is proposed of the form

$$u = \frac{\rho g}{2E}\left[2xa - x^2 - \nu y^2\right] \quad \text{and} \quad v = -\frac{\nu \rho g}{E}\,[(a - x)\,y],$$

(the bar is in a state of plane stress and is orientated with respect to the x and y-axes as indicated in the diagram). Determine the strains and the stresses in the bar. Is equilibrium satisfied (a) within the bar, and (b) on its boundary?

16. The radial and hoop displacements in an axisymmetric rotating body are $u_r = u_r(r)$ and $u_\theta = 0$ (r, θ and z are cylindrical polar coordinates). If a state of plane stress exists, show that the in-plane stresses σ_r, σ_θ and $\tau_{r\theta}$ are given by:

$$\sigma_r = \frac{E}{1 - \nu^2}\left[\frac{\partial u_r}{\partial r} + \frac{\nu u_r}{r}\right]$$

$$\sigma_\theta = \frac{E}{1 - \nu^2}\left[\nu\frac{\partial u_r}{\partial r} + \frac{u_r}{r}\right] \qquad \text{(i)}$$

$$\tau_{r\theta} = 0$$

Show also that the radial equation of stress equilibrium can be rewritten in terms of displacement as

$$\frac{\partial^2 u_r}{\partial r^2} + \frac{1}{r}\frac{\partial u_r}{\partial r} - \frac{u_r}{r^2} + (1 - \nu^2)\,\frac{\rho\omega^2 r}{E} = 0, \qquad \text{(ii)}$$

where ρ is the density of the material and ω the angular velocity of rotation about the axis of symmetry. Confirm that the displacement field

$$u_r = Ar + \frac{B}{r} - (1 - \nu^2)\,\frac{\rho\omega^2 r^3}{8E}, \qquad \text{(iii)}$$

satisfies the above equation for arbitrary constants A and B. By substituting (iii) into (i) and imposing the condition that σ_r is zero at $r = a$ and finite at $r = 0$, show that a solution for the stresses in a solid rotating disc of radius a is

$$\sigma_r = (3 + \nu)\,\frac{\rho\omega^2}{8}\,(a^2 - r^2)$$

and

$$\sigma_\theta = \frac{\rho\omega^2}{8}\left[(3 + \nu)\,a^2 - (1 + 3\nu)r^2\right].$$

17. The rotating body of problem 16 is subject to a temperature rise $T(r)^\circ$. Show that equation (ii) becomes

$$\frac{\partial^2 u_r}{\partial r^2}+\frac{1}{r}\frac{\partial u_r}{\partial r}-\frac{u_r}{r^2}-\alpha(1-\nu)\frac{\mathrm{d}T}{\mathrm{d}r}+(1-\nu^2)\frac{\rho\omega^2 r}{E}=0.$$

Solve to obtain expressions for the stress distribution in a solid rotating disc of radius a, in which the temperature varies linearly (with r) from a value T_c at the centre to T_r at the rim.

18. A prismatic bar undergoes a bending deformation of the type shown in Fig. 2.13(b). The deformation results from application of the forces and moments shown in Fig. 2.13(a) and also from a temperature change $T(x, z)^\circ$. Show that the axial stress in the beam (expression 2.33) must be modified, in the presence of the temperature change, to

$$\sigma_x=\sigma^0+\sigma^\mathrm{b}+\sigma^\mathrm{T}$$

where $\sigma^\mathrm{T}=-E\alpha T(x, z)$, and α denotes the coefficient of thermal expansion. Show also that equation 2.36 becomes

$$\frac{\mathrm{d}^2 w_0}{\mathrm{d}x^2}=-\frac{(M+M_\mathrm{T})}{EI} \tag{i}$$

where M is the applied moment and M_T is the 'thermal' moment given by

$$M_\mathrm{T}=\alpha E\int_A zT\,\mathrm{d}A.$$

19. A simply supported beam of length L has a rectangular cross-section of depth h and width b. It experiences a temperature distribution varying linearly with depth from zero at the lower surface to T° at the upper surface. Show that the thermal moment M_T (see problem 18) is

$$M_\mathrm{T}=-\alpha\frac{EIT}{h}$$

where $I=h^3b/12$.

If the entire beam is initially at ambient temperature and carries no load, is then loaded with a uniformly distributed load of magnitude w per unit length and at the same time the upper surface is heated to a temperature T°, determine the resulting central deflection in terms of α, w, E, I, L, and T (assume that the bottom surface remains at ambient temperature and that a linear variation of temperature exists from the lower to upper surface).

20. A flat plate is of sandwich construction with a layer of material of thickness t_A, Young's modulus E_A and Poisson's ratio ν_A bonded between two

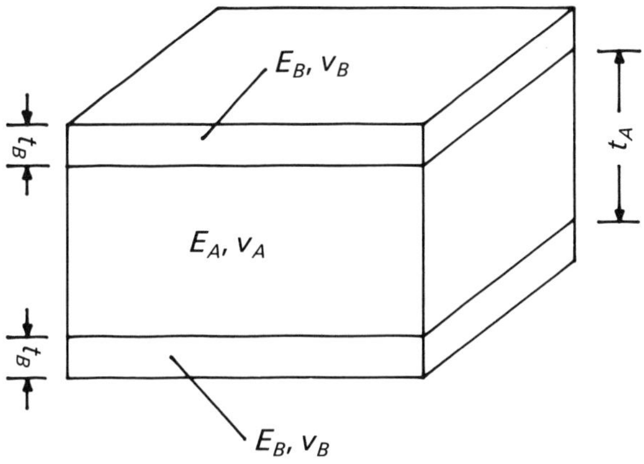

Problem 20.

layers of material of thickness t_B, Young's modulus E_B and Poisson's ratio ν_B (see figure). The assumptions made regarding the deformation of the plate are the same as those made in the analysis of an isotropic plate (see equation 2.37). Show that the moment–curvature relationship (equation 2.45) is unaltered if an 'effective' flexural rigidity, D, and an 'effective' Poisson's ratio, ν, are used, where

$$D = D_A + D_B \text{ and } \nu = \frac{D_A\nu_A + D_B\nu_B}{D_A + D_B}$$

and where D_A and D_B are given by

$$D_A = \frac{E_A t_A^3}{12\,(1-\nu_A)^2} \quad \text{and} \quad D_B = E_B\,\frac{[(t_A + 2t_B)^3 - (t_A)^3]}{12\,(1-\nu_B^2)} \,.$$

3

Energy formulations

3.1 INTRODUCTION

The calculation of stresses and displacements in an elastic continuum was defined in the preceding chapter as the solution of a coupled set of algebraic and differential equations. The problem can also be cast in 'variational' form. Explicit solutions of the equilibrium, compatibility and constitutive equations are then replaced by an assertion that the 'correct' distribution of stress or displacement is that which minimizes the energy of the system. The problem then becomes one of constrained optimization and lends itself to approximate solution. There are a number of variational statements which can be used. They differ in their definition of an 'energy functional' and in the parameters which are permitted to vary during its minimization. The two principal approaches form the bases of the 'displacement' and 'force' methods of analysis.

In the 'displacement method', the displaced state of the system is permitted to vary, subject only to the restriction that external kinematic constraints and internal compatibility requirements are satisfied. The associated state of stress does not, in general, satisfy equilibrium. The correct state both of stress and deformation (the one which *does* satisfy equilibrium) is then determined by minimizing an integral quantity termed the 'total energy' of the system. The minimization is performed with respect to compatible displacement configurations.

In the 'force method', no restriction is placed upon the displaced state of the system. Constraints are, however, imposed on stresses and internal forces to ensure that equilibrium is identically satisfied at all points. The 'correct' state of stress and deflection (that which also satisfies compatibility) is then obtained by minimizing a quantity termed the 'complementary energy' of the system. This minimization is performed with respect to variations of permissible (equilibrating) stress and redundant force states.

Displacements therefore form the unknown parameters in solutions based on the displacement approach, whereas stresses or internal forces perform this role in the force approach. A comprehensive discussion of both methods, their relationship to the continuum equations (through the methodology of variational calculus) and their relationship to a number of more general variational statements lies well beyond the scope of the current treatment but is to be found in Washizu's classic text [1] and also, in a more concise form, in the final chapter of Sokolnikoff's treatise on the theory of elasticity [2].

It is the displacement approach which forms a convenient theoretical basis for much finite element analysis. Paradoxically, it is the force method which is generally of more use in manual calculations, and in the years preceding the widespread use of digital computers, it was the force method which was generally favoured for matrix structural analysis. Even in the late sixties when the first versions of many of today's finite element programs were being commissioned, there was serious debate as to whether the future of computational methods lay in the force or displacement directions. Events since then have unequivocally favoured the latter, certainly in the more straightforward applications with which we are concerned here. It is with the displacement approach that the reminder of this chapter is exclusively concerned.

3.2 VARIATIONAL APPROACH, A SIMPLE DEMONSTRATION

The difference between the 'equilibrium solution' of an elastic problem (the explicit solution of the equilibrium, compatibility and constitutive equations) and an equivalent 'variational formulation' is illustrated by the following example. A bar of Young's modulus E, length L and cross-sectional area A, is fixed at one end and extends a distance δ at the other under the action an axial force F (Fig. 3.1). The state of stress in the bar is uniaxial, that is σ_x is the only non-zero component, and the axial displacement, $u(x)$, is uniform across each cross-section. If we approach this problem from a traditional standpoint, the calculation of stress and deflection reduces to an explicit solution of the equilibrium, compatibility and stress–strain equations. These find mathematical expression as follows:

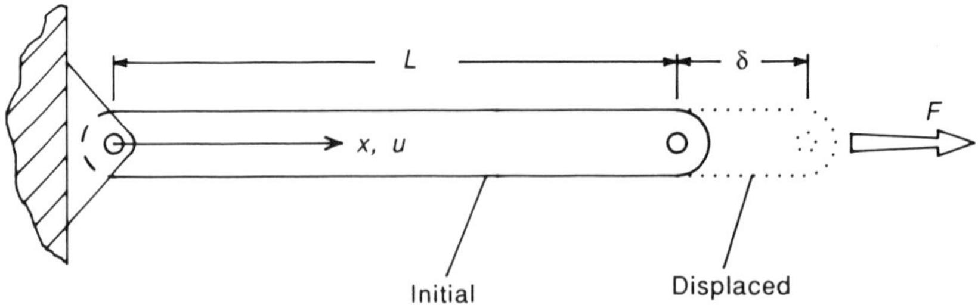

Fig. 3.1 *Initial and displaced state of a prismatic bar.*

1. *Equilibrium*
   Stress equilibrium within the bar:

$$\frac{d\sigma_x}{dx} = 0, 0 < x < L. \tag{3.1}$$

Boundary equilibrium at the free end:

$$A\sigma_x = F, \text{ at } x = L. \tag{3.2}$$

2. *Strain–displacement/compatibility*
Strain–displacement within the bar:

$$e_x = \frac{du}{dx}, 0 < x < L. \tag{3.3}$$

Compatibility within the bar:

$$u(x) \text{ continuous}, 0 < x < L. \tag{3.4a}$$

Compatibility at the fixed end:

$$u(0) = 0. \tag{3.4b}$$

3. *Stress–strain*
Hooke's law (uniaxial):

$$\sigma_x = Ee_x. \tag{3.5}$$

The solution is relatively straightforward. Integration of equation 3.1 yields the information that σ_x is a constant, its value being determined by the boundary condition in equation 3.2. That is

$$\sigma_x = F/A.$$

Equations 3.3 and 3.5 then yield

$$\frac{du}{dx} = \frac{\sigma_x}{E} = \frac{F}{EA},$$

which can be integrated to give

$$u(x) = \left(\frac{F}{EA}\right)x + C,$$

The constant of integration, C, is zero however by virtue of the compatibility boundary condition 3.4b. The final solution for $u(x)$ is therefore

$$u(x) = \left(\frac{F}{EA}\right)x.$$

The displacement at the right hand, δ say, is then

$$\delta = u(L) = \frac{FL}{EA},$$

or, rewriting in terms of an axial stiffness $k(=EA/L)$

$$\delta = \frac{F}{k}. \tag{3.6}$$

The steps involved in reaching this final relationship between load and displacement are presented here in rather more detail than they deserve to emphasize the basic structure of the traditional 'equilibrium' solution. They demonstrate that all three components of the original problem (equilibrium, compatibility and the constitutive relationship) play a part in, and are satisfied by, the final solution.

A 'displacement variational solution' of the same problem proceeds as follows. First a quantity termed the 'total energy' (or 'total potential') of the system is defined. The system in this case comprises both the bar *and* the external load. Its total energy is formed from two contributions; the energy stored in the elastic deformation of the bar, and the energy associated with the capacity of the external load to do work. The first of these is termed 'strain energy' and is denoted by U. The second is termed 'potential energy' and is denoted by V. The total energy, χ, is therefore

$$\chi = U + V. \tag{3.7a}$$

In the present instance the strain energy of the bar and the potential energy of the load are, $\frac{1}{2}k\delta^2$, and $-F\delta$, respectively. The manner in which these quantities are derived will be discussed later in this chapter and need not concern us here. If we assume their validity for the purposes of the present demonstration, equation 3.7a becomes

$$\chi = \tfrac{1}{2}k\delta^2 - F\delta. \tag{3.7b}$$

A variational principle which then applies, not only to this problem but to *all* elastic systems, states that among 'permissible' states of displacement, the 'correct' one is that which yields a stationary value of the total energy. Discussion of the precise meaning of 'permissible', the conditions under which the principal holds and proof of its validity, are deferred until later in this chapter. For the time being however, if we accept the principle as it stands, we can obtain a solution to the current problem by locating the displaced states (in this case values of δ) which give a stationary value of χ. These are readily identified if we sketch χ as a function of δ. This gives the parabola of Fig. 3.2 which has a stationary value (a minimum) at the point P. The value of δ at this point is obtained by imposing the condition that the slope of the curve is zero. In other words,

$$\frac{d\chi}{d\delta} = k\delta - F = 0,$$

which gives

$$\delta = \frac{F}{k}.$$

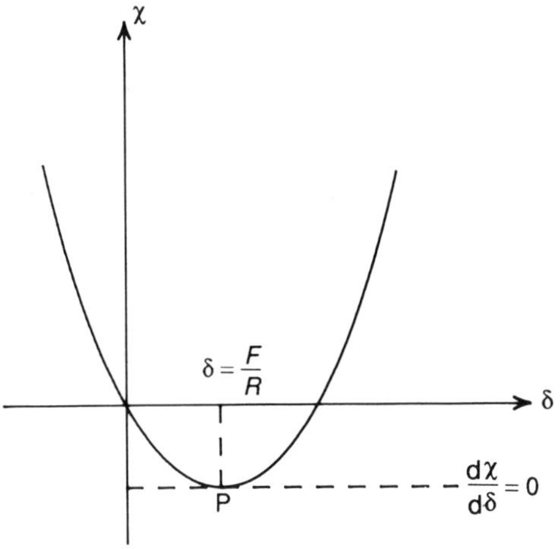

Fig. 3.2 *Total energy of a prismatic bar as a function of extension.*

This is indeed the 'correct' value of δ as predicted by the traditional equilibrium solution (*cf.* equation 3.6).

An important feature of the variational procedure, apparent even in this simple example, is its ability to produce a solution without direct reference to (or integration of) the differential equations of equilibrium. Location of the stationary value of the energy of the system, replaces, in effect, the equilibrium component of the original problem.

Consider now a modification to this example whereby the uniform bar is replaced by a stepped bar with submembers (a) and (b) of lengths L_a and L_b, areas A_a and A_b, and Young's moduli E_a and E_b (Fig. 3.3). A load F is again applied at the free end. The displaced state of the system is now defined by two parameters, δ_1 and δ_2, and the total energy may be written

$$\chi = U_a + U_b + V,$$

where U_a and U_b are the strain energies of the individual submembers. V is again the potential energy of the axial load F. Using the same expression as

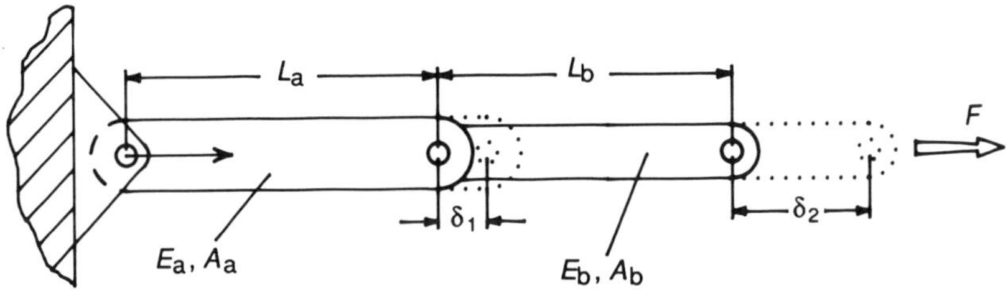

Fig. 3.3 *Initial and displaced state of a stepped bar.*

before for the strain energy of a single bar ($U = \frac{1}{2}k\delta^2$) and noting that the extension of bar (a) is δ_1, and that of bar (b) $\delta_2 - \delta_1$, we obtain

$$\chi = \tfrac{1}{2}k_a\delta_1^2 + \tfrac{1}{2}k_b(\delta_2 - \delta_1)^2 - F\delta_2,$$

where $k_a(= E_aA_a/L_a)$ and $k_b(= E_bA_b/L_b)$ are the axial stiffnesses of submembers (a) and (b), respectively. When plotted against δ_1 and δ_2, χ now forms the paraboloid surface shown in Fig. 3.4. By applying the same variational principle as in the previous example, we can obtain the 'correct' solution by locating the point on this surface where χ has a stationary value, that is, the point Q where it is a minimum. The values of δ_1 and δ_2 at this point are determined from the condition that the slopes $\partial\chi/\partial\delta_1$ and $\partial\chi/\partial\delta_2$ are zero. In other words, they are given by the solution of the equations

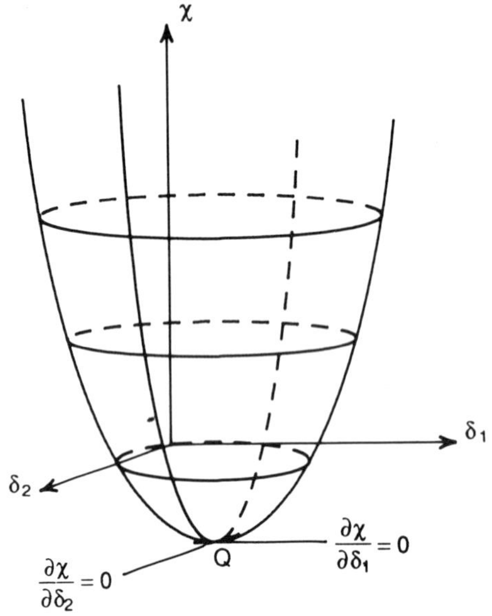

Fig. 3.4 *Total energy of a stepped bar as a function of extension.*

$$\frac{\partial\chi}{\partial\delta_1} = k_a\delta_1 - k_b(\delta_2 - \delta_1) = 0, \tag{3.8a}$$

and

$$\frac{\delta\chi}{\partial\delta_2} = k_b(\delta_2 - \delta_1) - F = 0. \tag{3.8b}$$

This gives

$$\delta_1 = \frac{F}{k_a}$$

and

$$\delta_2 = \frac{F}{k_a} + \frac{F}{k_b}.$$

As in the first example, these values are indeed 'correct' in that they correspond to those obtained from a full solution of the equilibrium, compatibility and constitutive equations in each submember (proof of this statement is left to the reader as an exercise).

We are given an intimation of the direction in which the variational approach is leading us if we rearrange equations 3.8 slightly to give

$$(k_a + k_b)\delta_1 + (-k_b)\delta_2 = 0$$

and

$$(-k_b)\delta_1 + (k_b)\delta_2 = F,$$

and then rewrite them in matrix form as

$$\begin{bmatrix} k_a + k_b & -k_b \\ -k_b & k_b \end{bmatrix} \begin{bmatrix} \delta_1 \\ \delta_2 \end{bmatrix} = \begin{bmatrix} 0 \\ F \end{bmatrix}. \tag{3.9}$$

The matrix equation produced in this way is identical in character to the matrix stiffness relationship discussed in Chapter 1 (see equation 1.6). The displacement energy approach has, in effect, assigned specific values to the components of a stiffness matrix for this simple system, bypassing in the process explicit consideration of the equilibrium equations. This falls somewhat short of the 'automatic' generation of such terms promised by the finite element method, but is sufficiently encouraging to warrant closer study of the variational principle which has produced it.

3.3 DISPLACEMENT ENERGY PRINCIPLES

The energy principles which are discussed in the remainder of this chapter can be stated in a multiplicity of ways. They are derived here in their most common forms and only in the context of 'displacement' (rather than 'force') methods of analysis. Proof of their validity may be approached along two distinct paths. The first starts with the equations of equilibrium and establishes their equivalence to a variational statement by means of formal mathematical proof. This is certainly the most rigorous approach but requires some familiarity with the calculus of variations. The second approach starts one step further back and derives the variational statement from the same physical arguments which produced the equilibrium equations in the first place. This latter approach is adopted here, variational statements being derived from physical rather than mathematical considerations. A somewhat fuller treatment based on the same general philosophy is to be found in reference [3]. Readers who wish to proceed along the more mathematical path should consult references [1] and [2]. Indeed, those wishing to come to terms with the mathematics involved may also wish to study the more general methodology of the calculus of variations which establishes the relationship between a variational functionals and an equivalent set of ordinary or partial differential equations [4, 5]. A useful review of many of these concepts in the specific context of finite element application is to be found in reference [6].

3.3.1 Principle of virtual work

The most general energy principle is that of 'virtual work'. When stated for a single particle it appears quite trivial. When extended to collections of particles, however, it provides the basis for much useful analysis.

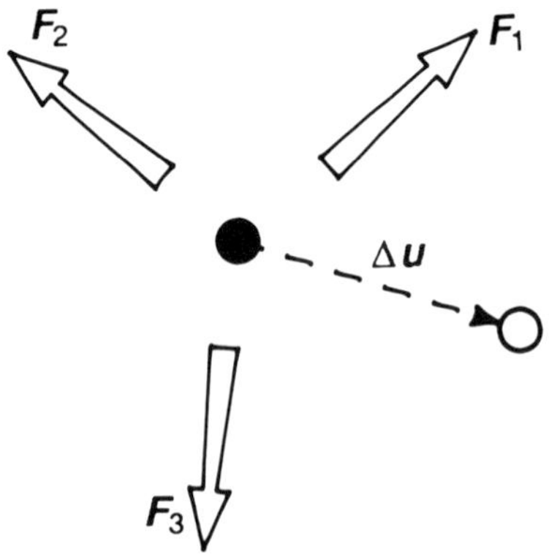

Fig. 3.5 *Virtual displacement of a particle.*

Consider first a single particle subject to loads $\boldsymbol{F}_1, \boldsymbol{F}_2, \ldots, \boldsymbol{F}_n$ (Fig. 3.5). Provided that the particle is in static equilibrium, the vector sum of the applied forces equates to zero, giving

$$\boldsymbol{F}_1 + \boldsymbol{F}_2 + \ldots + \boldsymbol{F}_n = 0. \tag{3.10}$$

Suppose now that the particle displaces through a distance $\Delta \boldsymbol{u}$ in such a way that the forces $\boldsymbol{F}_1, \boldsymbol{F}_2, \ldots \boldsymbol{F}_n$ are unaltered (and this will certainly be so if $\Delta \boldsymbol{u}$ is infinitesimally small). The work ΔW which is done by the forces is then

$$\begin{aligned}\Delta W &= \boldsymbol{F}_1 \,.\, \Delta \boldsymbol{u} + \boldsymbol{F}_2 \,.\, \Delta \boldsymbol{u} + \ldots + \boldsymbol{F}_n \,.\, \Delta \boldsymbol{u},\\ &= (\boldsymbol{F}_1 + \boldsymbol{F}_2 + \ldots + \boldsymbol{F}_n) \,.\, \Delta \boldsymbol{u}.\end{aligned}$$

From equation 3.10 however, the vector sum $(\boldsymbol{F}_1 + \boldsymbol{F}_2 + \ldots + \boldsymbol{F}_n)$ is identically zero and the above expression reduces to

$$\Delta W = 0. \tag{3.11}$$

A displacement which does not alter the forces acting on the particle is termed a 'virtual displacement', and equation 3.11 becomes the **principle of virtual work**. Formally, the principle is stated as follows:

> *A necessary and sufficient condition for the equilibrium of a particle is that the work done by the forces on the particle is zero for any virtual displacement.*

Consider next a system of n particles each of which exerts an internal force on every other. The force exerted by the jth particle on the ith particle is denoted by $\boldsymbol{F}_{ij}$. External loads are also applied, the resultant external force acting on the ith particle being denoted by $\boldsymbol{P}_i$. Such a system is illustrated in Fig. 3.6. The principle of virtual work may be applied to each particle in turn. In other words, the virtual work ΔW_i done on the ith particle as it displaces through a virtual displacement $\Delta \boldsymbol{u}_i$, is zero, giving

$$\begin{aligned} \Delta W_i &= \sum_{j=1}^{n} \boldsymbol{F}_{ij} \cdot \Delta \boldsymbol{u}_i + \boldsymbol{P}_i \cdot \Delta \boldsymbol{u}_i, \\ &= \Delta W_i^{\text{int}} + \Delta W_i^{\text{ext}} = 0, \end{aligned} \tag{3.12}$$

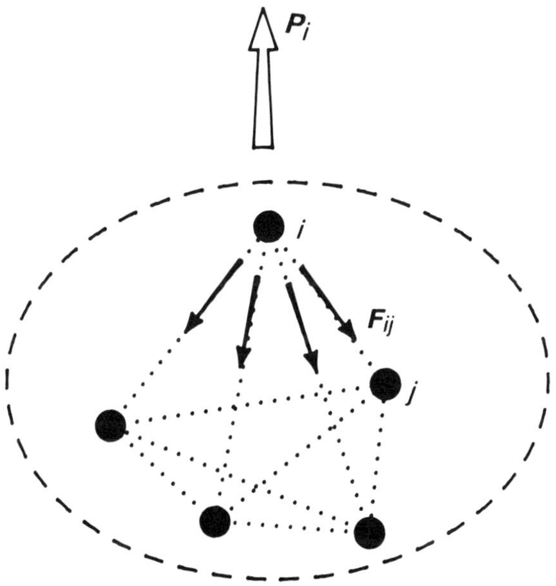

Fig. 3.6 *Equilibrium of a system of particles.*

where ΔW_i^{int} is the work done on the ith particle by the *internal* forces and ΔW_i^{ext} is the work done by the *external* force. The summation of equations 3.12 for all particles then gives

$$\Delta W = \sum_{i=1}^{n} \Delta W_i = \Delta W^{\text{int}} + \Delta W^{\text{ext}} = 0, \tag{3.13a}$$

where

$$\Delta W^{\text{int}} = \sum_{i=1}^{n} \Delta W_i^{\text{int}} \tag{3.13b}$$

and

$$\Delta W^{\text{ext}} = \sum_{i=1}^{n} \Delta W_i^{\text{ext}}. \tag{3.13c}$$

Equations 3.13a together with 3.13b and c are equivalent to the statement:

> *A necessary and sufficient condition for static equilibrium of a system of particles is that the work done by the internal forces plus the work done by the external loads during any virtual displacement is zero.*

This is the principle of virtual work in its most general form. Since any solid body may be regarded as a collection of particles, the phrase 'static equilibrium of a system of particles' may be replaced by the more useful 'static equilibrium of a solid body'. This statement holds irrespective of the nature of the internal forces.

3.3.2 Principle of stationary total potential

The principle of virtual work may be restated in a number of different ways. A particularly useful form is obtained by introducing the concept of internal and external energy to describe the work done by the internal and external forces.

'Energy' is defined as the capacity of a system to do work. As work is done *by* the system, energy is lost, and as work is done *on* it, energy is gained. The 'internal energy' of a solid body (also termed the 'strain energy') is defined incrementally in the following way. First, a reference state is chosen for which the internal energy is nominally zero (this is usually taken to be a state of zero stress). As the system undergoes an infinitesimal (and hence 'virtual' in the sense already defined) deformation, the strain energy *de*crement, $-\Delta U$, is defined to be the work done by the internal forces. Reversing signs, the strain energy *in*crement, $+\Delta U$, is given by

$$\Delta U = -\Delta W^{\text{int}}. \tag{3.14}$$

By using equation 3.13, which permits the substitution of ΔW^{ext} for $-\Delta W^{\text{int}}$, this can be rewritten

$$\Delta U = \Delta W^{\text{ext}}. \tag{3.15}$$

Equation 3.15 is then equivalent to the statement:

A necessary and sufficient condition for static equilibrium of a system of particles or a continuum is that the work done by the external loads during any virtual displacement is equal to the increase in internal energy of the system.

Turning now to the external loads, a 'potential energy' V can be defined in a similar way. That is to say, a *de*crement of energy, $-\Delta V$, is defined as the work done by the external forces during an incremental virtual displacement. In other words, using the notation of equation 3.13,

$$-\Delta V = \Delta W^{\text{ext}}. \tag{3.16a}$$

By eliminating ΔW^{ext} between equations 3.15 and 3.16, we then obtain

$$\Delta U = -\Delta V \quad \text{or} \quad \Delta(U+V) = 0. \tag{3.16b,c}$$

The quantity $(U+V)$ is termed the 'total energy' or 'total potential' of the system and will be denoted by the symbol χ. Equation 3.16 is therefore equivalent to the statement:

For a system in static equilibrium the increment in the total energy during any virtual displacement is zero.

A 'conservative system of forces' is one in which no energy is dissipated. In other words, it is a system in which all work done *on* the system is subsequently recovered as work done *by* it. Elastic forces are conservative by definition since elastic deformations are intrinsically reversible. No internal energy is therefore 'lost' when an elastic body is strained and subsequently returns to its initial state. External loads of fixed magnitude and direction are also conservative, since they do no net work as they move around closed paths returning to their original positions. An elastic body subject to fixed external loads therefore constitutes a conservative system of forces. The work done on and by such a system is dependent only on the current displacement configuration and not on the loading path which led to it. The internal and external energies of such a system are consequently single-valued 'functionals' of the current displaced state, a funct*ional*, as opposed to a funct*ion*, being a function of other functions, the displacement field in this case. Equation 3.16 can therefore be regarded as a statement that the single-valued, total energy of a conservative system neither increases nor decreases during any virtual displacement. This certainly includes infinitesmal displacements. In other words, the equilibrium state of such a system corresponds to a stationary value of the total energy. This is stated formally as the 'principal of stationary total potential' (or energy):

Of all compatible displacement states of a conservative system, those that satisfy the equations of equilibrium make the total energy stationary with respect to small variations of displacement.

It is not difficult to show that if the stationary value is a minimum, which is usually the case, the equilibrium state is stable. The principle is, for this reason, often referred to as the 'principle of minimum total potential'.

The principle of stationary/minimum total potential, as outlined above, forms a suitable basis for the development of finite element models for elastic systems. Before we can apply it to real problems, however, we must be able to calculate the total energy of an elastic system as a function of its displaced state.

3.4 STRAIN ENERGY OF AN ELASTIC CONTINUUM

The strain energy of a deformable body, as defined in the previous section, is the energy stored during the process of deformation. The calculation of this quantity for a complex elastic body is achieved by integrating the strain energy density (strain energy per unit volume) over the volume of the body. With this in mind, expressions are now derived for the strain energy of a small differential element, of side δx, δy and δz (Fig. 3.7) for some simple load states.

3.4.1 Strain energy due to direct stresses

Consider first the case of a normal stress σ acting in the x direction (Fig. 3.7). The resulting strain in the x direction is denoted by e. If both quantities are increased by small ammounts $\Delta\sigma$ and Δe, the work done by the stress on the face ABCD as it moves through an incremental distance $(\Delta e \delta x)$ is, to a first approximation, $(\sigma \delta y \delta z)(\Delta e \delta x)$. This quantity of work done *on* the element is equal to the increment in strain energy by virtue of equation 3.15. Note that the 'external' forces on the element in this case are the stresses applied to its surface by the rest of the continuum. Equation 3.15 therefore gives

$$\Delta U = (\sigma \Delta e)\, \delta x \delta y \delta z ,$$

or

$$\Delta U = (\sigma \Delta e)\, \delta V ,$$

where δV is the volume of the element. If the body deforms from an initial state of zero strain to a final value e', the above quantity can be integrated to give the strain energy per unit volume (*SE*/*vol*) in the final state of deformation as

$$SE/vol = \int_0^{e'} \sigma \, \mathrm{d}e . \tag{3.17a}$$

This quantity is uniquely defined for a given material by the area under the stress–strain curve, that is, by the shaded region of Fig. 3.8(a). In the case of

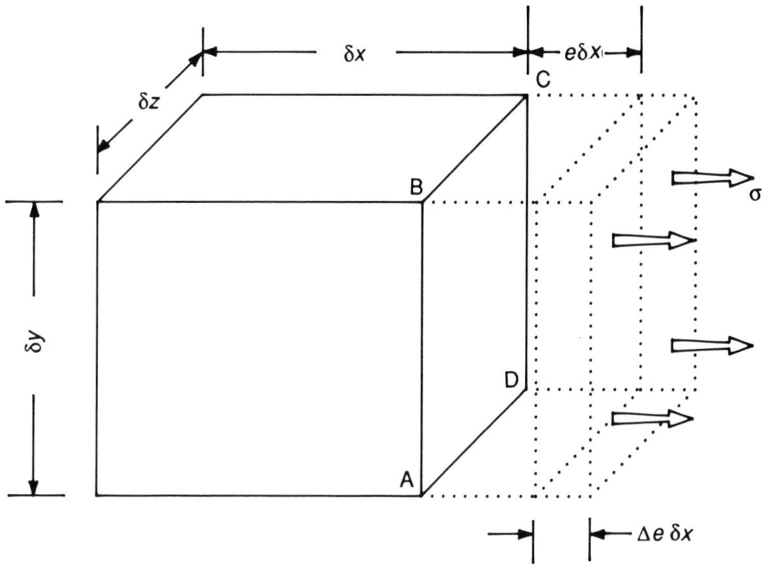

Fig. 3.7 *Incremental deformation of a differential element of elastic material subject to normal stress.*

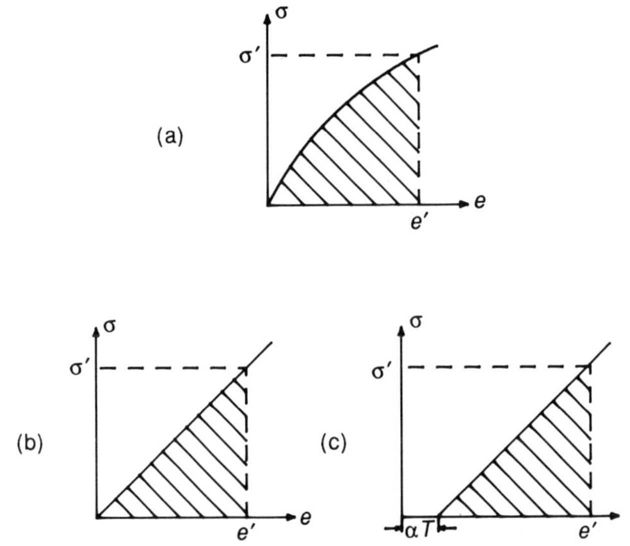

Fig. 3.8 *The strain energy integral for uniaxial stress. (a) general case, (b) linearly elastic (constant temperature), (c) linearly elastic (temperature rise T°).*

a linearly elastic material with no thermal strains, it becomes the shaded triangle of Fig. 3.8(b). If the prime (superscript ′) is now removed, so that σ and e denote the final values of stress and strain, the linearly elastic version of equation 3.17a is therefore

$$SE/vol = \tfrac{1}{2}\,\sigma e. \tag{3.17b}$$

This must be modified when thermal gradients are present. In the case of a temperature rise T, the material experiences an initial thermal strain $e^T = \alpha T$, and the stress–strain curve is then that of Fig. 3.8(c). Equation 3.17 is still valid, however, and the strain energy density is once again the shaded area under the curve, that is,

$$SE/vol = \tfrac{1}{2}\,\sigma(e - e^T). \tag{3.17c}$$

3.4.2 Strain energy in shear

The strain energy of a differential element in shear can be deduced using similar arguments. In a state of pure shear, the work done during an incremental deformation is

$$\Delta U = (\tau\,\Delta\gamma)\,\delta V \tag{3.18}$$

where τ, γ and $\Delta\gamma$ are the shear stress, shear strain and incremental shear strain, respectively. The summation of such contributions throughout the deformation process gives

$$SE/vol = \int_0^{\gamma'} \tau\,d\gamma, \tag{3.19a}$$

where γ' denotes the final value of shearing strain. In the linearly elastic case, the τ–γ curve is a straight line and this integral reduces to

$$SE/vol = \tfrac{1}{2}\,\tau\gamma. \tag{3.19b}$$

Here the prime (superscript ′) has again been removed so that τ and γ denote final values of stress and strain. No adjustment is required for thermal effects since these do not contribute to shearing deformation in an isotropic material.

3.4.3 Strain energy of a compound stress state

Expressions 3.17 and 3.19 give the strain energy per unit volume of an elastic material subject to direct stress and pure shear. They have been derived separately but can be applied to situations when both components act simultaneously simply by summing the two contributions. This procedure can be repeated for any number of orthogonal components. Thus, in the most general case of a stress state defined by orthogonal components $\sigma_1, \sigma_2, \sigma_3, \tau_{12}, \tau_{13}$, and τ_{23}, and corresponding strain components $e_1, e_2, \ldots$, expression 3.17a is replaced by

$$SE/vol = \int (\sigma_1 \mathrm{d}e_1 + \sigma_2 \mathrm{d}e_2 + \sigma_3 \mathrm{d}e_3 + \tau_{12}\mathrm{d}\gamma_{12} + \tau_{13}\mathrm{d}\gamma_{13} + \tau_{23}\mathrm{d}\gamma_{23}). \quad (3.20)$$

Moreover, if the material is linearly elastic and experiences a temperature rise T, this becomes

$$SE/vol = \tfrac{1}{2}\,(\sigma_1\,(e_1 - e_1^T) + \sigma_2\,(e_2 - e_2^T) + \sigma_3\,(e_3 - e_3^T) + \tau_{12}\gamma_{12} + \tau_{13}\gamma_{13} + \tau_{23}\gamma_{23}) \quad (3.21)$$

When written in the above form, the strain energy density appears to depend both upon the stresses ($\sigma_1, \sigma_2 \ldots$) and the strains ($e_1, e_2 \ldots$). In reality, we can always express the stresses in terms of the strains (or vice versa) using Hooke's law, and can therefore eliminate one or the other of these quantities from equation 3.21. This leaves an expression for the strain energy density which is either a function of stress, or a function of strain. As an example of an expression for the strain energy of a body in terms of strain only, we can now validate the formula used in section 3.2 for the strain energy of a uniform bar. The bar is of length L, area A, Young's modulus E, is fixed at one end and undergoes an axial displacement δ at the other (Fig. 3.1). The axial strain e, assumed uniform, is given by

$$e = \frac{\delta}{L}.$$

and from Hooke's law, the axial stress is

$$\sigma = Ee = E\frac{\delta}{L}.$$

This can be used to eliminate σ from expression 3.17b, giving

$$SE/vol = \tfrac{1}{2}\,\sigma e = \frac{1}{2}\left(E\frac{\delta}{L}\right)\left(\frac{\delta}{L}\right) = \frac{1}{2}E\left(\frac{\delta}{L}\right)^2.$$

Integrating this (constant) quantity over the volume of the bar, we finally obtain

$$U = (SE/vol) \times (vol.\ of\ bar) = \frac{1}{2}E\left(\frac{\delta}{L}\right)^2 \times (AL)$$

$$= \tfrac{1}{2}\,k\delta^2, \quad \text{where} \quad k = \frac{EA}{L},$$

confirming the expression used already in equation 3.7b.

3.5 STRAIN ENERGY OF BEAMS AND PLATES

The integration of strain energy over the volume of an elastic body is simplified in the case of a beam or a plate by our prior knowledge that stress and strain vary linearly from the upper to the lower surface. An expression for the strain energy per unit length (in the case of a beam) or the strain energy per unit area (in the case of a plate) can therefore be obtained by integrating the strain energy per unit volume explicitly over the area or thickness.

3.5.1 Strain energy of a beam

Consider first the case of a beam subject to an axial load P and bending moment M both of which act in the x–z plane. This corresponds to the configuration of Fig. 2.13(a). The axial stress and axial strain are then given by expressions 2.29 and 2.35, that is

$$e_x = \frac{\mathrm{d}u_0}{\mathrm{d}x} - z\left(\frac{\mathrm{d}^2 w_0}{\mathrm{d}x^2}\right) \tag{3.22a}$$

and

$$\sigma_x = \frac{P}{A} + z\left(\frac{M}{I}\right) \tag{3.22b}$$

Since these are the only non-zero components, the strain energy density is

$$SE/vol = \tfrac{1}{2}\,\sigma_x e_x = \frac{1}{2}\left[\frac{P}{A} + z\left(\frac{M}{I}\right)\right]\left[\frac{\mathrm{d}u_0}{\mathrm{d}x} - z\left(\frac{\mathrm{d}^2 w_0}{\mathrm{d}x^2}\right)\right].$$

By expanding this expression as a quadratic function in z and integrating over the cross-section, we obtain

$$SE/length = \frac{1}{2}\left\{\frac{P}{A}\frac{\mathrm{d}u_0}{\mathrm{d}x}\int_A \mathrm{d}A + \left[\frac{M}{I}\frac{\mathrm{d}u_0}{\mathrm{d}x} - \frac{P}{A}\frac{\mathrm{d}^2 w_0}{\mathrm{d}x^2}\right]\int_A z\,\mathrm{d}A - \left[\frac{M}{I}\frac{\mathrm{d}^2 w_0}{\mathrm{d}x^2}\right]\int_A z^2\,\mathrm{d}A\right\},$$

which upon substitution of the identities

$$\int_A \mathrm{d}A = A, \quad \int_A z\,\mathrm{d}A = 0 \text{ and } \int_A z^2\,\mathrm{d}A = I$$

reduces to

$$SE/length = \frac{1}{2}\left[P\left(\frac{\mathrm{d}u_0}{\mathrm{d}x}\right) + M\left(-\frac{\mathrm{d}^2 w_0}{\mathrm{d}x^2}\right)\right]. \tag{3.23}$$

This can be written in a number of different forms, since the axial force P and bending moment M can be expressed in terms of u_0 and w_0 (or vice versa) through the stress–strain and moment–curvature relationships (equations 2.34 and 2.36). Eliminating P and M in favour of u_0 and w_0, for example, gives

$$SE/length = \frac{1}{2}\left[EA\left(\frac{\mathrm{d}u_0}{\mathrm{d}x}\right)^2 + EI\left(\frac{\mathrm{d}^2w_0}{\mathrm{d}x^2}\right)^2\right]. \tag{3.24}$$

Conversely, eliminating $\mathrm{d}u_0/\mathrm{d}x$ and $\mathrm{d}^2w_0/\mathrm{d}x^2$ in favour of P and M, we obtain

$$SE/length = \frac{1}{2}\left[\frac{P^2}{EA} + \frac{M^2}{EI}\right] \tag{3.25}$$

A third variant is derived by reintroducing the quantities σ^0 and e^0 used in section 2.5 to denote centroidal stress and strain (see equations 2.30 and 2.33). Equation 3.23 then becomes

$$SE/length = \frac{1}{2}\left[A\sigma^0 e^0 + M\left(-\frac{\mathrm{d}^2w_0}{\mathrm{d}x^2}\right)\right]. \tag{3.26}$$

This is a useful precursor of an analogous, but more complex, expression which will be derived shortly for the strain energy of a plate.

3.5.2 Strain energy of structural members

Expressions 3.23 and 3.24 can easily be extended to prismatic members which are simultaneously bent about two principal axes. Moments M_y and M_z then act at any cross-section, as shown in Fig. 3.9, and expression 3.23 must be amended to include the additional bending contribution. It becomes

$$SE/length = \frac{1}{2}\left[P\frac{\mathrm{d}u_0}{\mathrm{d}x} + M_y\left(-\frac{\mathrm{d}^2w_0}{\mathrm{d}x^2}\right) + M_z\left(-\frac{\mathrm{d}^2v_0}{\mathrm{d}x^2}\right)\right], \tag{3.27}$$

where $v_0(x)$ is the centroidal deflection in the y direction. The appropriate modification of expression 3.24 is

$$SE/length = \frac{1}{2}\left[EA\left(\frac{\mathrm{d}u_0}{\mathrm{d}x}\right)^2 + EI_y\left(\frac{\mathrm{d}^2w_0}{\mathrm{d}x^2}\right)^2 + EI_z\left(\frac{\mathrm{d}^2v_0}{\mathrm{d}x^2}\right)^2\right], \tag{3.28}$$

where I_y and I_z are the second moments of area about the y and z-axes, respectively. For completeness, the effect of a torsional moment T acting about the centroidal axis will also be included. This is accomodated by the addition of a term $\frac{1}{2}T(\mathrm{d}\phi/\mathrm{d}x)$ in expression 3.27 or $\frac{1}{2}(C)(\mathrm{d}\phi/\mathrm{d}x)^2$ in expression 3.28,

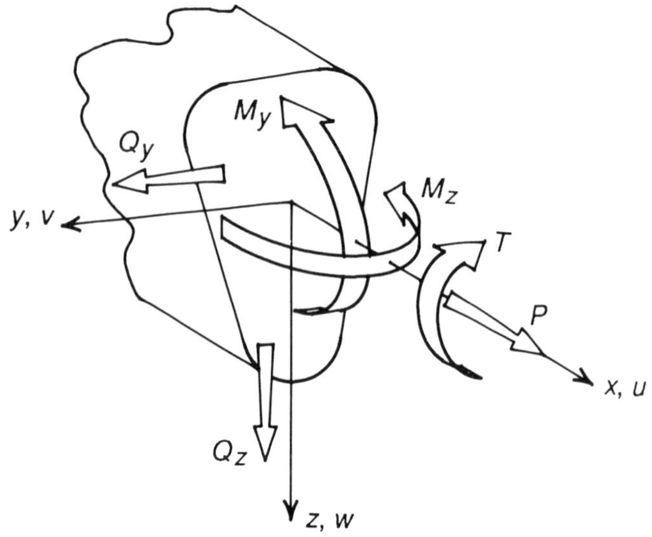

Fig. 3.9 *Forces and moments acting on a structural member.*

where $\varphi(x)$ is the angle of twist and C is the torsional rigidity of the member. In the case of a circular section, C is simply equal to the product of the shear modulus G and the polar moment of area J. For noncircular sections, it must be determined from a more complex analysis (see Chapter 6 of [7]).

Our expressions for the strain energy per unit length include as yet no contribution from the bending shear stresses associated with shear forces Q_y and Q_z which are inevitably present unless M_y and M_z are constant (that is, unless *pure* bending occurs in both planes). Their omission is, in fact, consistent with our neglect of shear deformation in beam theory itself (refer to the comments at the end of section 2.5.1). In most instances, this is quite acceptable, since the overall effect of such terms is insignificant unless the beam is very short. These effects can, however, be included if required. The consequent modifications to the strain energy expressions are discussed further in chapter nine when Timoshenko's theory is introduced.

3.5.3 Strain energy of plates

The strain energy per unit area of a plate is obtained in much the same way as that of a beam, by integrating the strain energy per unit volume through the thickness of the plate. The stress state within the plate is defined by the components σ_x, σ_y and τ_{xy} (the x–y plane forms the centroidal plane of the plate, see Fig. 2.14), and the strain energy per unit area is obtained by integrating expression 3.21 with respect to z. This gives

$$SE/area = \int_{-t/2}^{t/2} \tfrac{1}{2} \{ \sigma_x e_x + \sigma_y e_y + \tau_{xy}\gamma_{xy} \} \, dz. \tag{3.29}$$

The stress and strain components occuring in the integrand are formed from membrane and bending contributions as detailed in section 2.5.2. It is convenient to rewrite them here in a slightly modified form as

$$e_x = e_x^0 - z \frac{\partial^2 w_0}{\partial x^2} \tag{3.30a}$$

$$e_y = e_y^0 - z \frac{\partial^2 w_0}{\partial y^2} \tag{3.30b}$$

$$\gamma_{xy} = \gamma_{xy}^0 - 2z \frac{\partial^2 w_0}{\partial x \partial y} \tag{3.30c}$$

and

$$\sigma_x = \sigma_x^0 + z \frac{12M_x}{t^3} \tag{3.31a}$$

$$\sigma_y = \sigma_y^0 + z \frac{12M_y}{t^3} \tag{3.31b}$$

$$\tau_{xy} = \tau_{xy}^0 + z \frac{12M_{xy}}{t^3} \tag{3.31c}$$

where expressions 3.30 are obtained by combining elements of equations 2.38 and 2.39, and expressions 3.31 by combining elements of equations 2.40–2.46. In both cases, the terms with the zero superscript (^{0}) represent in-plane *membrane* contributions which are constant with z (the notation is the same as that used in section 2.5.2). The remaining terms represent bending contributions which vary linearly with z. Substitution of equations 3.30 and 3.31 into the integrand of equation 3.29 followed by integration with respect to z (this may be performed explicitly) then gives the following expression for the strain energy per unit area

$$SE/area = \frac{1}{2} t \{ \sigma_x^0 e_x^0 + \sigma_y^0 e_y^0 + \tau_{xy}^0 \gamma_{xy}^0 \} + \frac{1}{2} \left\{ M_x \left(-\frac{\partial^2 w_0}{\partial x^2} \right) + M_y \left(-\frac{\partial^2 w_0}{\partial y^2} \right) + 2M_{xy} \left(-\frac{\partial^2 w_0}{\partial x \partial y} \right) \right\}. \tag{3.32}$$

Note that the 'cross' terms in the original integrand, products of membrane stresses and bending strains and vice versa, disappear during the integration so that the the strain energies due to membrane and bending effects are 'uncoupled' in the final expression. This means that the first three terms on the right hand side of equation 3.32 represent the strain energy due to in-plane loads, and the remaining three terms represent the strain energy due to bending. When both effects occur simultaneously the net strain energy is obtained simply by summing the two contributions (since strain energy is a quadratic function of stress and strain, this is not in fact as obvious a result as it sounds!).

Expression 3.32 must now be integrated over area to obtain the total strain energy of a particular plate. An example of such a calculation is included later in this chapter (worked example, section 3.8.3).

3.6 POTENTIAL ENERGY OF EXTERNAL LOADS

In stating the principle of stationary total energy, the energy associated with external loads was defined by the incremental relationship

$$\Delta V = -\Delta W^{\text{ext}}. \tag{3.33}$$

In the case of a concentrated load, $\boldsymbol{P}$, acting at a point which experiences a displacement $\boldsymbol{u}$, the increase in potential energy during an incremental displacement $\Delta \boldsymbol{u}$ is

$$\Delta V = -\boldsymbol{P} \cdot \Delta \boldsymbol{u}. \tag{3.34}$$

A suitable potential function V is therefore

$$V = -\boldsymbol{P} \cdot \boldsymbol{u}, \tag{3.35}$$

this being equivalent to equation 3.34 provided that the applied load $\boldsymbol{P}$ is independent of $\boldsymbol{u}$. If the applied load *does* depend on $\boldsymbol{u}$ then a stiffness relationship of some sort exists — a sprung support, for example — and the work associated with this mechanism must be included as part of the internal energy of the system.

Expression 3.35 can clearly be generalized to include any number of loads by summing individual contributions. The potential energy associated with external loads $\boldsymbol{P}_1, \boldsymbol{P}_2 \ldots \boldsymbol{P}_n$ acting at points with displacements $\boldsymbol{u}_1, \boldsymbol{u}_2, \ldots, \boldsymbol{u}_n$ is therefore

$$V = -\boldsymbol{P}_1 \cdot \boldsymbol{u}_1 - \boldsymbol{P}_2 \cdot \boldsymbol{u}_2 \ldots - \boldsymbol{P}_n \cdot \boldsymbol{u}_n. \tag{3.36}$$

In the case of a continuously distributed load, the summation is replaced by an integral. The potential energy associated with an external traction $\boldsymbol{t}$ acting over a boundary S on which the displacement is $\boldsymbol{u}$, is therefore

$$V = -\int_S (\boldsymbol{t} \cdot \boldsymbol{u})\, \mathrm{d}S. \tag{3.37}$$

Similarly, the potential energy associated with a body force $\boldsymbol{g}$ per unit volume acting over a volume V is

$$V = -\int_V (\boldsymbol{g} \,.\, \boldsymbol{u})\, \mathrm{dV}. \tag{3.38}$$

3.7 TOTAL ENERGY FUNCTIONAL

The total energy of an elastic system can now be calculated by combining one or more of the potential energy expressions of the previous section with a strain energy expression from sections 3.4 or 3.5. In the general case of a body subject to external concentrated loads $\boldsymbol{P}_i$ ($i = 1, \ldots n$), surface loads $\boldsymbol{t}$ per unit area and body forces $\boldsymbol{g}$ per unit volume, the total energy χ is therefore

$$\chi(\boldsymbol{u}) = U(\boldsymbol{u}) + V(\boldsymbol{u}) = U(\boldsymbol{u}) - \sum_{i=1}^{n} \boldsymbol{P}_i \,.\, \boldsymbol{u}_i - \int_S (\boldsymbol{t} \,.\, \boldsymbol{u})\, \mathrm{dS} - \int_V (\boldsymbol{g} \,.\, \boldsymbol{u})\, \mathrm{dV}, \tag{3.39}$$

where $\boldsymbol{u}(\boldsymbol{x})$ is the displacement field within the body. The principle of stationary total energy, when applied to expression 3.39, tells us that of all compatible displacement fields $\boldsymbol{u}(\boldsymbol{x})$, the correct one gives a stationary value of χ .

To conclude, let us revisit briefly the example which was used in section 3.2 to introduce the variational approach. We can now justify with hindsight the energy function χ which was used at that time. The expression for the strain energy has already been discussed (section 3.4) and is now known to be $\frac{1}{2}k\delta^2$ for a uniform bar. The potential energy in this case is due to the concentrated load F at the free end (Fig. 3.1). Since the displacement at this point is δ, the potential energy is therefore (using equation 3.35)

$$V = -F\delta.$$

Combining these two expressions, the total energy of the system is

$$\chi = U + V = \frac{1}{2}k\delta^2 - F\delta,$$

as previously stated.

3.8 APPROXIMATE SOLUTIONS, THE RAYLEIGH–RITZ METHOD

Although the principle of stationary total potential may be used as a statement of the full continuum problem, its main application is as a basis for approximate, rather than exact, solutions. By considering *all* permissible displacement fields, as the principle advocates, the number of displaced states from which the correct ones — those which yield a stationary value of the energy functional — must be selected, is infinite. It is more practicable in many instances to consider a finite class of displacements which may or may not include the exact solution but which, if judiciously chosen, will provide a reasonable approximation to it. The stationary values of the total energy are then located with respect to a 'trial' displacement field defined by a discrete set of unknown parameters. The more comprehensive the trial field, the better the correspondence between the approximate and the exact solution.

This procedure is termed the Rayleigh–Ritz approach. In practice, it involves trial displacements which are linear combinations of known 'basis' functions (typically polynomial or trigonometric expressions) and unknown coefficients, $a_1, a_2 \ldots a_n$ say. These are substituted into an appropriate expression for the strain energy density and integrated over the volume of the body to give its strain energy as an algebraic function of the unknown Rayleigh–Ritz coefficients. The potential energy is similarly obtained as a function of the same parameters. By summing these quantities, we then obtain the total energy of the system as function of the coefficients a_i, rather than as a function*al* of the displacement field. The approximate expression for χ can therefore be written in the form

$$\chi = \chi\,(a_1, a_2, \ldots, a_n),$$

and the stationary values of the discrete system obtained by solving the simultaneous equations

$$\frac{\partial \chi}{\partial a_1} = 0, \quad \frac{\partial \chi}{\partial a_2} = 0, \ldots, \frac{\partial \chi}{\partial a_n} = 0 \tag{3.40}$$

In the case of a linearly elastic system, these equations are linear in the unknown coefficients a_i and their solution is therefore relatively straightforward.

To illustrate the Rayleigh–Ritz procedure, we conclude this chapter with three worked examples in which it is applied to elastic systems with one and two degrees of freedom. The first problem involves an assembly of pin-jointed bars, the second a simply supported beam and the third a rectangular plate.

3.8.1 Rayleigh–Ritz worked example 1, a plane framework

A simple framework consists of two bars, each of cross-sectional area A and Young's modulus E, which are pinned to supports at B and C and pinned together at D. A horizontal load P is applied at D (Fig. 3.10). The horizontal and vertical displacements at this point are a_1 and a_2. If a state of uniform strain exists in each bar, show that the total energy of the system is

$$\chi = \frac{EA}{L}\left[\frac{\sqrt{3}}{4}\left(\frac{a_2\sqrt{3}}{2} + \frac{a_1}{2}\right)^2 + \frac{a_2^2}{2}\right] - Pa_1.$$

Apply the Rayleigh–Ritz procedure to determine a_1 and a_2 in terms of P and determine also the stresses and forces in each bar.

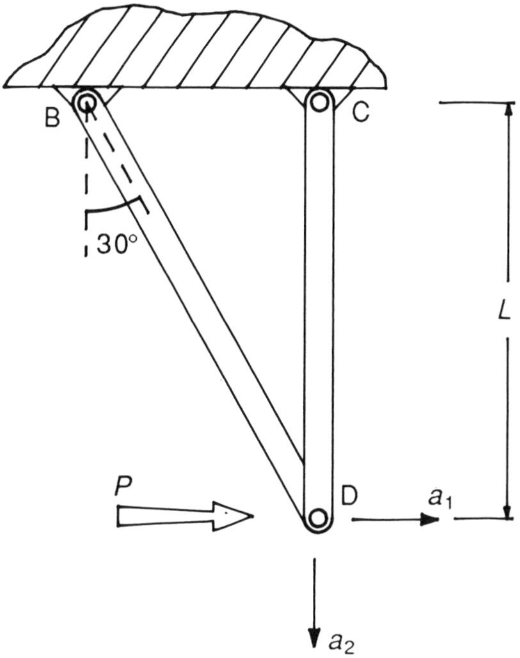

Fig. 3.10 *Worked example 1.*

Solution

The total energy χ of the system can be written

$$\chi = U_{BD} + U_{CD} + V, \tag{i}$$

where U_{BD} and U_{CD} are the strain energies of the bars BD and CD, and V is the potential energy of the concentrated load P. The horizontal displacement at the point of application of P is a_1, so V is given by

$$V = -Pa_1 \tag{ii}$$

The strain energies U_{BD} and U_{CD} can be calculated using the general expression derived at the end of section 3.4.3 which gives the strain energy of a uniform bar which undergoes an extension δ as $\frac{1}{2}(EA/L)\delta^2$. In the case of bar BD, the extension, δ_{BD}, is

$$\delta_{BD} = a_2 \cos 30^\circ + a_1 \sin 30^\circ = \frac{a_2\sqrt{3}}{2} + \frac{a_1}{2},$$

so that

$$U_{BD} = \frac{1}{2}\left(\frac{EA}{L_{BD}}\right)\left(\frac{a_2\sqrt{3}}{2} + \frac{a_1}{2}\right)^2 \tag{iii}$$

where L_{BD} is the length BD. In the case of bar CD, the extension is a_2 itself, giving

$$U_{CD} = \frac{1}{2}\left(\frac{EA}{L_{CD}}\right)a_2^2\,, \tag{iv}$$

combining expressions (i), (ii), (iii) and (iv), and setting L_{BD} and L_{CD} equal to $2L/\sqrt{3}$ and L, respectively, we obtain

$$\chi = \frac{EA\sqrt{3}}{4L}\left(\frac{a_2\sqrt{3}}{2} + \frac{a_1}{2}\right)^2 + \frac{EA}{2L}a_2^2 - Pa_1$$

or

$$\chi = \frac{EA}{L}\left[\frac{\sqrt{3}}{4}\left(\frac{a_2\sqrt{3}}{2} + \frac{a_1}{2}\right)^2 + \frac{a_2^2}{2}\right] - Pa_1\,,$$

as required.

The Rayleigh–Ritz solution is now obtained by locating the stationary values of χ with respect to a_1 and a_2, in other words, by setting $\partial\chi/\partial a_1$ and $\partial\chi/\partial a_2$ to zero. This gives

$$\frac{\partial\chi}{\partial a_1} = \frac{\sqrt{3}}{4}\left(\frac{EA}{L}\right)\left(\frac{a_2\sqrt{3}}{2} + \frac{a_1}{2}\right) - P = 0, \tag{v}$$

$$\frac{\partial\chi}{\partial a_2} = \frac{EA}{L}\left[\frac{3}{4}\left(\frac{a_2\sqrt{3}}{2} + \frac{a_1}{2}\right) + a_2\right] = 0, \tag{vi}$$

which can be solved to give

$$a_1 = \left(\frac{8}{\sqrt{3}} + 3\right)\left(\frac{PL}{EA}\right)$$

and

$$a_2 = -\sqrt{3}\left(\frac{PL}{EA}\right). \tag{vii}$$

The axial stresses σ_{BD} and σ_{CD} are then calculated using Hooke's law in each bar, that is

$$\sigma_{BD} = Ee_{BD} = E\frac{\delta_{BD}}{L_{BD}} = E\frac{(a_2\sqrt{3}/2 + a_1/2)}{2L/\sqrt{3}} = \frac{2P}{A}$$

and

$$\sigma_{CD} = Ee_{CD} = E\frac{\delta_{CD}}{L_{CD}} = E\frac{a_2}{L} = -\sqrt{3}\frac{P}{A} \tag{viii}$$

when multiplied by the cross-sectional area, these convert to to axial forces $2P$ (tensile) and $-\sqrt{3}P$ (compressive) in BD and CD, respectively. This completes the Rayleigh–Ritz solution.

Readers can easily confirm for themselves that the member forces calculated in this way are indeed 'correct', in the sense that they agree with the traditional equilibrium solution. This is true also of the displacements a_1 and a_2. The precise correspondence between the Rayleigh–Ritz and equilibrium solutions is not coincidental. Although not stated explicitly, the Rayleigh–Ritz formulation has incorporated in its evaluation of strain energy an assumption that the axial deformation varies linearly along each bar. This follows from the assumption that the strain is uniform. It imposes a severe restriction on the class of displacement fields from which the Rayleigh–Ritz solution is drawn but does not exclude the 'true' equilibrium solution for which stress and strain are indeed constant within each bar. The minimum energy state of the discrete system is therefore the minimum energy state of the continuous one and the Rayleigh–Ritz procedure captures the exact solution without approximation. This is not the case in the two examples to follow.

3.8.2 Rayleigh–Ritz worked example 2, deflection of a simply supported beam

An elastic beam of flexural rigidity EI and length L is simply supported at each end and carries a uniform load p_0 per unit length on its upper surface (Fig. 3.11). It carries no axial load, being free to move horizontally at the right hand end. If the vertical displacement of the beam is approximated by the trial function

$$w_0(x) = a_1 \sin\frac{\pi x}{L}, \tag{i}$$

show that the total energy of the system is

$$\chi = \chi(a_1) = \frac{EI\pi^4 a_1^2}{4L^3} - \frac{2p_0 a_1 L}{\pi}.$$

Use the Rayleigh–Ritz procedure to obtain an approximate expression for the central deflection and compare with the exact value. Compare also the exact and Rayleigh–Ritz values for the maximum bending moment.

Solution

The strain energy per unit length along the beam is given by expression 3.24 with the first term omitted, that is,

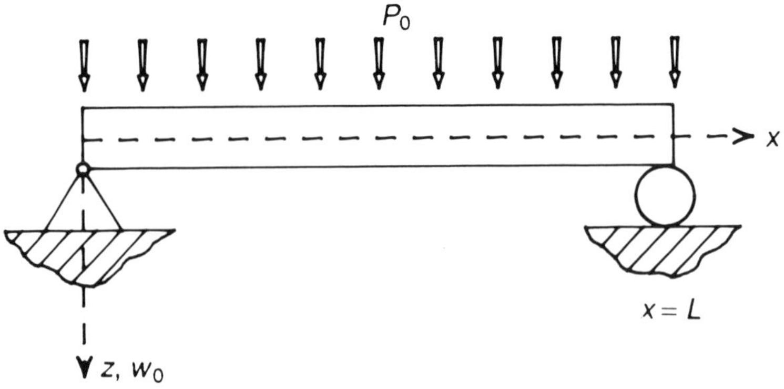

Fig. 3.11 *Worked example 2.*

$$SE/length = \frac{1}{2} EI \left(\frac{d^2 w_0}{dx^2} \right)^2 .$$

Substitution of trial function (i) into the above expression, followed by integration from $x = 0$ to $x = L$, gives the strain energy of the beam as

$$U = \int_0^L \frac{1}{2} EI \left(\left(\frac{a_1 \pi^2}{L^2} \right) \sin \frac{\pi x}{L} \right)^2 dx$$

$$= \frac{EI a_1^2 \pi^4}{4L^3} . \tag{ii}$$

The potential energy of the distributed load is (see expression 3.37)

$$V = - \int_0^L p_0 w_0 (x) \, dx = - \int_0^L p_0 \left(a_1 \sin \frac{\pi x}{L} \right) dx$$

$$= - \frac{2 p_0 a_1 L}{\pi} , \tag{iii}$$

and the total energy of the system is therefore

$$\chi = U + V = \frac{EI \pi^4 a_1^2}{4L^3} - \frac{2 p_0 a_1 L}{\pi} , \tag{iv}$$

as required.

The Rayleigh–Ritz solution is now obtained by locating the stationary value(s) of χ. This is equivalent in the current instance to locating the value of a_1 for which $d\chi/da_1$ is zero. In other words, the value of a_1 for which,

$$\frac{\partial \chi}{\partial a_1} = \frac{\partial}{\partial a_1}\left\{\frac{EI\pi^4 a_1^2}{4L^3} - \frac{2p_0 a_1 L}{\pi}\right\}$$

$$= \left(\frac{EI\pi^4}{4L^3}\right)a_1 - \frac{2p_0 L}{\pi} = 0.$$

This gives

$$a_1 = \frac{4p_0 L^4}{EI\pi^5} = \frac{0.0131 p_0 L^4}{EI}. \tag{v}$$

Given that $w_0(L/2)$ is identically equal to a_1, expression (v) can also be regarded as the Rayleigh–Ritz estimate of the central deflection of the beam. An 'exact' solution for this problem is easily obtained by integrating the deflection equation (equation 2.32) to give a central deflection, $(5/384)\, p_0 L^4/EI$. This is within 1% of the Rayleigh–Ritz estimate (equation (v)). The closeness of the two results tells us, not surprisingly, that the exact deflection — a quartic polynomial in x — is approximated well by the sine function used for the trial displacement.

The errors in the bending moment are somewhat larger. A consistent, Rayleigh–Ritz expression for the bending moment is obtained by substituting the trial displacement into the moment–curvature relationship to give

$$M(x) = -EI\frac{\mathrm{d}^2 w_0}{\mathrm{d}x^2} = EIa_1\left(\frac{\pi}{L}\right)^2 \sin\frac{\pi x}{L}.$$

This has a maximum value at midspan ($x = L/2$) of

$$M_{\max} = EIa_1\left(\frac{\pi}{L}\right)^2.$$

Substitution of expression (v) for a_1 then yields

$$M_{\max} = 0.129\, p_0 L^2. \tag{vi}$$

The 'exact' value of the bending moment at the midpoint of a uniformly loaded, simply supported beam is readily obtained from a free-body diagram and is given by

$$M_{\max} = \tfrac{1}{8}\, p_0 L^2 = 0.125\, p_0 L^2. \tag{vii}$$

The error in the Rayleigh–Ritz estimate is therefore of the order of 3.0–4.0%. Once again, the size of the error is measure of the ability of the trial function (for the bending moment in this case) to accurately represent the true solution.

Although the results obtained with the current trial function are reasonably good, they can be improved by including additional terms in the original expansion. A more accurate, trial function of this type is suggested in problem seven at the end of this chapter.

3.8.3 Rayleigh–Ritz worked example 3, deflection of a square plate

A square plate of side L is simply supported at its edges and carries a concentrated load P at its centre. The plate is of flexural rigidity D and lies in the x–y plane of a cartesian coordinate system (Fig. 3.12). No in-plane loads are applied and it is assumed that in-plane stresses are negligibly small. Calculate the strain energy of the plate using the trial function

$$w_0(x, y) = a_1 \sin\left(\frac{\pi x}{L}\right) \sin\left(\frac{\pi y}{L}\right), \qquad \text{(i)}$$

and obtain a Rayleigh–Ritz estimate for the central deflection. Compare with an exact result and comment.

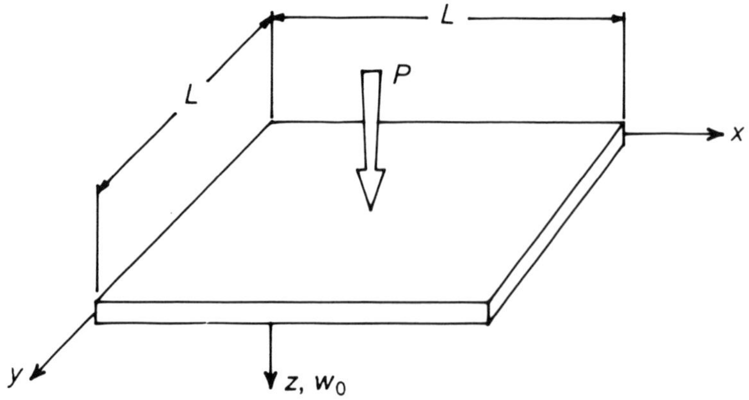

Fig. 3.12 *Worked example 3.*

(*Note*: the calculation of an exact result for the deflection of the plate lies well beyond the limited treatment of plate theory offered in Chapter 2. For the purposes of comparison, however, the exact value of the central displacement can be taken to be $0.0114PL^2/D$. Full details of this solution are to be found in Chapter 13 of [7]).

Solution

The strain energy of the plate is obtained by integrating the strain energy per unit area (equation 3.32) over the area of the plate. Since there are no in-plane stresses, the first three terms in equation 3.32 disappear and the integral simplifies to

$$U = \int_0^L \int_0^L \frac{1}{2} \left\{ M_x \left(-\frac{\partial^2 w_0}{\partial x^2} \right) + M_y \left(-\frac{\partial^2 w_0}{\partial y^2} \right) + 2\, M_{xy} \left(-\frac{\partial^2 w_0}{\partial x \partial y} \right) \right\} \mathrm{d}x\mathrm{d}y \qquad \text{(ii)}$$

The partial derivatives of w_0 which occur in the integrand are obtained by differentiating the trial solution to give

$$\begin{aligned} \frac{\partial^2 w_0}{\partial x^2} &= -\, a_1 \left(\frac{\pi}{L} \right)^2 \sin\left(\frac{\pi x}{L} \right) \sin\left(\frac{\pi y}{L} \right), \\ \frac{\partial^2 w_0}{\partial y^2} &= -\, a_1 \left(\frac{\pi}{L} \right)^2 \sin\left(\frac{\pi x}{L} \right) \sin\left(\frac{\pi y}{L} \right), \\ \frac{\partial^2 w_0}{\partial x \partial y} &= \quad a_1 \left(\frac{\pi}{L} \right)^2 \cos\left(\frac{\pi x}{L} \right) \cos\left(\frac{\pi y}{L} \right). \end{aligned} \qquad \text{(iii)}$$

The moments M_x, M_y and M_{xy} are then obtained from the moment–curvature relationship of Chapter 2 (equations 2.41), and are given by

$$M_x = -\, D \left\{ \frac{\partial^2 w_0}{\partial x^2} + \nu \frac{\partial^2 w_0}{\partial y^2} \right\} = D a_1 \left\{ \left(\frac{\pi}{L} \right)^2 + \nu \left(\frac{\pi}{L} \right)^2 \right\} \sin\left(\frac{\pi x}{L} \right) \sin\left(\frac{\pi y}{L} \right),$$

$$M_y = -\, D \left\{ \frac{\partial^2 w_0}{\partial y^2} + \nu \frac{\partial^2 w_0}{\partial x^2} \right\} = D a_1 \left\{ \left(\frac{\pi}{L} \right)^2 + \nu \left(\frac{\pi}{L} \right)^2 \right\} \sin\left(\frac{\pi x}{L} \right) \sin\left(\frac{\pi y}{L} \right),$$

$$M_{xy} = -\, D(1 - \nu) \frac{\partial^2 w_0}{\partial x \partial y} = -\, D a_1 (1 - \nu) \left(\frac{\pi}{L} \right) \left(\frac{\pi}{L} \right) \cos\left(\frac{\pi x}{L} \right) \cos\left(\frac{\pi y}{L} \right) \qquad \text{(iv)}$$

Substitution of (iii) and (iv) into (ii) followed by integration over the area of the plate then gives, after some rearrangement of terms,

$$\begin{aligned} U &= \frac{1}{2} D a_1^2 \int_0^L \int_0^L \left[2\,(1 + \nu) \left(\frac{\pi}{L} \right)^4 \sin^2\left(\frac{\pi x}{L} \right) \sin^2\left(\frac{\pi y}{L} \right) \right. \\ &\qquad \left. + \, 2\,(1 - \nu) \left(\frac{\pi}{L} \right)^4 \cos^2\left(\frac{\pi x}{L} \right) \cos^2\left(\frac{\pi y}{L} \right) \right] \mathrm{d}y\mathrm{d}x \\ &= \frac{1}{2} D L^2 a_1^2 \left(\frac{\pi}{L} \right)^4 \end{aligned} \qquad \text{(v)}$$

This completes the calculation of strain energy. The potential energy of the concentrated load is

$$V = -P\,w_0\left(\frac{L}{2}, \frac{L}{2}\right) = -Pa_1, \tag{vi}$$

and the total energy of the system is therefore

$$\chi = U + V = \left(\frac{D\pi^4}{2L^2}\right)a_1^2 - Pa_1. \tag{vii}$$

The Rayleigh–Ritz solution is now obtained by locating the stationary values of χ with respect to a_1, that is, by writing

$$\frac{\mathrm{d}\chi}{\mathrm{d}a_1} = \left(\frac{D\pi^4}{L^2}\right)a_1 - P = 0,$$

and solving for a_1 to give

$$a_1 = \frac{PL^2}{D\pi^4} = 0.0103\frac{PL^2}{D}. \tag{viii}$$

Consistent with the original trial function, the Rayleigh–Ritz estimate of the central deflection is simply a_1 itself, since $w_0\,(\frac{L}{2}, \frac{L}{2}) \equiv a_1$. Expression (viii) can therefore be compared directly to the 'exact' result, $0.0114\,PL^2/D$, noted at the beginning of this example. The error is of the order of 10%. As in the previous example this may be reduced by using a more complex initial trial function (see problem 9 the end of this chapter).

3.9 CONCLUDING COMMENTS

We have now reached the point where it is appropriate to discuss general strategies for the selection of Rayleigh–Ritz trial solutions for solid and structural problems. In the worked examples proceding these remarks, solutions were obtained in an *ad hoc* way using trial functions defined by no more than two unknown coefficients. The same approach can be extended to more complex systems but requires a larger number of unknown coefficients. The number and complexity of the strain energy integrals then increase rapidly, as does the algebra involved in keeping track of the resulting terms and differentiating them correctly to give the final equation set. Moreover, the solution of the equation set itself becomes a major exercise as the number of unknowns becomes large. The prospect therefore of applying the method as it stands to systems which require a large number of degrees of freedom is largely impractical. What is needed in such situations is a structured, automatic approach in which robust algorithms keep track of things for us, irrespective of the complexity of the problem or the number of unknown coefficients. This is precisely how matters are organized in a finite element model. In the next chapter the basic concepts of the finite element idea are introduced as a particular application of the Rayleigh–Ritz procedure.

REFERENCES

[1] Washizu, K. (1975) *Variational Methods in Elasticity and Plasticity* 2nd ed., Pergammon, London.
[2] Sokolnikoff, I. S., (1956), *Mathematical Theory of Elasticity*, McGraw-Hill, New York, Chapter 7.
[3] Rivello, R. M., (1969) *Theory and Analysis of Flight Structures*, McGraw-Hill, New York, Chapter 6.
[4] Finlayson, B. A., (1972) *The Method of Weighted Residuals and Variational Principles*, Academic, New York.
[5] Pars, L. A., (1962) *An Introduction to the Calculus of Variations*, Heinemann, London.
[6] Zienkiewicz, O. C. and Taylor, R. L., (1990) *The Finite Element Method* (4th ed.), vol. 1, *Basic Formulation and Linear Problems*, McGraw-Hill, New York, Chapter 9.
[7] Ugural, A. C and Fenster, S. K. (1987) *Advanced Strength and Applied Elasticity* 2nd ed., Elsevier, New York.

PROBLEMS

1. A non-linear, elastic material has a uniaxial stress–strain curve defined by

$$\sigma = \alpha e^{\beta} \quad (\alpha, \beta \text{ constant}).$$

Show that the strain energy per unit volume in an element of this material subject to a uniaxial stress σ and corresponding strain e is

$$\frac{\sigma e}{1+\beta}.$$

A bar of cross-section A and length L is constructed of the above material. One end of the bar is fixed and the other extends axially through a distance δ. The stress and strain are assumed uniform throughout the bar. Show that the strain energy of the bar is

$$U = \frac{\alpha A \delta^{(\beta+1)}}{(\beta+1)L^{\beta}}.$$

2. Show that the total energy of the bar of problem one when extended by a concentrated load F applied at its free end is

$$\chi(\delta) = \frac{\alpha A \delta^{(\beta+1)}}{(\beta+1)L^{\beta}} - F\delta.$$

Use the principle of stationary total energy to obtain the extension δ in terms of the applied load F. Does this value agree with the *equilibrium* solution of the problem?

3. Repeat the worked example of section 3.8.1 with the horizontal load P replaced by a vertical load P' acting in the direction of the displacement a_2.

4. Three bars of cross-sectional area A and Youngs modulus E are pinned together at **F** and pinned to fixed supports at B, C and D as shown. A concentrated load P acts vertically downwards at **F**. Show that the total energy of the system is

$$\frac{1}{2}\frac{EA}{L}\left\{a_1^2+\frac{1}{2\sqrt{2}}(a_1+a_2)^2+\frac{1}{2}\left(\frac{a_1}{2}+\frac{a_2\sqrt{3}}{2}\right)^2\right\}+Pa_2$$

where a_1 and a_2 are the displacements at F (assume a state of uniform strain in each bar). Obtain expressions for a_1 and a_2 using the Rayleigh–Ritz procedure and show that the resulting forces in the bars satisfy equilibrium at F.

5. Use the Rayleigh–Ritz method to estimate the maximum deflection and maximum bending moment in a cantilever beam (shown below) subject to a concentrated end load P. Use the trial function

$$W_0(x)=a_1\left[1-\cos\left(\frac{\pi x}{2L}\right)\right],$$

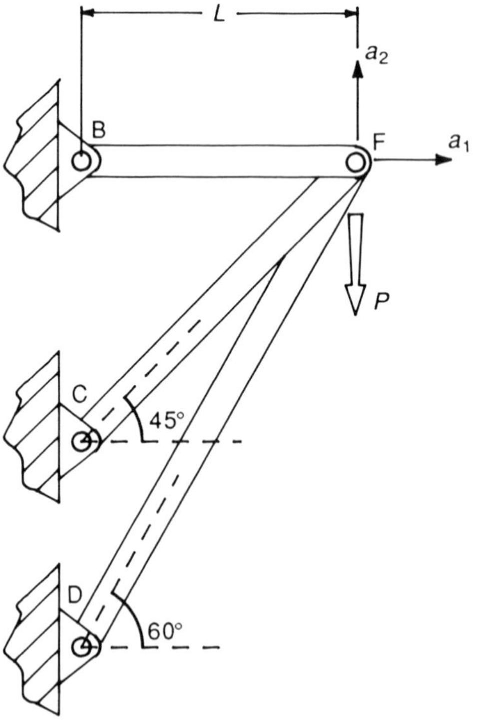

Problem 4.

and show that the total energy of the system is given by

$$\chi = \left(\frac{\pi^4 EI}{64L^3}\right) a_1^2 - P a_1.$$

Compare the Rayleigh–Ritz result with an exact solution for the same problem.

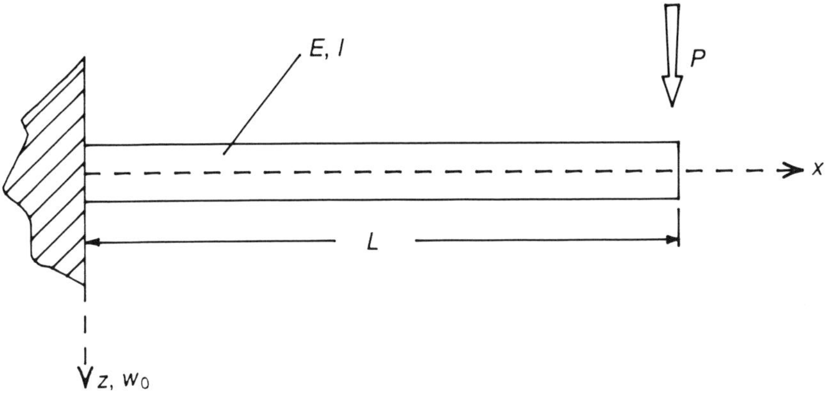

Problem 5.

6. Repeat problem five using a trial function

$$w_0(x) = a_1 x^2 + a_2 x^3.$$

Show that this produces the exact solution and explain carefully why this is so.

7. Use a trial function

$$w_0(x) = a_1 \sin\frac{\pi x}{L} + a_2 \sin\frac{3\pi x}{L}.$$

for the worked example of section 3.8.2 and show that the total energy χ is then given by

$$\chi = \chi\,(a_1, a_2) = \frac{EI\pi^4}{4L^3}\left[a_1^2 + 81 a_2^2\right] - \frac{2p_0}{\pi} L \left[a_1 + \frac{a_2}{3}\right].$$

Obtain a Rayleigh–Ritz estimate for the central deflection of the beam and compare with the exact solution.

8. A square plate of side L is clamped on two adjacent sides and subject to a uniformly distributed pressure p_0 on its upper surface (see figure). Use a Rayleigh–Ritz trial function

$$w_0(x, y) = a_1 x^2 y^2,$$

to show that the total energy of the system can be approximated by the expression

$$Da_1^2\left(\frac{116}{45} - \frac{4}{3}\nu\right)L^6 - \frac{p_0 a_1 L^6}{9} \quad (D \text{ is the flexural rigidity}).$$

Obtain an estimate for the deflection at the unconstrained corner of the plate in terms of L, D, ν and p_0.

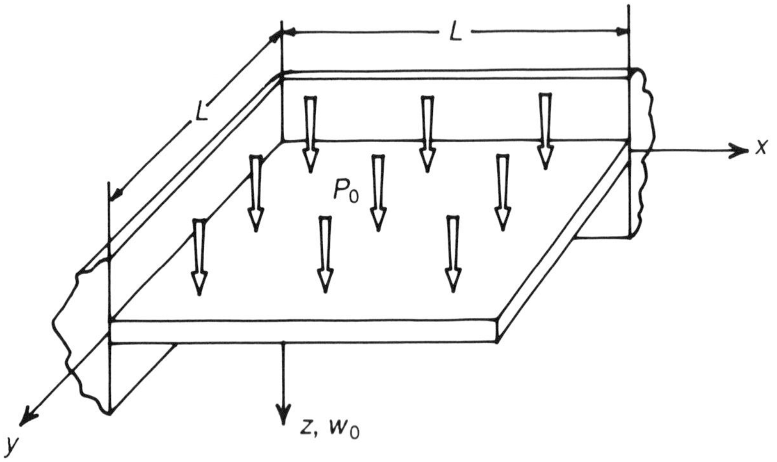

Problem 8.

9. Repeat the worked example of section 3.8.3 using a trial function

$$w_0(x, y) = a_1 \sin\frac{\pi x}{L}\sin\frac{\pi y}{L} + a_2 \sin\frac{3\pi x}{L}\sin\frac{\pi y}{L} + a_3 \sin\frac{\pi x}{L}\sin\frac{3\pi y}{L}.$$

Confirm that the resulting estimate for the central deflection is more accurate than that of section 3.8.3.

10. A square plate of side L and flexural rigidity D is clamped along one side and free on the others (see figure). A concentrated load P acts at one of the corners as shown. If the vertical deflection $w_0(x, y)$ of the centroidal plane is approximated by the trial solution

$$w_0(x, y) = \left[a_1 + (a_2 - a_1)\left(\frac{y}{L}\right)\right]\left[1 - \cos\frac{\pi x}{2L}\right],$$

show that the total energy of the system is

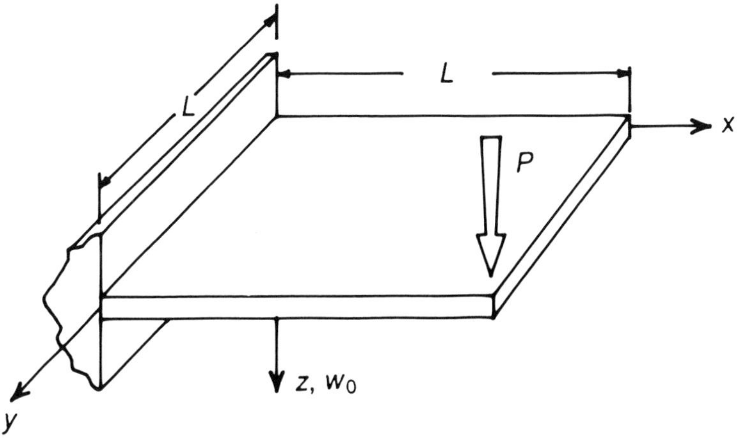

Problem 10.

$$\chi(a_1, a_2) = \frac{D}{L^2}\left\{ \frac{\pi^4}{192}\left[a_1^2 + a_2^2 + a_1 a_2\right] + \frac{\pi^2}{8}(1-\nu)\left[a_1 - a_2\right]^2 \right\} - Pa_2.$$

Use the Rayleigh–Ritz procedure to obtain an estimate for the deflection of the plate at the point where the load is applied.

4

Displacement finite elements, the basic approach

4.1 INTRODUCTORY COMMENTS

The finite element method is developed, in the chapters that follow, as an application of the principle of stationary total energy. This is one of several possible interpretations. The most straightforward is the **direct stiffness approach** [1]. This is somewhat simpler to apply in the case of a framed structure where members are joined together at well defined nodes, but runs into conceptual difficulties when extended to a continuum. The Galerkin or **weighted residual approach** [2–4], on the other hand, is of much broader application but is generally less intuitive and relates less directly to the physical problem.

In the present treatment, the finite element method is presented as a systematic application of the Rayleigh–Ritz procedure. The distinctive feature of the finite element approach is that the trial displacement field is generated 'automatically' from a representation of the body as an assembly of discrete elements. The interpolation functions within each element then become the basis functions of the Rayleigh–Ritz expansion and the values of displacement at discrete nodes the unknown coefficients.

The elements themselves are of simple geometrical form (triangular, rectangular, tetrahedral, et cetera) but may be combined to form shapes of great complexity. A typical finite element model is shown in Fig. 4.1, and will serve to demonstrate the basic steps in the finite element procedure. These are summarized as follows:

1. *Element definition.* The body to be analysed is subdivided into discrete elements. These are of finite size but do not necessarily correspond to distinct physical submembers. In Fig. 4.1, for example, the component is divided into three-dimensional blocks whose size and location are defined by a sequence of 'nodes' on its periphery (the 'topology' of the element). In the case of the block element shown in Fig. 4.1, there are eight nodes located at the corners. Material attributes such as Young's modulus, Poisson's ratio, density, are associated with each element, and may vary between elements as required.
2. *Element interpolation.* The displacement field within the element is interpolated using values of displacement at the nodes. These are ordered

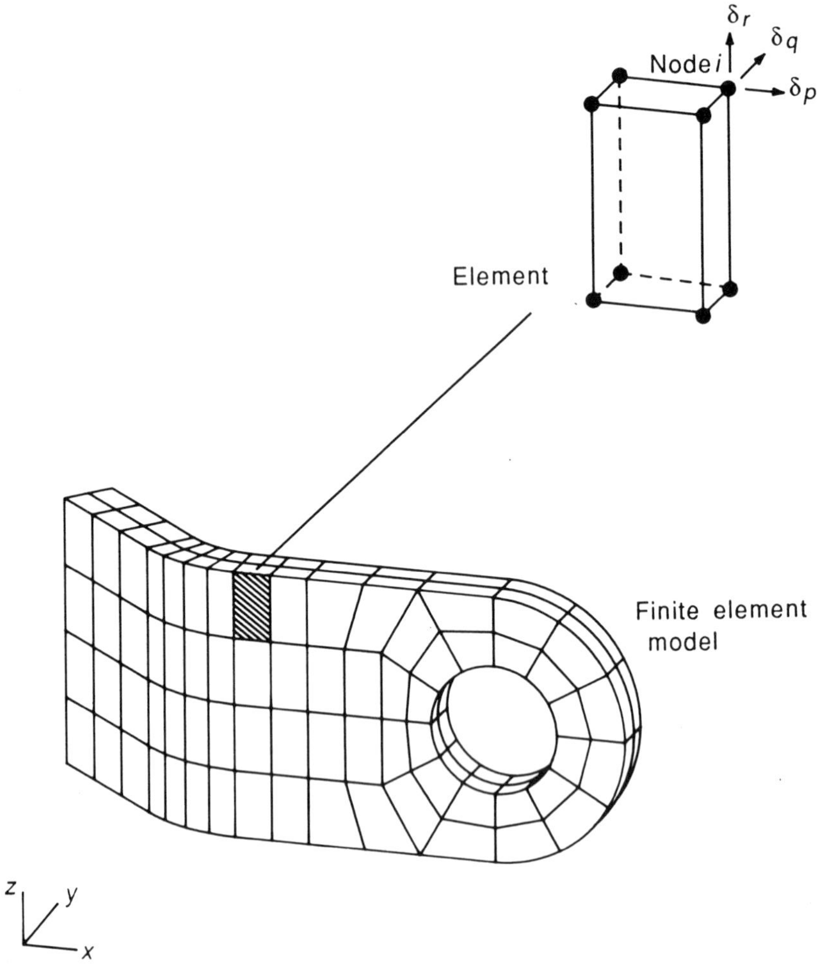

Fig. 4.1 *A finite element model.*

sequentially for the whole model. In the current notation, they are labelled, $\delta_1, \delta_2, \ldots, \delta_n$. In the case of Fig. 4.1, the displacements at a typical node, i, are shown as δ_p, δ_q and δ_r.

3. *Calculation of element energy.* The strain energy, U^e, of each element is determined by calculating the strain energy density in terms of the interpolated displacement field, and then integrating this quantity over the volume of the element. The potential energy V^e is evaluated in the same way by integrating body and surface potentials over the element. Both quantities are obtained as functions of the nodal displacements of the element.

4. *Assembly*. Element contributions to the strain and potential energies are summed over all elements to give the total energy of the assembled system as a function of its nodal displacements. Implicitly, χ can be written in the form

$$\chi = \chi(\delta_1, \delta_2, \ldots, \delta_n) \tag{4.1}$$

5. *Minimization and solution*. χ is minimized with respect to the nodal displacements. This yields a set of simultaneous equations of the form

$$\frac{\partial \chi}{\partial \delta_i} = 0, \quad i \in (1, 2, \ldots, n) \tag{4.2}$$

   Some nodal displacements are known (for example, at points of support where they are zero) and are excluded from this process. The number of equations, however, is always equal to the number of unknowns irrespective of the nature of the support. If the system is linearly elastic, the equations are linear in the variables δ_i and can be solved relatively simply.

The above process is largely a restatement of the general Rayleigh–Ritz procedure outlined in the preceding chapter. The distinguishing feature of the current formulation is that the Rayleigh–Ritz coefficients a_i have been replaced by the unknown nodal displacements δ_i.

Upon first sight, there seems little likelihood of implementing such a procedure in a problem-independent way. This is not, in fact, the case by virtue of the following considerations:

(a) Step 3 is intensely repetitive. In computational terms, a single subprogram may be used to process all elements of the same type irrespective of their size, shape or location.
(b) Steps 4 and 5 are not performed explicitly. In the finite element solution, a single algorithm both sums the energies of the individual elements and assembles equations 4.2 without forming an intermediate expression of type 4.1, and without differentiating it explicitly with respect to the nodal displacements.

As the finite element procedure is developed, it will become clear how a methodical implementation along these lines is not only possible but relatively straightforward.

4.2 ELEMENT FORMULATION

The basic steps of the finite element solution are described in the remainder of this chapter with reference to the simplest of elements, the one-dimensional bar. While concentrating on this element, we will, however, develop the relationships which govern the finite element model in a form which makes them more generally applicable.

We look first at steps 1, 2 and 3 of the general procedure, that is, at element definition, element interpolation, and the evaluation of the strain and potential energies of a single element.

4.2.1 Element definition

A bar 'element' (also termed a 'rod' or a 'truss') is a finite segment of a prismatic bar. For the purposes of the present discussion, it will be assumed that the element lies parallel to the x-axis of a cartesian coordinate system (Fig. 4.2), is of length L, cross-sectional area A and Young's modulus E. It has nodes at each end, labelled '1' and '2'. The element may be regarded as an entity in its own right or as part of a larger structure. In either event, it displaces under the action of external loads (unspecified at this stage) so that node 1 displaces through an axial distance u_1 and node 2 through a distance u_2. Lateral displacements representing rigid-body translations and rotations may also occur but do not contribute to the strain energy of the element and will be incorporated at a later stage.

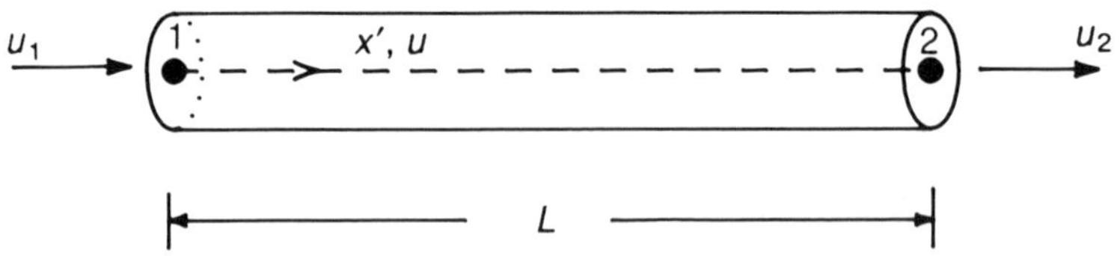

Fig. 4.2 *The uniaxial bar element, geometry and topology.*

4.2.2 Interpolation of displacement, element shape functions

Before defining an interpolation within the element, it is convenient to define a 'local' coordinate system. This is done by selecting an origin at the left hand end of the element and defining an axial coordinate x' which takes values 0 and L at nodes 1 and 2, respectively. An interpolation is now sought for the axial displacement, u, in terms of x'. Although the nature of the variation is not known in advance, it *is* known that u takes values u_1 and u_2 at nodes 1 and 2. An appropriate interpolation must therefore be chosen which satisfies this requirement. In the present instance, we will assume a linear variation, that is, an interpolation of the form

$$u(x') = \alpha_1 + \alpha_2 x'. \qquad (4.3)$$

The values of the constants α_1 and α_2 are determined from the conditions $u(0) = u_1$ and $u(L) = u_2$. This gives

$$\alpha_1 = u_1 \text{ and } \alpha_2 = (u_2 - u_1)/L,$$

which, upon substitution into equation 4.3 and some rearrangement of terms, gives

$$u = n_1(x')u_1 + n_2(x')u_2, \tag{4.4}$$

where $n_1(x') = 1 - x'/L$, and $n_2(x') = x'/L$. (4.5)

$n_1(x')$ and $n_2(x')$ are the 'shape functions' of the element. They define the displacement at any point within it in terms of the displacements of the nodes. The bar element has two nodes and hence two shape functions. In more complex elements, the number can be quite large. In all cases, however, the shape functions define an interpolation which is of the same general form as equation 4.4. The universal nature of this relationship can be emphasized by rewriting it in matrix form as,

$$\boldsymbol{u} = \boldsymbol{N}^{\mathrm{e}}\,\boldsymbol{d}^{\mathrm{e}}, \tag{4.6}$$

where $\boldsymbol{u} = [u]$, $\boldsymbol{N}^{\mathrm{e}} = [n_1(x'), n_2(x')]$, and $\boldsymbol{d}^{\mathrm{e}} = \begin{bmatrix} u_1 \\ u_2 \end{bmatrix}$ (4.7)

In this notation, $\boldsymbol{u}$ is a column vector which contains the displacement component(s) at a point (in the bar element there is only one of these but in other elements there may be two or three) $\boldsymbol{N}^{\mathrm{e}}$ is a 'shape matrix' whose components are the shape functions of the element, and $\boldsymbol{d}^{\mathrm{e}}$ is a column vector which contains the nodal displacements. The superscript 'e' is used here as elsewhere to signify a matrix or vector quantity which relates to a particular element rather than to the system as a whole. Although equation 4.6 has been derived specifically for a bar element, for which $\boldsymbol{N}^{\mathrm{e}}$ and $\boldsymbol{d}^{\mathrm{e}}$ are of order 1×2 and 2×1, the general form of the equation applies equally well to more complex elements. That is to say, although the size of $\boldsymbol{d}^{\mathrm{e}}$ and $\boldsymbol{N}^{\mathrm{e}}$ may vary, the expression of the vector of displacements within the element as the product of a 'shape matrix' and a 'nodal displacement vector' remains the same.

4.2.3 Interpolation of strain, strain–displacement matrix

An interpolated strain field within the element is implicit in the interpolation of displacement. It is obtained by substituting the shape relationship into the strain–displacement equations. In the case of the bar element, this gives

$$e_x = \frac{\mathrm{d}u}{\mathrm{d}x} = \frac{\mathrm{d}n_1}{\mathrm{d}x'}u_1 + \frac{\mathrm{d}n_2}{\mathrm{d}x'}u_2 \tag{4.8}$$

which, after substitution of the expressions for $n_1(x')$ and $n_2(x')$, using equation 4.5, reduces to

$$e_x = \left(-\frac{1}{L}\right)u_1 + \left(\frac{1}{L}\right)u_2 = \left[-\frac{1}{L}, \frac{1}{L}\right]\begin{bmatrix} u_1 \\ u_2 \end{bmatrix}. \tag{4.9}$$

This also may be written in matrix notation as,

$$\boldsymbol{e} = \boldsymbol{B}^{\mathrm{e}}\boldsymbol{d}^{\mathrm{e}}, \tag{4.10a}$$

where

$$\boldsymbol{e} = [e_x] \text{ and } \boldsymbol{B}^{\mathrm{e}} = \left[-\frac{1}{L}, \frac{1}{L}\right]. \tag{4.10b}$$

In a more general context, $\boldsymbol{e}$ is a column vector which contains the component(s) of strain (e_x in the case of the bar element) and $\boldsymbol{B}^{\mathrm{e}}$ is the 'strain–displacement matrix' whose components are derivatives of the shape functions. Although derived here specifically for a bar element, a version of equation 4.10(a) holds for all elements.

If we assume that Hooke's law holds within each bar, a stress field consistent with the above expression, is given by

$$\alpha_x = Ee_x = E\left[\left(-\frac{1}{L}\right)u_1 + \left(\frac{1}{L}\right)u_2\right] = [E]\left[-\frac{1}{L}, \frac{1}{L}\right]\begin{bmatrix} u_1 \\ u_2 \end{bmatrix}. \tag{4.11}$$

This also can be written in matrix form as

$$\boldsymbol{s} = \boldsymbol{D}\,\boldsymbol{B}^{\mathrm{e}}\,\boldsymbol{d}^{\mathrm{e}}, \tag{4.12}$$

where

$$\boldsymbol{s} = [\sigma_x] \text{ and } \boldsymbol{D} = [E]. \tag{4.13}$$

In the general case, $\boldsymbol{s}$ is a column vector of the stress components at a point, and $\boldsymbol{D}$ is a stress–strain matrix. For the bar element, $\boldsymbol{D}$ is simply a 1×1 matrix which has E as its only component. In two- and three-dimensional formulations, it contains elastic constants extracted from the appropriate stress–strain equations. Irrespective of the order of $\boldsymbol{D}$, the stresses within an element will always be given by a version of equation 4.12.

4.2.4 Strain energy of an element, element stiffness matrix

Provided that no temperature change occurs (thermal effects are considered at later stage) the strain energy density of a uniform bar is, $\frac{1}{2}\sigma_x e_x$. The strain energy of the element, U^{e}, is therefore

$$U^{\mathrm{e}} = \int_0^L \left(\tfrac{1}{2}\sigma_x e_x\right) A\,\mathrm{d}x'. \tag{4.14}$$

Consistent with our interpolation of the displacement field within the element, we can replace e_x and σ_x by scalar versions of expressions 4.9 and 4.11 to give, after some rearrangement of terms,

$$U^{\mathrm{e}} = \int_0^L \tfrac{1}{2} E \left(\frac{u_2 - u_1}{L} \right)^2 A\,\mathrm{d}x' = \tfrac{1}{2} k(u_2 - u_1)^2,$$

where $k = EA/L$. This is the strain energy of the element expressed in terms of its nodal displacements. When expanded as a quadratic expression in u_1 and u_2, it becomes

$$U^{\mathrm{e}} = \tfrac{1}{2}(ku_1^2 + (-k)u_1u_2 + (-k)u_2u_1 + ku_2^2),$$

which in turn can be written as a matrix triple product,

$$U^{\mathrm{e}} = \tfrac{1}{2}[u_1\ u_2] \begin{bmatrix} k & -k \\ -k & k \end{bmatrix} \begin{bmatrix} u_1 \\ u_2 \end{bmatrix}. \tag{4.15}$$

The square matrix at the centre of this expression is the **element stiffness matrix**. It is denoted by the symbol $\boldsymbol{K}^{\mathrm{e}}$, and is an entity central to all finite element analysis. Adopting again a more general, matrix notation, equation 4.15 can be written

$$U^{\mathrm{e}} = \tfrac{1}{2} \boldsymbol{d}^{\mathrm{eT}} \boldsymbol{K}^{\mathrm{e}} \boldsymbol{d}^{\mathrm{e}}, \tag{4.16}$$

and in this form, holds for all elements. That is to say, although the components of the above stiffness matrix are peculiar to the bar element, the idea of expressing the strain energy of the element as a matrix product of this form extends to all elements.

A general, integral expression for the stiffness matrix of an element can be obtained by returning to equation 4.14 and re-writing the strain energy density, $\frac{1}{2}\sigma_x e_x$, as the inner product of stress and strain vectors, that is, as $\frac{1}{2}(\boldsymbol{s}^{\mathrm{T}}\boldsymbol{e})$. In the uniaxial case, this is a trivial restatement, since $\boldsymbol{s}$ and $\boldsymbol{e}$ contain only one component each. The expression is valid however for all stress states. In the case of plane-stress, for example, when $\boldsymbol{s}$ and $\boldsymbol{e}$ have components $(\sigma_x, \sigma_y, \tau_{xy})$ and (e_x, e_y, γ_{xy}), it is still true that the strain energy density, now given by the scalar quantity, $\frac{1}{2}(\sigma_x e_x + \sigma_y e_y + \tau_{xy}\gamma_{xy})$, is equal to the inner product $\frac{1}{2}(\boldsymbol{s}^{\mathrm{T}}\boldsymbol{e})$.

Returning to integral in equation 4.14, the strain energy of an element can now be written quite generally as

$$U^{\mathrm{e}} = \int_V \tfrac{1}{2}(\boldsymbol{s}^{\mathrm{T}}\boldsymbol{e})\,\mathrm{d}V \tag{4.17}$$

where V denotes the volume of the element and $\mathrm{d}V$ $(= A\mathrm{d}x'$ in the case of the bar element) is a volume differential. Substitution of equations 4.10a and 4.12 into equation 4.17 then gives

$$U^{\mathrm{e}} = \tfrac{1}{2} \boldsymbol{d}^{\mathrm{eT}} \boldsymbol{K}^{\mathrm{e}} \boldsymbol{d}^{\mathrm{e}}, \text{ where } \boldsymbol{K}^{\mathrm{e}} = \int_V (\boldsymbol{B}^{\mathrm{eT}} \boldsymbol{D}^{\mathrm{T}} \boldsymbol{B}^{\mathrm{e}})\,\mathrm{d}V. \tag{4.18}$$

This is a restatement of expression 4.16 but with a general, integral expression for the element stiffness matrix. In the case of the bar element, this yields the same result as before. That is to say, substituting $\boldsymbol{B}^{\mathrm{e}} = [-1/L, 1/L]$ and $\boldsymbol{D} = [E]$ into the above integral for $\boldsymbol{K}^{\mathrm{e}}$, we obtain

$$\int_V (\boldsymbol{B}^{\mathrm{eT}}\boldsymbol{D}^{\mathrm{T}}\boldsymbol{B}^{\mathrm{e}})\,\mathrm{d}V = \int_0^L \begin{bmatrix} -1/L \\ 1/L \end{bmatrix} [E]\,[-1/L, 1/L]\,A\,\mathrm{d}x'$$

$$= \int_0^L \begin{bmatrix} E/L^2, & -E/L^2 \\ -E/L^2, & E/L^2 \end{bmatrix} A\,\mathrm{d}x', \tag{4.19}$$

which, after integration of the individual (constant) terms, reduces to

$$\boldsymbol{K}^{\mathrm{e}} = \begin{bmatrix} k, & -k \\ -k, & k \end{bmatrix} \tag{4.20}$$

where $k = EA/L$. This is in agreement with the previous version (equation 4.15).

The integral in equation 4.18 also establishes symmetry as a general property of the element stiffness matrix. Since the stress–strain matrix $\boldsymbol{D}$ is symmetric for isotropic materials, the entire integrand is invariant to transposition and $\boldsymbol{K}^{\mathrm{e}}$ is therefore symmetric by definition. Note also that since $\boldsymbol{D}$ is symmetric, $\boldsymbol{D}$ and $\boldsymbol{D}^{\mathrm{T}}$ are interchangeable. The transpose superscript can therefore be dropped from $\boldsymbol{D}$ in equation 4.18 to give the more common form of this integral,

$$\boldsymbol{K}^{\mathrm{e}} = \int_V (\boldsymbol{B}^{\mathrm{eT}}\boldsymbol{D}\boldsymbol{B}^{\mathrm{e}})\,\mathrm{d}V. \tag{4.21}$$

4.2.5 Potential energy of an element, equivalent nodal forces

The principle of minimum total energy requires that we evaluate not only the strain energy of the element but also its potential energy. Discussion of the potential energy due to concentrated loads is deferred until the assembly of element energies in section 4.3. That due to distributed loads is considered now.

In the case of the bar element, suppose that a body force, g per unit volume, acts in the axial direction. The potential energy of this load is given by (see equation 3.38)

$$V^{\mathrm{e}} = -\int_V (ug)\,\mathrm{d}V. \tag{4.22a}$$

Substitution of equation 4.4 and the replacement of the differential volume $\mathrm{d}V$ by $A\mathrm{d}x'$ then gives

$$V^e = -\int_0^L (n_1(x')u_1 + n_2(x')u_2)\, gA\,dx' \tag{4.22b}$$

which can be written

$$V^e = -u_1 g_1 - u_2 g_2 = -[u_1, u_2]\begin{bmatrix} g_1 \\ g_2 \end{bmatrix}, \tag{4.23}$$

where the quantities g_1 and g_2 have the dimensions of force and are given by

$$g_1 = \int_0^L n_1(x')gA\,dx', \text{ and } g_2 = \int_0^L n_2(x')gA\,dx'. \tag{4.24}$$

Expression 4.23 has an interesting physical interpretation. It tells us that the potential energy of the distributed load is exactly equivalent to the potential energy of concentrated loads g_1 and g_2 acting at nodes 1 and 2. In other words, the distributed load can be replaced by strategically placed 'equivalent nodal forces' g_1 and g_2 whose magnitudes are determined by equations 4.24. In the case of a constant body force ($g(x') = g_0$) this gives the unsurprising result that

$$g_1 = \int_0^L g_0 n_1(x')A\,dx' = \int_0^L g_0\left(1 - \frac{x'}{L}\right)A\,dx' = \tfrac{1}{2}(g_0 AL) \tag{4.25a}$$

and

$$g_2 = \int_0^L g_0 n_2(x')A\,dx' = \int_0^L g_0\left(\frac{x'}{L}\right)A\,dx' = \tfrac{1}{2}(g_0 AL). \tag{4.25b}$$

In other words, the body force is modelled by assigning one half of the net resultant load ($g_0 AL$) to each end of the element.

A more general expression for these 'equivalent nodal forces' is obtained by returning to the integrand of equation 4.22a and rewriting it as a vector product, ($\boldsymbol{u}^T\boldsymbol{g}$), where $\boldsymbol{g}$ is a vector of body force components. In the case of the bar element, $\boldsymbol{u}$ and $\boldsymbol{g}$ have only one component each (u and g) and the difference between the vector and scalar products is somewhat academic. This is not so in two and three-dimensional problems, since $\boldsymbol{u}$ and $\boldsymbol{g}$ then contain several components each, but the expression of potential energy density as the inner product of $\boldsymbol{u}$ and $\boldsymbol{g}$ remains valid. Its substitution into equation 4.22 yields

$$V^e = -\int_V \boldsymbol{u}^T\boldsymbol{g}\,dV, \tag{4.26}$$

which upon substitution of equation 4.6, gives

$$V^e = -\mathbf{d}^{eT}\boldsymbol{f}_g^e, \text{ where } \boldsymbol{f}_g^e = \int_V \boldsymbol{N}^{eT}\boldsymbol{g}\,dV. \tag{4.27}$$

The first part of this expression is now identical to the latter part of equation 4.23. The second provides a general definition for the equivalent nodal force vector, f_g^e. We can confirm that this gives the same result as before for the bar element by replacing the volume differential dV by Adx' and substituting the components of N^{eT} into equation 4.27 to give

$$f_g^e = \int_0^L \begin{bmatrix} n_1(x') \\ n_2(x') \end{bmatrix} [g]\, A\mathrm{d}x' = \begin{bmatrix} g_1 \\ g_2 \end{bmatrix},$$

where g_1 and g_2 are again given by equation 4.24.

4.2.6 Thermal effects, equivalent thermal loads

In calculating the strain energy of the element, we assumed that no temperature variation occurred. For completeness, this restriction is now removed.

Suppose that the bar experiences a temperature rise, $T(x')^\circ$. The stress–strain relationship, previously given by $\sigma_x = Ee_x$, then becomes

$$\sigma_x = E(e_x - \alpha T), \tag{4.28}$$

and the strain energy density is (see equation 3.21)

$$\tfrac{1}{2}\sigma_x(e_x - \alpha T).$$

Combining these expressions, the strain energy of the element is

$$U^e = \int_0^L \tfrac{1}{2} E(e_x - \alpha T)^2 A\mathrm{d}x'. \tag{4.29}$$

Substitution of equation 4.9 and evaluation of the integral, then gives, after some manipulation,

$$U^e = \tfrac{1}{2} k(u_1 - u_2)^2 - (u_1 h_1 + u_2 h_2) + \int_0^L \tfrac{1}{2} E(\alpha T)^2 A\mathrm{d}x' \tag{4.30}$$

where $h_1 = -\int_0^L \left(\frac{E\alpha \mathrm{T}}{L}\right) A\mathrm{d}x'$ and $h_2 = \int_0^L \left(\frac{E\alpha \mathrm{T}}{L}\right) A\mathrm{d}x'$. (4.31)

Expression 4.30 contains three contributions. The first is the quadratic expression in u_1 and u_2 which is identical to that obtained in the absence of temperature effects (equation 4.15). The second involves products of u_1 and u_2 with the quantities h_1 and h_2 defined by the integrals in equation 4.31. These have the units of force, and contribute a quantity of strain energy which is equal to the potential energy which would result if they were applied as concentrated loads at each end of the element, that is $-u_1h_1 - u_2h_2$. They are in this sense 'equivalent thermal forces' whose effect is identical to that of the distributed temperature rise. In the case of a constant temperature rise T_0, their values are $h_1 = -EA\alpha T_0$ and $h_2 = +EA\alpha T_0$. These are simply the tensile forces required to produce an elastic deformation equal to the the thermal expansion.

The third contribution to equation 4.30 is an integral which does not depend upon the deformation of the element, but only on the temperature distribution within it. Since the ultimate object of the Rayleigh–Ritz procedure is to minimize the total energy of the system with respect to the displacement field, this term can be regarded as constant during the minimization process and will play no further part in the analysis. In fact, it can be removed entirely from the calculation at this stage by redefining the original datum energy level of the element to include this term. If this is done, equation 4.30 can be rewritten as

$$U^e = \frac{1}{2}[u_1, u_2]\begin{bmatrix} k & -k \\ -k & k \end{bmatrix}\begin{bmatrix} u_1 \\ u_2 \end{bmatrix} - [u_1, u_2]\begin{bmatrix} h_1 \\ h_2 \end{bmatrix}, \tag{4.32}$$

and then becomes a clear extension of equation 4.15.

A more general version of equation 4.32 is obtained by returning to equation 4.29 and rewriting the integral for U^e as

$$U^e = \int_V \tfrac{1}{2}\boldsymbol{s}^T[\boldsymbol{e} - \boldsymbol{e}_T]\,dV \tag{4.33}$$

where $\boldsymbol{e}_T$ is a vector of thermal strains. In the case of the bar element, it has only one component, αT. In two and three-dimensional problems, it contains strains αT in each row corresponding to a direct strain. The modified stress–strain relationship is then, $\boldsymbol{s} = \boldsymbol{D}[\boldsymbol{e} - \boldsymbol{e}_T]$ (replacing $\boldsymbol{s} = \boldsymbol{De}$) and equation 4.33 therefore becomes

$$U^e = \int_V \tfrac{1}{2}[\boldsymbol{e}^T - \boldsymbol{e}_T]^T\boldsymbol{D}^T[\boldsymbol{e} - \boldsymbol{e}_T]\,dV. \tag{4.34}$$

After substitution of the discrete strain–displacement relationship (equation 4.10) and further manipulation, this reduces to

$$U^e = \tfrac{1}{2}\boldsymbol{d}^{eT}\boldsymbol{K}^e\boldsymbol{d}^e - \boldsymbol{d}^{eT}\boldsymbol{f}_T^e, \tag{4.35}$$

where the stiffness matrix, $\boldsymbol{K}^e$, is as previously defined, and the thermal load vector $\boldsymbol{f}_T^e$ is given by

$$\boldsymbol{f}_T^e = \int_V (\boldsymbol{B}^{eT}\boldsymbol{D}\boldsymbol{e}_T)\,dV. \tag{4.36}$$

An additional term which depends only on the temperature distribution and not on the deformation has again been absorbed into the original datum strain energy of the element. Expression 4.36 now provides a general expression for the equivalent thermal loads associated with temperature variation over an element.

4.2.7 Summary. Total energy of an element

The total energy associated with the two-noded bar element can now be written, combining equations 4.23 and 4.32, as

$$\chi^{e} = U^{e} + V^{e} = \frac{1}{2}[u_1, u_2]\begin{bmatrix} k & -k \\ -k & k \end{bmatrix}\begin{bmatrix} u_1 \\ u_2 \end{bmatrix} - [u_1, u_2]\left[\begin{bmatrix} g_1 \\ g_2 \end{bmatrix} + \begin{bmatrix} h_1 \\ h_2 \end{bmatrix}\right], \qquad (4.37)$$

where g_i and $h_i (i = 1, 2)$ are equivalent nodal forces arising from body forces and temperature variations respectively and where $k(=EA/L)$ is the axial stiffness of the bar.

An equivalent, general statement is

$$\chi^{e} = \tfrac{1}{2}\boldsymbol{d}^{eT}\boldsymbol{K}^{e}\boldsymbol{d}^{e} - \boldsymbol{d}^{eT}[\boldsymbol{f}_{g}^{e} + \boldsymbol{f}_{T}^{e}], \qquad (4.38)$$

where $\boldsymbol{K}^{e}, \boldsymbol{f}_{g}^{e}$ and $\boldsymbol{f}_{T}^{e}$ are given by equations 4.21, 4.27, and 4.36. These are repeated here for ease of future reference, they are:

$$\boldsymbol{K}^{e} = \int_{V} (\boldsymbol{B}^{eT}\boldsymbol{D}\boldsymbol{B}^{e})\,dV,$$

$$\boldsymbol{f}_{T}^{e} = \int_{V} (\boldsymbol{B}^{eT}\boldsymbol{D}\boldsymbol{e}_{T})\,dV, \qquad (4.39)$$

$$\boldsymbol{f}_{g}^{e} = \int_{V} (\boldsymbol{N}^{eT}\boldsymbol{g})\,dV.$$

For the particular case of the bar element, the components of $\boldsymbol{K}^{e}, \boldsymbol{f}_{g}^{e}$ and $\boldsymbol{f}_{T}^{e}$ are given explicitly by expressions 4.15, 4.24 and 4.31. In the general case, they are obtained by integrating expressions 4.39 after defining appropriate element matrices, $\boldsymbol{N}^{e}$, $\boldsymbol{B}^{e}$ and $\boldsymbol{D}$.

4.3 ASSEMBLY

The total energy of the finite element model is obtained by summing the element energies χ^{e} given by equation 4.38. In theory, these can be expanded as scalar algebraic expressions and simply added together. In reality, this is impractical except in the simplest of cases. It is more sensible to retain the matrix notation of expression 4.38 and to extend it to the whole structure. A simple algorithm then emerges which performs the required summations. This process is termed 'assembly' and is illustrated first with reference to a structure which consists entirely of bar elements of the type described in the previous section.

4.3.1 Assembly, a one-dimensional illustration

Bar elements of varying stiffnesses are joined end to end as shown in Fig. 4.3. They are numbered $1, 2 \ldots, m$, element 'e' being a typical member. The node points at which elements are joined are numbered, $1, 2, \ldots, n$. The ordering of both nodes and elements is arbitrary in the sense that neither need follow any particular sequence. Element 'e', for example, is defined by nodes i_1 and i_2, where i_1 and i_2 lie in the range $1, \ldots, n$ but are not necessarily adjacent numbers.

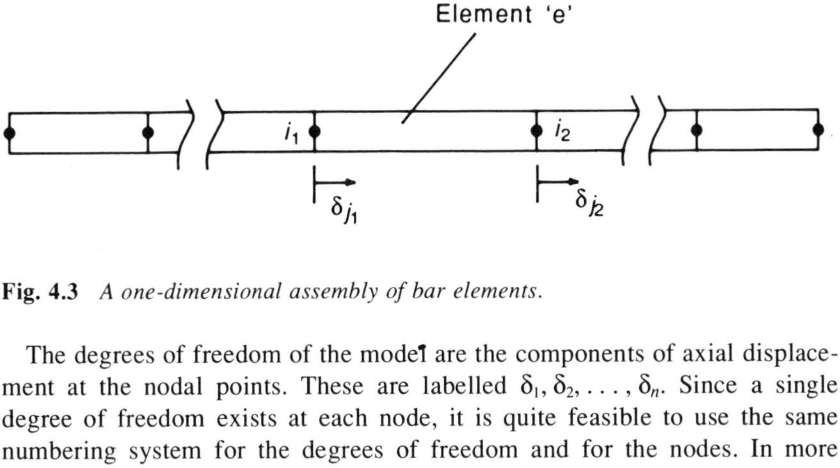

Fig. 4.3 *A one-dimensional assembly of bar elements.*

The degrees of freedom of the model are the components of axial displacement at the nodal points. These are labelled $\delta_1, \delta_2, \ldots, \delta_n$. Since a single degree of freedom exists at each node, it is quite feasible to use the same numbering system for the degrees of freedom and for the nodes. In more complex formulations, however, where there are two or more degrees of freedom at each node, this is no longer possible. We will therefore assume, for generality, that the numbering system used for the degrees of freedom is independent of that used for the nodes. The nodal displacements at nodes i_1 and i_2 are therefore denoted by δj_1 and δj_2 where j_2 and j_2 are integers in the range $1, \ldots, n$, but are not necessarily equal to i_1 and i_2.

A simple illustration of these three numbering systems (element, node and degree of freedom) is contained in Fig. 4.4 which shows a two-element model of a uniform bar. The elements and nodes are numbered from left to right and the degrees of freedom from right to left.

Before proceeding further, the material and geometric properties of each element must be specified. We will assume here that our typical element has a Young's modulus E, cross-sectional area A and length L. An axial body force g_0 per unit volume and a temperature rises T_0 are also assumed. All of these quantities can vary from element to element but are taken to be constant within each.

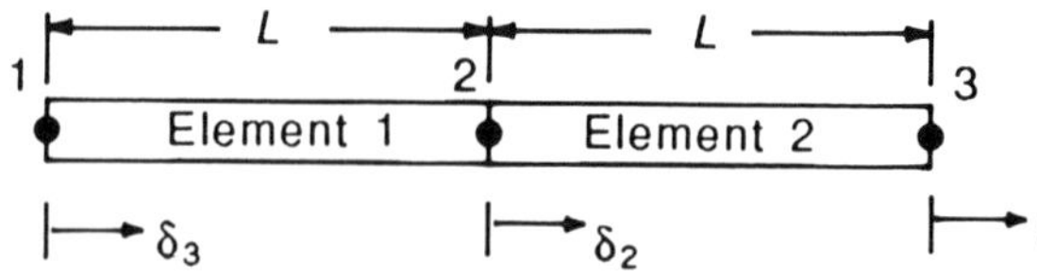

Fig. 4.4 *A two-bar assembly.*

The contribution from the eth element to the total energy of the system is then given by expression 4.38, with u_1 and u_2 replaced by δj_1 and δj_2. Written in component form, this gives

$$\chi^e = \tfrac{1}{2}\,[\delta j_1\ \delta j_2]\begin{bmatrix} k_{11}^e, & k_{12}^e \\ k_{21}^e, & k_{22}^e \end{bmatrix}\begin{bmatrix} \delta j_1 \\ \delta j_2 \end{bmatrix} - [\delta j_1\ \delta j_2]\left[\begin{bmatrix} g_1^e \\ g_2^e \end{bmatrix} + \begin{bmatrix} h_1^e \\ h_2^e \end{bmatrix}\right], \qquad (4.40a)$$

where $k_{11}^e = k_{22}^e = -k_{12}^e = -k_{21}^e = (EA/L)$, $h_1^e = -h_2^e = -EA\alpha T_0$, and $g_1^e = g_2^e = (\frac{1}{2}g_0AL)$. Note that the superscript e has been attached to the components of the element stiffness matrix and load vectors to distinguish them from their 'assembled' counterparts which will be encountered shortly. Expression 4.40a is now rewritten in an expanded form so that it involves *all* of the degrees of freedom of the model and not simply those relating to this particular element. This is done by increasing to n the order of the element stiffness matrix and load vectors while 'padding' them with zeroes in all but the j_1th and j_2th rows and columns. Expression 4.40(a) then becomes

$$\chi^e = \tfrac{1}{2}[\delta_1, \ldots \delta_n]\begin{bmatrix} 0 & 0 & 0 & 0 \\ 0 & k_{11}^e & k_{12}^e & 0 \\ 0 & k_{21}^e & k_{22} & 0 \\ \vdots & & & \\ 0 & 0 & 0 & 0 \end{bmatrix}\begin{bmatrix} \delta_1 \\ \delta_{j1} \\ \delta_{j2} \\ \vdots \\ \delta_n \end{bmatrix} - [\delta_1 \ldots \delta_n]\left\{\begin{bmatrix} 0 \\ g_1^e \\ g_2^e \\ \vdots \\ 0 \end{bmatrix} + \begin{bmatrix} 0 \\ h_1^e \\ h_2^e \\ \vdots \\ 0 \end{bmatrix}\right\} \qquad (4.40b)$$

In this form, contributions from different elements can be added together to give a homogeneous expression for the entire structure of the same form, that is:

$$\sum_{e=1}^{m} \chi^{e} = \tfrac{1}{2}[\delta_1 \ldots \delta n] \begin{bmatrix} k_{11} & k_{12} & k_{1n} \\ k_{21} & k_{22} & k_{2n} \\ \vdots & \vdots & \vdots \\ k_{n1} & k_{n2} & k_{nn} \end{bmatrix} \begin{bmatrix} \delta_1 \\ \delta_2 \\ \vdots \\ \delta_n \end{bmatrix} - [\delta_1 \ldots \delta_n] \left\{ \begin{bmatrix} g_1 \\ g_2 \\ \vdots \\ g_n \end{bmatrix} + \begin{bmatrix} h_1 \\ h_2 \\ \vdots \\ h_n \end{bmatrix} \right\},$$

or,

$$\sum_{e=1}^{m} \chi^{e} = \tfrac{1}{2} \boldsymbol{d}^{\mathrm{T}} \boldsymbol{K} \boldsymbol{d} - \boldsymbol{d}^{\mathrm{T}} [\boldsymbol{f}_{\mathrm{g}} + \boldsymbol{f}_{\mathrm{T}}]. \tag{4.41}$$

$\boldsymbol{K}$, $\boldsymbol{f}_{\mathrm{g}}$, $\boldsymbol{f}_{\mathrm{T}}$ and $\boldsymbol{d}$ are now the stiffness matrix, nodal force vectors and nodal displacement vector for the assembled system.

The terms within $\boldsymbol{K}$, $\boldsymbol{f}_{\mathrm{g}}$ and $\boldsymbol{f}_{\mathrm{T}}$ are obtained by incrementing them one element at a time, adding contributions of the type in equation 4.40(b). This involves the the following sequence of operations:

1. Initialize $\boldsymbol{K}$, $\boldsymbol{f}_{\mathrm{g}}$ and $\boldsymbol{f}_{\mathrm{T}}$ by equating all components to zero.
2. Select an element.
3. Evaluate the components of the element stiffness matrix $\boldsymbol{K}^{\mathrm{e}}$ and nodal force vectors $\boldsymbol{f}_{\mathrm{g}}^{\mathrm{e}}$ and $\boldsymbol{f}_{\mathrm{T}}^{\mathrm{e}}$, noting the degree of freedom numbers j_1 and j_2.
4. Insert each component of $\boldsymbol{K}^{\mathrm{e}}$ into the appropriate location in $\boldsymbol{K}$, adding to any number already there. That is:
   add k_{11}^{e} to $k_{j1\,j1}$,
   add k_{12}^{e} to $k_{j1\,j2}$,
   add k_{21}^{e} to $k_{j2\,j1}$,
   add k_{22}^{e} to $k_{j2\,j2}$.
5. Insert each component of $\boldsymbol{f}_{\mathrm{g}}^{\mathrm{e}}$ and $\boldsymbol{f}_{\mathrm{T}}^{\mathrm{e}}$ into the appropriate row of $\boldsymbol{f}_{\mathrm{g}}$ and $\boldsymbol{f}_{\mathrm{T}}$ adding to any number already there. That is:
   add g_1^{e} to g_{j1}, and h_1^{e} to h_{j1},
   add g_2^{e} to g_{j2} and h_2^{e} to h_{j2}.
6. Return to step 2, select a new element and repeat steps 3, 4 and 5 until all elements have been processed.

At the end of this sequence, $\boldsymbol{K}$, $\boldsymbol{f}_{\mathrm{g}}$ and $\boldsymbol{f}_{\mathrm{T}}$ are complete. The above procedure is termed 'assembly'.

Example

The assembly procedure is demonstrated by forming the stiffness matrix, body-force load vector and thermal load vector for the finite element model of Fig. 4.4. This is a uniform bar of length $2L$, modelled by two bar elements joined end to end. Each element is of length L, cross-sectional area A and Young's modulus E, and experiences a constant body force g_0 per unit volume and a uniform temperature rise $\mathring{T}_0$. The assembly of element contributions proceeds as follows.

First, we initialize a 3×3 matrix $\boldsymbol{K}$ and 3×1 vectors $\boldsymbol{f}_{\mathrm{g}}$ and $\boldsymbol{f}_{\mathrm{T}}$ to zero (step 1 of the assembly procedure) giving:

$$\boldsymbol{K} = \begin{bmatrix} 0 & 0 & 0 \\ 0 & 0 & 0 \\ 0 & 0 & 0 \end{bmatrix}, \quad \boldsymbol{f}_{\mathrm{g}} = \begin{bmatrix} 0 \\ 0 \\ 0 \end{bmatrix}, \quad \boldsymbol{f}_{\mathrm{T}} = \begin{bmatrix} 0 \\ 0 \\ 0 \end{bmatrix}.$$

Next we take element number one and calculate the components of its stiffness matrix and load vectors (steps 2 and 3). These are:

$$\boldsymbol{K}^{\mathrm{e}} = \begin{bmatrix} k, & -k \\ -k, & k \end{bmatrix}, \quad \boldsymbol{f}_{\mathrm{T}}^{\mathrm{e}} = \begin{bmatrix} -EA\alpha T_0 \\ EA\alpha T_0 \end{bmatrix}, \text{ and } \boldsymbol{f}_{\mathrm{g}}^{\mathrm{e}} = \begin{bmatrix} \frac{1}{2} g_0 AL \\ \frac{1}{2} g_0 AL \end{bmatrix}, \quad \left(k = \frac{EA}{L}\right).$$

The nodal displacements, δ_{j1} and d_{j2}, are δ_3 and δ_2, respectively. Step 4 of the assembly procedure therefore requires us to place stiffness terms, k, $-k$, $-k$ and k into locations 3–3, 3–2, 2–3 and 2–2 of the stiffness matrix $\boldsymbol{K}$. Similarly, step 5 requires us to place terms $\frac{1}{2} g_0 AL$ in the third and second rows of $\boldsymbol{f}_{\mathrm{g}}$, and terms $-E\alpha AT_0$ and $+E\alpha AT$ in the third and second rows of $\boldsymbol{f}_{\mathrm{T}}$. When this is done, $\boldsymbol{K}$, $\boldsymbol{f}_{\mathrm{g}}$ and $\boldsymbol{f}_{\mathrm{T}}$ look like

$$\boldsymbol{K} = \begin{bmatrix} 0 & 0 & 0 \\ 0 & k & -k \\ 0 & -k & k \end{bmatrix}, \quad \boldsymbol{f}_{\mathrm{g}} = \begin{bmatrix} 0 \\ \frac{1}{2} g_0 AL \\ \frac{1}{2} g_0 AL \end{bmatrix}, \quad \boldsymbol{f}_{\mathrm{T}} = \begin{bmatrix} 0 \\ E\alpha AT_0 \\ -E\alpha AT_0 \end{bmatrix}.$$

The procedure is now repeated for the element two, for which δ_{j1} and δ_{j2} are δ_2 is δ_1, respectively. The element itself is identical to the first one, and has the same stiffness matrix and load vectors. Steps 4 and 5 therefore require insertion of the same terms as before but in rows and columns 2 and 1 instead of 2 and 3. This gives

$$\boldsymbol{K} = \begin{bmatrix} k & -k & 0 \\ -k & 2k & -k \\ 0 & -k & k \end{bmatrix}, \quad \boldsymbol{f}_{\mathrm{g}} = \begin{bmatrix} \frac{1}{2} g_0 AL \\ g_0 AL \\ \frac{1}{2} g_0 AL \end{bmatrix}, \quad \boldsymbol{f}_{\mathrm{T}} = \begin{bmatrix} E\alpha AT_0 \\ 0 \\ -E\alpha AT_0 \end{bmatrix} \tag{4.42}$$

Since the current model has only two elements, the assembly of $\boldsymbol{K}$, $\boldsymbol{f}_{\mathrm{g}}$ and $\boldsymbol{f}_{\mathrm{T}}$ is now complete.

4.3.2 Assembly, the general case

Although the assembly procedure has been introduced here using two-noded bar elements, the assembly algorithm outlined above can be applied with little modification to any type of element. The only difference in the general case is that the element has p degrees of freedom, say, rather than two. Within the assembly procedure, the element stiffness matrix and load vectors will then be of order $p \times p$ and $p \times 1$, instead of 2×2 and 2×1. The general algorithm can be summarized formally as follows:

1. Initialize $\boldsymbol{K}, \boldsymbol{f}_{\mathrm{b}}$ and $\boldsymbol{f}_{\mathrm{T}}$ by equating all terms to zero.
2. Select an element.
3. Calculate $\boldsymbol{K}^{\mathrm{e}}, \boldsymbol{f}_{\mathrm{b}}^{\mathrm{e}}$ and $\boldsymbol{f}_{\mathrm{T}}^{\mathrm{e}}$ using equations 4.39 and note the degree of freedom numbers of the element; $j_1, j_2, \ldots, j_p$.
4. Add $k^{\mathrm{e}}_{\alpha\beta}$ to $k_{j\alpha j\beta}$ for all indices $\alpha = 1, \ldots, p$ and $\beta = 1, \ldots, p$.
5. Add $(f_{\mathrm{g}}^{\mathrm{e}})_{\alpha}$ to $(f_{\mathrm{g}})_{j\alpha}$ and $(f_{\mathrm{T}}^{\mathrm{e}})_{\alpha}$ to $(f_T)_{j\alpha}$ for all indices $\alpha = 1, \ldots, p$.
6. Return to step 2 and select another element. When all elements have been processed the assembly of $\boldsymbol{K}, \boldsymbol{f}_{\mathrm{b}}$ and $\boldsymbol{f}_{\mathrm{T}}$ is complete.

Further to comments made at the end of section 4.2.4, note that, since the element stiffness matrices are individually symmetric, step 4 ensures that the assembled stiffness matrix $\boldsymbol{K}$ is also symmetric. This is an important characteristic and one which we will make use of shortly when we attempt to minimize the total energy of the assembled system.

4.3.3 Inclusion of concentrated loads, the final form of the energy function

We are almost in a position to apply the Rayleigh–Ritz procedure to the assembled model. There is, however, one contribution to the total energy which has not yet been accounted for. This is the potential energy due to concentrated loads applied at the nodes. Such contributions are not associated with any particular element and have not therefore been included in the summation of element energies.

A concentrated, nodal load, P, acting in the direction of a nodal displacement δ_i, contributes a quantity $-\delta_i P$ to the energy of the system. This is equivalent to the 'inner' product

$$-\boldsymbol{d}^{\mathrm{T}}\boldsymbol{f}_P \tag{4.43}$$

where $\boldsymbol{f}_P$ is a vector containing P as its ith component and with zeros else where. When a number of concentrated loads are applied at different nodes, the value of each load can be placed in the appropriate row of $\boldsymbol{f}_P$. The addition of this final contribution to the existing expression for $\Sigma\chi^{\mathrm{e}}$ (equation 4.41) yields an expression for the total energy of the system, of the form

$$\chi = \sum_{\mathrm{e}=1}^{m} \chi^{\mathrm{e}} - \boldsymbol{d}^{\mathrm{T}}\boldsymbol{f}_P = \tfrac{1}{2}\boldsymbol{d}^{\mathrm{T}}\boldsymbol{K}\boldsymbol{d} - \boldsymbol{d}^{\mathrm{T}}\boldsymbol{f}, \tag{4.44}$$

where $\boldsymbol{f}$ is a nodal force vector which includes distributed, thermal and point loads. It is given by

$$\boldsymbol{f} = \boldsymbol{f}_g + \boldsymbol{f}_T + \boldsymbol{f}_P. \qquad (4.45)$$

With expression 4.44 in place, the next step is to minimize this quantity with respect to the components of $\boldsymbol{d}$.

4.4 MINIMIZATION AND SOLUTION

4.4.1 Differentiation of the energy functional

A nodal displacement is free to vary within the Rayleigh–Ritz scheme provided that it is unconstrained by support conditions. The derivative of the total energy with respect to such displacements must be zero. With a view to applying this constraint to expression 4.44, we now form a vector, $\boldsymbol{d}\chi$, which contains the derivatives of χ with respect to the nodal displacements of the system. The vector $\boldsymbol{d}\chi$ has components given by

$$\boldsymbol{d}\chi = \begin{bmatrix} \partial\chi/\partial\delta_1 \\ \partial\chi/\partial\delta_2 \\ \vdots \\ \delta\chi/\partial\delta_n \end{bmatrix}. \qquad (4.46)$$

Each can be evaluated — in theory at least — by writing equation 4.44 as a summation of scalar terms and then differentiating the resulting expression. In the general case, χ can be written

$$\chi(\delta_1\, \delta_2 \ldots \delta_n) = \frac{1}{2}\left(\sum_{j=1}^{n}\sum_{i=1}^{n} k_{ij}\,\delta_i\delta_j\right) - \sum_{j=1}^{n} f_j\delta_j, \qquad (4.47)$$

which, upon partial differentiation with respect to δ_k, yields

$$\frac{\partial\chi}{\partial\delta_k} = \frac{1}{2}\left\{\sum_{j=1}^{n}\sum_{i=1}^{n} k_{ij}\left[\left(\frac{\partial\delta_i}{\partial\delta_k}\right)\delta_j + \left(\frac{\partial\delta_j}{\partial\delta_k}\right)\delta_i\right]\right\} - \sum_{j=1}^{n} f_j\left(\frac{\partial\delta_j}{\partial\delta_k}\right). \qquad (4.48)$$

By definition, however, $\dfrac{\partial\delta_i}{\partial\delta_k} = 1 \text{ if } i = k,$

$$= 0 \text{ if } i \neq k,$$

and equation 4.48 reduces to

$$\frac{\partial \chi}{\partial \delta_k} = \frac{1}{2}\left\{\sum_{j=1}^{n} k_{kj}\,\delta_j + \sum_{i=1}^{n} k_{ik}\,\delta_i\right\} - f_k .$$

This is further simplified by the symmetry of $\boldsymbol{K}$ which permits the replacement of k_{ik} by k_{ki} in the second summation. After replacement also of the *dummy* index i by j, we finally obtain

$$\frac{\partial \chi}{\partial \delta_k} = \sum_{j=1}^{n} k_{kj}\,\delta_j - f_k. \tag{4.49}$$

In matrix notation, this becomes

$$\begin{bmatrix} \partial\chi/\partial\delta_1 \\ \partial\chi/\partial\delta_2 \\ \vdots \\ \delta\chi/\partial\delta_n \end{bmatrix} = \begin{bmatrix} k_{11} & k_{12} & \dots & k_{1n} \\ k_{21} & k_{22} & \dots & k_{2n} \\ \vdots & \vdots & \dots & \vdots \\ k_{n1} & k_{n2} & \dots & k_{nn} \end{bmatrix} \begin{bmatrix} \delta_1 \\ \delta_2 \\ \vdots \\ \delta_n \end{bmatrix} - \begin{bmatrix} f_1 \\ f_2 \\ \vdots \\ f_n \end{bmatrix} \tag{4.50}$$

or

$$\boldsymbol{d}\chi = \boldsymbol{K}\,\boldsymbol{d} - \boldsymbol{f}. \tag{4.51}$$

4.4.2 Imposition of constraints, the partitioned equation

The Rayleigh–Ritz solution now requires that the partial derivatives of χ are zero for those degrees of freedom δ_i which are free to move. In order to distinguish more easily between constrained and unconstrained degrees of freedom, let us adopt the convention that the unconstrained displacements are numbered first and constrained ones last. In the general case, suppose that there are r unconstrained degrees of freedom, $\delta_1, \delta_2, \dots, \delta_r$, and that the remainder, $\delta_{r+1}, \delta_{r+2}, \dots, \delta_n$, are constrained. The nature of the constraint varies from problem to problem. The simplest type of constraint is that of zero displacement in a particular direction at a point of support ($\delta_k = 0$). In some instances, however, we may wish to impose displacements which are known but nonzero (where a point comes to rest against a 'stop' for example). A general set of nodal constraints is therefore

$$\delta_r = \delta_r^0,\ \delta_{r+1} = \delta_{r+1}^0, \dots, \delta_n = \delta_n^0 \tag{4.52}$$

where $\delta_{r+1}^0 \dots \delta_n^0$ are known values (frequently zero).

The derivatives $\partial\chi/\partial\delta_1, \partial\chi/\partial\delta_2, \dots, \partial\chi/\partial\delta_r$ must now equate to zero. Writing equation 4.50 in 'partitioned' form, this gives

$$\begin{bmatrix} 0 \\ \vdots \\ 0 \\ \hline \partial\chi/\partial\delta_{r+1} \\ \vdots \\ \partial\chi/\partial\delta_n \end{bmatrix} = \left[\begin{array}{c|c} \boldsymbol{K}_{11} & \boldsymbol{K}_{12} \\ \hline \boldsymbol{K}_{21} & \boldsymbol{K}_{22} \end{array}\right] \begin{bmatrix} \delta_1 \\ \vdots \\ \delta_r \\ \hline \delta^0_{r+1} \\ \vdots \\ \delta^0_n \end{bmatrix} - \begin{bmatrix} f_1 \\ \vdots \\ f_r \\ \hline f_{r+1} \\ \vdots \\ f_n \end{bmatrix}, \tag{4.53}$$

where $\boldsymbol{K}_{11}, \boldsymbol{K}_{12}, \boldsymbol{K}_{21}$ and $\boldsymbol{K}_{22}$ are submatrices of $\boldsymbol{K}$, of order $r \times r$ and $r \times (n-r)$, $(n-r) \times r$ and $(n-r) \times (n-r)$ respectively. The first r rows of matrix 4.53 can then be written separately as,

$$\boldsymbol{K}_{11}\boldsymbol{d}' = \boldsymbol{f}' - \boldsymbol{K}_{12}\boldsymbol{d}_0 \tag{4.54}$$

where $\boldsymbol{d}' = \begin{bmatrix} \delta_1 \\ \delta_2 \\ \vdots \\ \delta_r \end{bmatrix}$, $\boldsymbol{f}' = \begin{bmatrix} f_1 \\ f_2 \\ \vdots \\ f_r \end{bmatrix}$ and $\boldsymbol{d}_0 = \begin{bmatrix} \delta^0_{r+1} \\ \vdots \\ \delta^0_n \end{bmatrix}$.

This reduced, equation set forms a complete system of linear equations for the unknown displacements $\delta_1, \delta_2, \ldots, \delta_r$, and can be solved to give the displacements at all points. The finite element solution is then complete.

4.4.3 Partition and solution, a simple example

The above procedure is now illustrated using as an example a finite element model formed from two bar elements. The problem to be solved is that of a uniform bar of length $2L$ and cross-sectional area A, which is fixed at one end and subject to an axial load P at the other (Fig. 4.5). It is subject also to an axial body force g_0 per unit volume. A traditional 'equilibrium' solution, obtained from a free body diagram, gives

$$\sigma_x = \frac{P + G(1 - x/2L)}{A} \tag{4.55}$$

where $G = 2ALg_0$ (this is the net body force acting on the bar). The corresponding axial displacement $u(x)$ is

$$u(x) = \frac{Px + Gx(1 - x/4L)}{EA}. \tag{4.56}$$

Consider now a finite element solution for the same problem, obtained by subdividing the bar into two equal elements to give a model identical to that of Fig. 4.4. The stiffness matrix and load vectors have already been assembled for this configuration of elements (see expressions 4.42, since no thermal loads

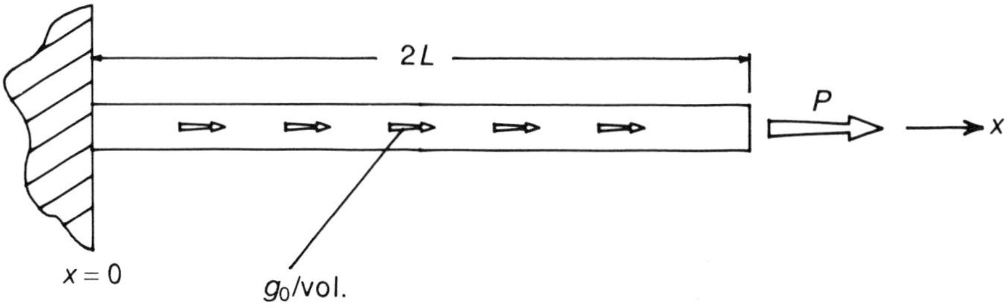

Fig. 4.5 *Worked example, geometry and loads.*

are present in the current example, f_T can be ignored). It remains to add the external concentrated load acting at the right hand end in the direction of δ_1 . The load vector f_P, therefore has components,

$$f_P = \begin{bmatrix} P \\ 0 \\ 0 \end{bmatrix}.$$

Adding this to the load vector f_g in expression 4.42 and replacing $2ALg_0$ by G, following the notation of the exact solution, we obtain

$$\chi = \frac{1}{2}[\delta_1, \delta_2, \delta_3]\begin{bmatrix} k, & -k, & 0 \\ -k, & 2k, & -k \\ 0, & -k, & k \end{bmatrix}\begin{bmatrix} \delta_1 \\ \delta_2 \\ \delta_3 \end{bmatrix} - [\delta_1, \delta_2, \delta_3]\begin{bmatrix} P + G/4 \\ G/2 \\ G/4 \end{bmatrix}.$$

Note that the unconstrained degrees of freedom (δ_1 and δ_2) have been numbered first and the constrained one at the fixed end (δ_3) last. The integers r and n are therefore two and three, respectively, and the partitioned stiffness matrix and load vector of expression 4.53 look like:

$$\left[\begin{array}{cc|c} k & -k & 0 \\ -k & 2k & -k \\ \hline 0 & -k & k \end{array}\right] \text{ and } \left[\begin{array}{c} P + G/4 \\ G/2 \\ \hline G/4 \end{array}\right].$$

The reduced equation 4.54 then becomes

$$\begin{bmatrix} k, & -k \\ -k & 2k \end{bmatrix}\begin{bmatrix} \delta_1 \\ \delta_2 \end{bmatrix} = \begin{bmatrix} P + G/4 \\ G/2 \end{bmatrix},$$

giving a solution

$$\delta_1 = (2P + G)/k \text{ and } \delta_2 = (P + 3G/4)/k. \tag{4.57}$$

These values of displacement are identical to those of the equilibrium solution in equation 4.56 evaluated at $x = L$ and $x = 2L$. The finite element solution is therefore exact at the nodal points. This is not the case within each element, however, since the interpolated displacement varies linearly, whereas the exact solution varies quadratically along the length of the bar. A comparison of computed and exact displacement throughout the bar is plotted in Fig. 4.6(a).

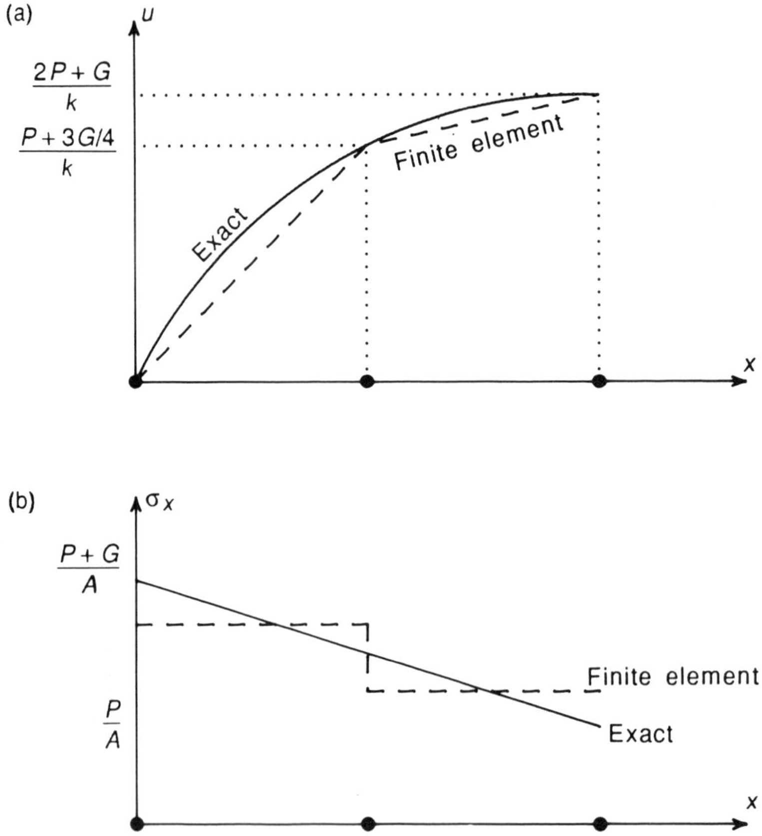

Fig. 4.6 *Worked example, comparison of computed and exact solutions: (a) the axial displacement u, (b) the axial stress* σ_x.

The stresses in the finite element model are obtained using equation 4.11 to calculate the stress within each element. In the first element, this gives

$$\sigma_x = [E]\,[-1/L, +1/L]\begin{bmatrix} u_1 \\ u_2 \end{bmatrix}, \text{ where } u_1 = \delta_3 \text{ and } u_2 = \delta_2.$$

Substituting $\delta_3 = 0$, and $\delta_2 = (P + 3G/4)/k$, we obtain

$$\sigma_x = [E]\,[-1/L, +1/L]\begin{bmatrix} 0 \\ (P+3G/4)/k \end{bmatrix} = \frac{(P+3G/4)}{A}\,. \tag{4.58}$$

In the second element which has a nodal displacement vector,

$$\begin{bmatrix} u_1 \\ u_2 \end{bmatrix} = \begin{bmatrix} \delta_2 \\ \delta_1 \end{bmatrix} = \begin{bmatrix} (P+3G/4)/k \\ (2P+G)/k \end{bmatrix},$$

the same procedure yields

$$\sigma_x = \frac{P+G/4}{A}\,.$$

These values, constant within each element, are plotted against axial distance in Fig. 4.6(b). The 'exact' solution, given by equation 4.55, is also shown. The finite element model approximates the exact solution by two constant 'steps', modelling in effect the average value of stress in each element.

A feature of Fig. 4.6(b) common to most finite element models is the presence of a discontinuity in stress between elements. At node two, for example, the axial stress changes abruptly from $(P+3G/4)/A$ in the first element to $(P+G/4)/A$ in the second. Such behaviour clearly violates equilibrium at the boundary between these elements. This is not in itself a cause for concern since equilibrium is modelled only in an approximate sense within the Rayleigh–Ritz procedure. Intuitively, however, the size of such discontinuities should decrease as the mesh is refined. This can, in fact, be used as a measure of the extent to which the finite element model has converged to the true solution (this will be discussed in greater detail in Chapter 7).

4.5 THE TWO-DIMENSIONAL BAR ELEMENT

The bar element discussed so far lies parallel to the x-axis of a global coordinate system (the term 'global' is used here to indicate that the coordinate system is universal to all elements). Any assembly of such elements forms a linear structure of the type shown in Fig. 4.3. A more useful bar element is one which is arbitrarily orientated in the x–y plane. This can be used to model plane structures of pin-jointed members.

4.5.1 Local-global transformation

Consider now a two-dimensional bar element which is orientated at an angle θ to the x-axis of a global (x, y) coordinate system. Typically such an element forms part of a pin-jointed, plane structure such as that indicated in Fig. 4.7. Apart from its orientation, the element is identical to the one-dimensional element of Fig. 4.2. Its nodes, however, displace in two directions and both components of displacement contribute to the elongation or contraction of the bar.

It is convenient to introduce a 'local' axial coordinate, x', which coincides with the axial coordinate used in the preceding section (it lies along the centroidal axis irrespective of the orientation of the element). The nodal displacements in the direction of x' are denoted by u_1 and u_2, those in the global directions by $\delta_1, \ldots, \delta_4$ (Fig. 4.7). Resolving the global displacements along the bar, we obtain

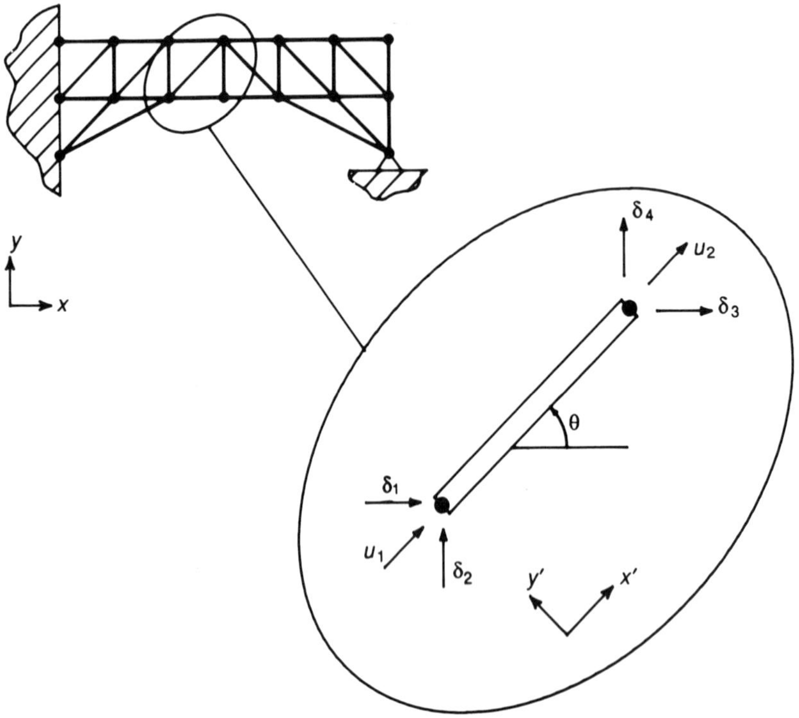

Fig. 4.7 *The two-dimensional bar element, local and global coordinate systems.*

$$u_1 = \delta_1 \cos\theta + \delta_2 \sin\theta$$

and

$$u_2 = \delta_3 \cos\theta + \delta_4 \sin\theta,$$

or

$$\begin{bmatrix} u_1 \\ u_2 \end{bmatrix} = \begin{bmatrix} \cos\theta & \sin\theta & 0 & 0 \\ 0 & 0 & \cos\theta & \sin\theta \end{bmatrix} \begin{bmatrix} \delta_1 \\ \delta_2 \\ \delta_3 \\ \delta_4 \end{bmatrix}. \qquad (4.60a)$$

This expresses the 'local' displacements, u_1 and u_2 in terms of the global displacements $\delta_1, \ldots, \delta_4$. In a more concise notation, equation 4.60 can be written as a transformation

$$\boldsymbol{d}^{e'} = \boldsymbol{T}^{e}\,\boldsymbol{d}^{e}, \tag{4.60b}$$

where $\boldsymbol{d}^{e'} = [u_1, u_2]^{T}$ is a vector of local, nodal displacements, $\boldsymbol{d}^{e} = [\delta_1, \delta_2, \delta_3, \delta_4]^{T}$ is the corresponding vector of global displacements, and $\boldsymbol{T}^{e}$ is a 2×4 matrix whose components are given in equation (4.60a).

4.5.2 Element stiffness matrix

From equation 4.16, the strain energy of the element (in the absence of thermal effects) is given by

$$U^{e} = \tfrac{1}{2}\,\boldsymbol{d}^{e'T}\,\boldsymbol{K}^{e'}\,\boldsymbol{d}^{e'}, \tag{4.61}$$

where $\boldsymbol{K}^{e'}$ is the local 2×2 stiffness matrix given by expression 4.20, that is

$$\boldsymbol{K}^{e'} = \begin{bmatrix} k & -k \\ -k & k \end{bmatrix},$$

where k is the axial stiffness $(=EA/L)$. U^{e} can be rewritten in terms of global displacements by substituting equation (4.60b) into equation 4.61 to give

$$U^{e} = \tfrac{1}{2}\,\boldsymbol{d}^{eT}\,\boldsymbol{K}^{e}\,\boldsymbol{d}^{e}, \tag{4.62}$$

where

$$\boldsymbol{K}^{e} = \boldsymbol{T}^{eT}\,\boldsymbol{K}^{e'}\,\boldsymbol{T}^{e}. \tag{4.63}$$

$\boldsymbol{K}^{e}$ is now the global stiffness matrix for the two-dimensional element with degrees of freedom $\delta_1, \ldots, \delta_4$. The components of $\boldsymbol{K}^{e}$ are obtained by evaluating explicitly the triple matrix product of equation 4.63. This gives, after some manipulation,

$$\boldsymbol{K}^{e} = k \begin{bmatrix} c^2 & sc & -c^2 & -sc \\ sc & s^2 & -sc & -s^2 \\ -c^2 & -sc & c^2 & sc \\ -sc & -s^2 & sc & s^2 \end{bmatrix}, \tag{4.64}$$

where $c = \cos\theta$ and $s = \sin\theta$. The assembly of such elements to model plane, pin-jointed structures then follows the general procedure described in section 4.3.2.

4.5.3 A two-dimensional assembly, worked example

A finite element solution for a planar assembly of bars is now illustrated using the structure shown in Fig. 4.8. This is formed from three bars of length L and $\sqrt{2}L$. Each is of cross-section A and Young's modulus E. A vertical load, F, is applied as shown.

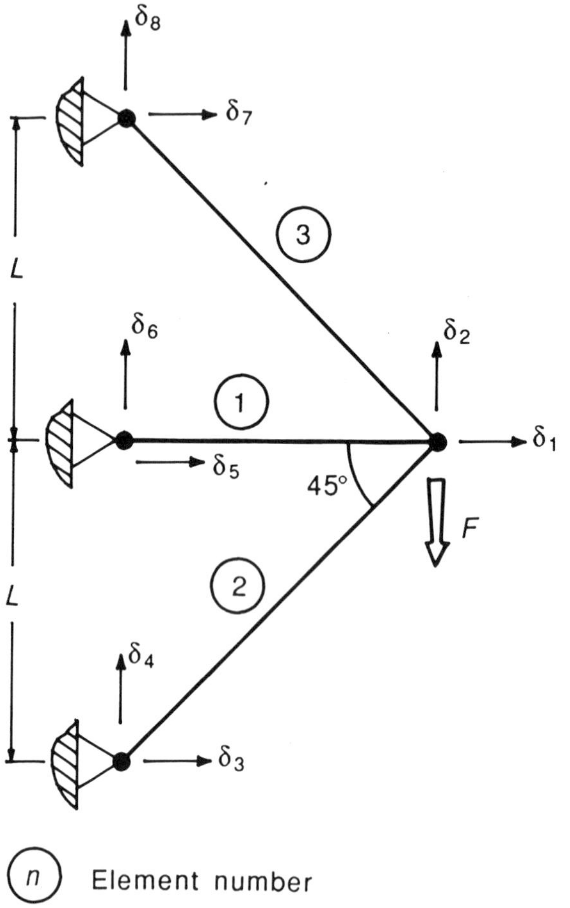

Fig. 4.8 *A finite element model for a three-bar assembly.*

The structure is modelled using three bar elements (numbered one, two and three). Node and degree of freedom numbers are assigned as shown. Note that in numbering the degrees of freedom, care is taken to number the unconstrained displacement components (δ_1 and δ_2) first and the constrained ones ($\delta_3, \ldots, \delta_8$) last.

The stiffness matrices of the elements are formed by substituting appropriate values for θ and k into equation 4.64. In the case of element one, with $\theta = 0$, this gives

$$\boldsymbol{K}^{(1)} = k^{(1)} \begin{bmatrix} 1 & 0 & -1 & 0 \\ 0 & 0 & 0 & 0 \\ -1 & 0 & 1 & 0 \\ 0 & 0 & 0 & 0 \end{bmatrix}, \text{ where } k^{(1)} = EA/L \ (=k, \text{ say}).$$

In the case of element two, with $\theta = 45^\circ$

$$\boldsymbol{K}^{(2)} = k^{(2)} \begin{bmatrix} 1/2 & 1/2 & -1/2 & -1/2 \\ 1/2 & 1/2 & -1/2 & -1/2 \\ -1/2 & -1/2 & 1/2 & 1/2 \\ -1/2 & -1/2 & 1/2 & 1/2 \end{bmatrix}, \text{ where } k^{(2)} = EA/\sqrt{2}L = k/\sqrt{2},$$

and in the case of element three, with $\theta = 45^\circ$,

$$\boldsymbol{K}^{(3)} = k^{(3)} \begin{bmatrix} 1/2 & -1/2 & -1/2 & 1/2 \\ -1/2 & 1/2 & 1/2 & -1/2 \\ -1/2 & 1/2 & 1/2 & -1/2 \\ 1/2 & -1/2 & -1/2 & 1/2 \end{bmatrix}, \text{ where } k^{(3)} = k/\sqrt{2}.$$

The nodal displacement vectors for the three elements are

$$\boldsymbol{d}^{(1)} = \begin{bmatrix} \delta_5 \\ \delta_6 \\ \delta_1 \\ \delta_2 \end{bmatrix}, \ \boldsymbol{d}^{(2)} = \begin{bmatrix} \delta_3 \\ \delta_4 \\ \delta_1 \\ \delta_2 \end{bmatrix} \text{ and } \boldsymbol{d}^{(3)} = \begin{bmatrix} \delta_7 \\ \delta_8 \\ \delta_1 \\ \delta_2 \end{bmatrix}.$$

Assembly of $\boldsymbol{K}^{(1)}, \boldsymbol{K}^{(2)}$ and $\boldsymbol{K}^{(3)}$ into an initially zero 8×8 matrix then gives:
(i) after assembly of element one

Column

$$\boldsymbol{K} = \begin{bmatrix} k & & & & -k & & & \\ & & & & & & & \\ & & & & & & & \\ & & & & & & & \\ -k & & & & k & & & \\ & & & & & & & \\ & & & & & & & \\ & & & & & & & \end{bmatrix}$$

Row	1	2	3	4	5	6	7	8
1	k				$-k$			
2								
3								
4								
5	$-k$				k			
6								
7								
8								

(ii) after assembly of element two

$$
\begin{array}{c}
\text{Column} \\
\begin{array}{cc}
\begin{array}{cccccccc}
1 & 2 & 3 & 4 & 5 & 6 & 7 & 8
\end{array} & \\
\mathbf{K} = \left[\begin{array}{cccccccc}
k + k/2\sqrt{2} & k/2\sqrt{2} & -k/2\sqrt{2} & -k/2\sqrt{2} & -k & & & \\
k/2\sqrt{2} & k/2\sqrt{2} & -k/2\sqrt{2} & -k/2\sqrt{2} & & & & \\
-k/2\sqrt{2} & -k/2\sqrt{2} & k/2\sqrt{2} & k/2\sqrt{2} & & & & \\
-k/2\sqrt{2} & -k/2\sqrt{2} & k/2\sqrt{2} & k/2\sqrt{2} & & & & \\
-k & & & & k & & & \\
 & & & & & & & \\
 & & & & & & & \\
 & & & & & & &
\end{array}\right] & \begin{array}{c} 1 \\ 2 \\ 3 \\ 4 \\ 5 \\ 6 \\ 7 \\ 8 \end{array} \text{Row}
\end{array}
\end{array}
$$

(iii) after assembly of element three

$$
\begin{array}{c}
\text{Column} \\
\begin{array}{cc}
\begin{array}{cccccccc}
1 & 2 & 3 & 4 & 5 & 6 & 7 & 8
\end{array} & \\
\mathbf{K} = \left[\begin{array}{cccccccc}
k + k/\sqrt{2} & 0 & -k/2\sqrt{2} & -k/2\sqrt{2} & -k & & -k/2\sqrt{2} & k/2\sqrt{2} \\
0 & k/\sqrt{2} & -k/2\sqrt{2} & -k/2\sqrt{2} & & & k/2\sqrt{2} & -k/2\sqrt{2} \\
-k/2\sqrt{2} & -k/2\sqrt{2} & k/2\sqrt{2} & k/2\sqrt{2} & & & & \\
-k/2\sqrt{2} & -k/2\sqrt{2} & k/2\sqrt{2} & k/2\sqrt{2} & & & & \\
-k & & & & k & & & \\
 & & & & & & & \\
-k/2\sqrt{2} & k/2\sqrt{2} & & & & & k/2\sqrt{2} & -k/2\sqrt{2} \\
k/2\sqrt{2} & -k/2\sqrt{2} & & & & & -k/2\sqrt{2} & k/2\sqrt{2}
\end{array}\right] & \begin{array}{c} 1 \\ 2 \\ 3 \\ 4 \\ 5 \\ 6 \\ 7 \\ 8 \end{array} \text{Row}
\end{array}
\end{array}
$$

Since there are only three elements in the current model, this completes the assembly process.

The only external load is the concentrated force at node four acting to oppose degree of freedom δ_2. The force vector $\boldsymbol{f}$ is therefore

$$
\boldsymbol{f} = \begin{bmatrix} 0 \\ -F \\ 0 \\ 0 \\ 0 \\ 0 \\ 0 \\ 0 \end{bmatrix}.
$$

Fixed supports at nodes one, two and three give zero displacements at these points. In other words, degrees of freedom $\delta_3, \delta_4 \ldots \delta_8$ are constrained to zero. Partitioning the stiffness equation (with $n = 8$ and $r = 2$, see equation 4.53) we obtain

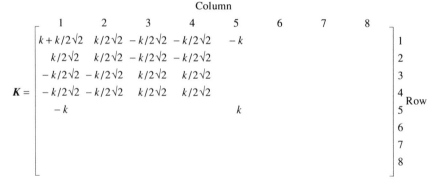

$$\begin{bmatrix} k + k/\sqrt{2} & 0 \\ 0 & k/\sqrt{2} \end{bmatrix} \begin{bmatrix} \delta_1 \\ \delta_2 \end{bmatrix} = \begin{bmatrix} 0 \\ -F \end{bmatrix}, \tag{4.65}$$

which can be solved (trivially) to give

$$\delta_1 = 0 \text{ and } \delta_2 = -F\sqrt{2}/k .$$

By resolving these nodal displacements along the axis of each bar, it is then simple to calculate the strain in each element. These are $F/(\sqrt{2}EA)$, 0 and $-F/(\sqrt{2}EA)$ in elements three, one and two, respectively, giving axial forces $F/\sqrt{2}$, 0 and $-F/\sqrt{2}$, in agreement with the equilibrium solution (confirmation of this last statement is left to the reader).

4.5.4 Concluding remarks

Although the preceding example involves a relatively simple structure, the same approach may be applied to pin-jointed frameworks of virtually unlimited complexity. Manual assembly and solution of the stiffness equations, time-consuming even in this simple example, becomes extremely tedious as the number of elements increases. For more realistic problems, it soon becomes necessary to write a computer program to perform these operations.

In doing so, an obvious way to reduce the computational effort is to assemble and process only that portion of the stiffness matrix which is required for the final reduced equation. In the preceding problem, for example, a substantial proportion of the effort involved in assembling the stiffness matrix is wasted in the sense that only four of the sixty four stiffness components (those in the first two rows and columns) are ultimately required to form equation 4.65. In the general case, components of element stiffness need only be assembled for those degrees of freedom which correspond to unconstrained displacements. In the context of the previous example it was only strictly necessary to assemble and store those terms in the first two rows and columns of the final matrix. With hindsight it is unnecessary even to number the constrained degrees of freedom. They can simply be 'tagged' and neglected altogether when they are encountered in the assembly process.

When the reduced matrix has been formed the next step is to solve the resulting set of linear equations. This is the most time-consuming part of the process for most linear problems and must be performed as efficiently as possible. Some simple steps which can be taken to minimize the computational effort involved are described in the next chapter.

REFERENCES.

[1] Stasa, F. L. (1985) *Applied Finite Element Analysis for Engineers*, CBS Publishing, Tokyo, pp. 45–62.

[2] Fletcher, C. A. J. (1984) *Computational Galerkin Methods*, Springer-Verlag, New York.
[3] Zienkiewicz, O. C. and Taylor R. L. (1990) *The Finite Element Method*, 4th edn, Vol. 1, McGraw-Hill, London, Chapter 9.
[4] Cook, R. D., Malkus D. S. and Plesha, M. E. (1989) *Concepts and Applications of Finite Element Analysis*, 3rd edn, John Wiley & Sons, New York, Chapter 15.

PROBLEMS

1. A bar element of Young's modulus E and length L has a variable cross-sectional area, $A(x')$, where x' is a local axial coordinate, (see figure). Nodes 1 and 2 are defined as shown with axial displacements u_1 and u_2. Linear interpolation of the axial displacement is assumed within the element. Show that the element stiffness matrix is

$$\boldsymbol{K}^{\mathrm{e}} = \bar{k}\begin{bmatrix} 1 & -1 \\ -1 & 1 \end{bmatrix}$$

where $\bar{k} = E\bar{A}/L$ and $\bar{A} = (1/L)\int_0^L A(x')\,\mathrm{d}x'$.

[*Note*: Since $u(x')$ varies linearly with x', the shape functions for this element are the same as for an element of constant cross-sectional area. See section 4.2]

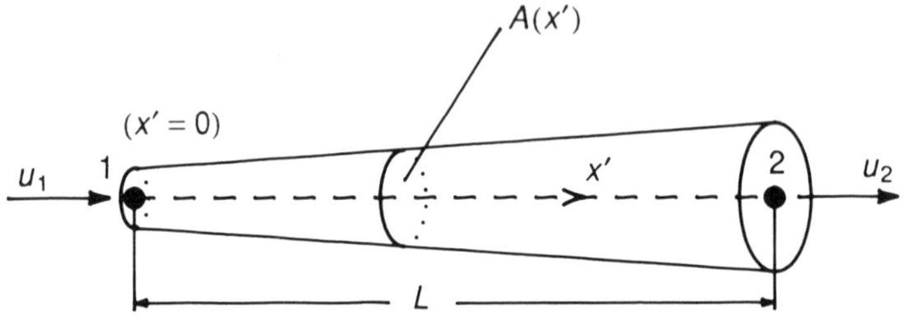

Problem 1.

2. A three-noded bar element is defined with one node at the centre and one at each end as shown. The element is of cross-sectional area A, Young's modulus E and length L. It has nodal displacements u_1, u_2 and u_3. A local axial coordinate, x', has its origin at the central node.

The axial displacement, u, is assumed to vary quadratically within the element so that, $u(x') = \alpha_1 + \alpha_2 x' + \alpha_3 x'^2$. Evaluate α_1, α_2 and α_3 in terms of u_1, u_2 and u_3, and show that u may also be written

$$u(x') = n_1(x')\,u_1 + n_2(x')\,u_2 + n_3(x')\,u_3$$

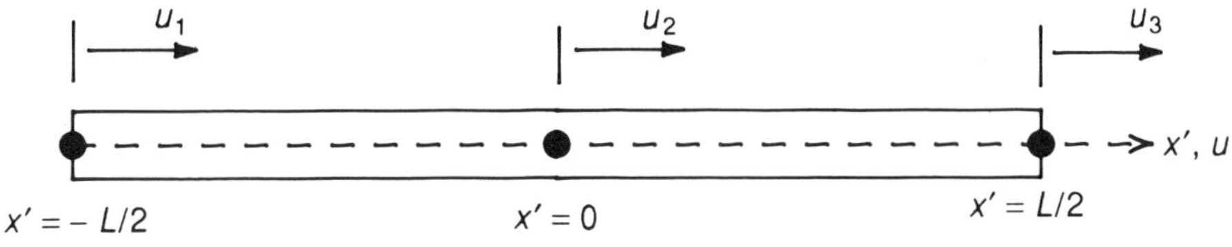

Problem 2.

where

$$n_1(x') = -\left(\frac{x'}{L}\right)\left(1 - \frac{2x'}{L}\right),$$

$$n_2(x') = \left(1 - \frac{4x'^2}{L^2}\right),$$

$$n_3(x') = \left(\frac{x}{L}\right)\left(1 + \frac{2x'}{L}\right).$$

If the nodal displacement vector for the element is defined to be

$$\boldsymbol{d}^{\mathrm{e}} = \begin{bmatrix} u_1 \\ u_2 \\ u_3 \end{bmatrix},$$

show that the strain displacement matrix $\boldsymbol{B}^{\mathrm{e}}$ has components

$$\boldsymbol{B}^{\mathrm{e}} = \left[\frac{1}{L}\left(\frac{4x'}{L} - 1\right), -\frac{8x'}{L^2}, \frac{1}{L}\left(1 + \frac{4x'}{L}\right)\right]$$

and that the stiffness matrix is

$$\boldsymbol{K}^{\mathrm{e}} = k\begin{bmatrix} 7/3 & -8/3 & 1/3 \\ -8/3 & 16/3 & -8/3 \\ 1/3 & -8/3 & 7/3 \end{bmatrix},$$

where $k = EA/L$

3. A two-noded bar element has the dimensions and properties shown in Fig. 4.2 and is subject to an axial body force which varies linearly from zero at node 1 to g_0 at node 2, that is

$$g(x') = g_0(x'/L).$$

Show that the equivalent nodal forces which must be applied to the element are $g_0AL/6$ at node 1 and $g_0AL/3$ at node 2.

4. A constant axial body force g_0 per unit volume acts on the three-noded element of problem two. Show that the resulting equivalent nodal forces are $2G/3$ at the centre and $G/6$ at each end ($G = g_0AL$).

5. A two-noded bar element has the dimensions and properties shown in Fig. 4.2 and is subject to an axially varying temperature rise $T(x')^\circ$ which varies linearly from 0 at node one to T_0 at node two, that is

$$T(x') = T_0\left(\frac{x'}{L}\right).$$

Show that the equivalent nodal forces are $\pm\frac{1}{2}EA\alpha T_0$ where α is the coefficient of thermal expansion.

6. Obtain a finite element solution for the problem of Fig. 4.5 by modelling the bar with a single, three-noded element of the type discussed in problem two. Show that the resulting solution is identical to the exact solution at all points and comment on this agreement.

7. Repeat problem 6 using a model formed by two two-noded elements of unequal length, as shown below. Compare the resulting displacements and stresses with those predicted by an exact solution.

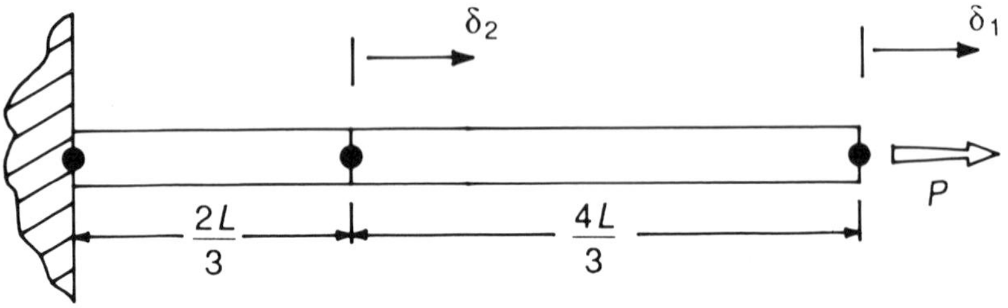

Problem 6.

8. Repeat problem six using three two-noded elements of equal length. Compare the accuracy of the resulting solution with that presented in the text.

9. A tapered bar of variable cross-sectional area shown below is subject to an axial load P at its free end. The bar has a Young's modulus E and of length $2L$. The cross-sectional area tapers linearly from $3A_0$, at the left hand end, to A_0 at the right.

   A finite element model is formed by dividing the bar into two elements of the type discussed in problem one. Numbering the degrees of freedom δ_1, δ_2 and δ_3 as shown, show that the stiffness matrix for the system is

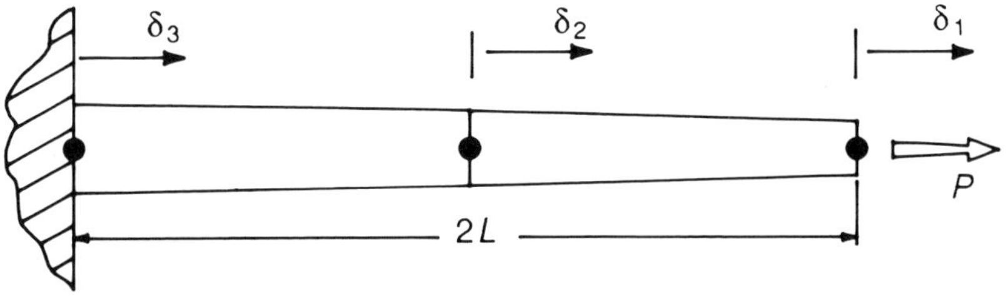

Problem 9.

$$\begin{bmatrix} 3k/2 & -3k/2 & 0 \\ -3k/2 & 4k & -5k/2 \\ 0 & -5k/2 & 5k/2 \end{bmatrix}$$

where $k = EA_0/L$. Obtain a solution by imposing suitable constraints and compare with the exact solution. [*Note*: The exact displacement field is obtained by integrating $\sigma_x = E\,(\mathrm{d}u/\mathrm{d}x) = P/A\,(x)$, subject to the boundary condition $u\,(0) = 0$]

10. Show that the thermal load vector f_T^e for a two-dimensional bar element (see Fig. 4.7) which experiences a uniform temperature rise T° is

$$f_T^e = \begin{bmatrix} -E\alpha AT\cos\theta \\ -E\alpha AT\sin\theta \\ E\alpha AT\cos\theta \\ E\alpha AT\sin\theta \end{bmatrix}$$

where α is the coefficient of thermal expansion. [*Note*: To obtain this result, transform the thermal load vector of the one-dimensional element using a local-global transformation for nodal forces similar to that used for nodal displacements in section 4.5.1]

Element number three of the three bar assembly of Fig. 4.8 is subject to a uniform temperature rise T_0°. No other loads are applied to the structure and the remaining elements are insulated from temperature change. Retaining the same degree of freedom numbering system as in Fig. 4.8, show that δ_1 and δ_2 are given by

$$\delta_1 = \frac{\alpha T_0 L}{1+\sqrt{2}} \quad \text{and} \quad \delta_2 = -\alpha T_0 L.$$

11. The plane structure (see figure overleaf) is formed by two, pin-jointed members of length L and $(\sqrt{5}L/2)$. Each is of Young's modulus E and

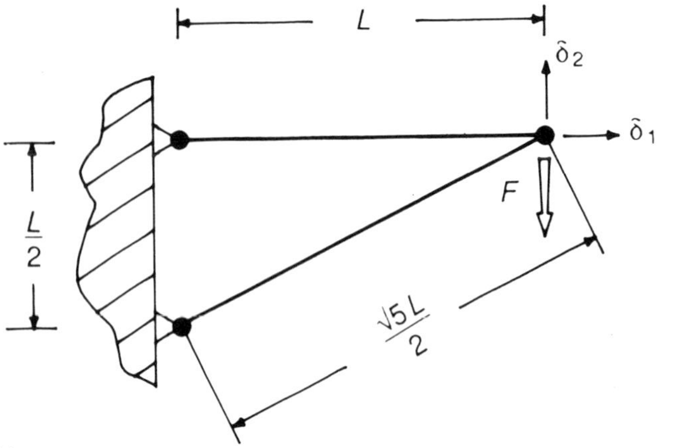

Problem 11.

cross-sectional area A. The framework is loaded with a vertical force F as shown. A finite element model is constructed by representing each member as a two-dimensional bar element. Show that the stiffness relationship for the model is

$$\begin{bmatrix} k+4k'/5 & 2k'/5 \\ 2k'/5 & k'/5 \end{bmatrix} \begin{bmatrix} \delta_1 \\ \delta_2 \end{bmatrix} = \begin{bmatrix} 0 \\ -F \end{bmatrix},$$

where $k = EA/L$ and $k' = 2EA/\sqrt{5}L$. Determine the displacements δ_1 and δ_2 and compare with an exact solution.

12. The framework shown in the figure is formed by four pin-jointed bars of Young's modulus E and cross-sectional area A. It is loaded by a horizontal force F. Treating each member as a bar element and numbering the unconstrained degrees of freedom δ_1, δ_2, δ_3 and δ_4 as shown, show that the stiffness relationship for the structure is

$$k\begin{bmatrix} (1+\sqrt{2}/4) & \sqrt{2}/4 & -1 & 0 \\ \sqrt{2}/4 & (1+\sqrt{2}/4) & 0 & 0 \\ -1 & 0 & 1 & 0 \\ 0 & 0 & 0 & 1 \end{bmatrix} \begin{bmatrix} \delta_1 \\ \delta_2 \\ \delta_3 \\ \delta_4 \end{bmatrix} = \begin{bmatrix} 0 \\ 0 \\ F \\ 0 \end{bmatrix},$$

where $k = EA/L$. Solve for $\delta_1, \ldots, \delta_4$ and show that the force in each member is the same as that predicted by a traditional, equilibrium analysis.

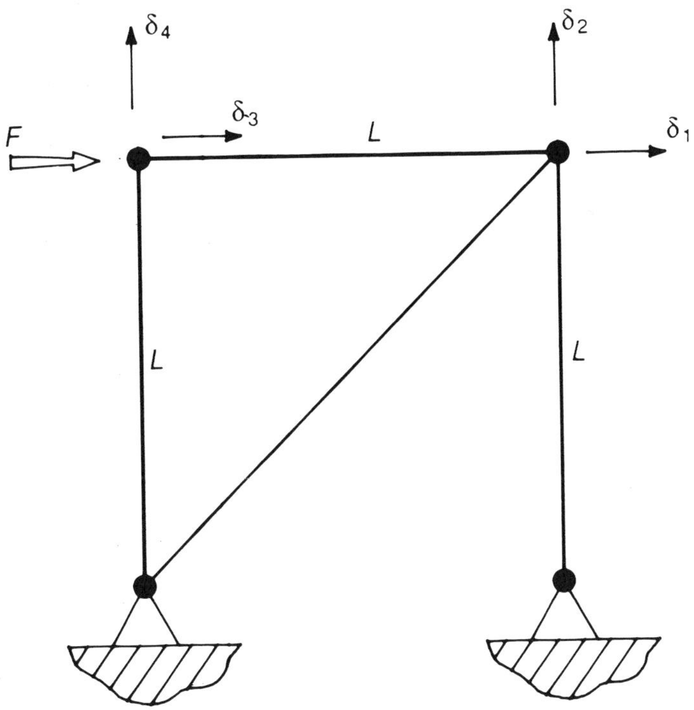

Problem 12.

5

Solutions of the finite element equations, some practical considerations

5.1 INTRODUCTION

In computing solutions to the finite element equations, the order in which the unconstrained degrees of freedom are numbered and the order in which the elements are assembled can be of considerable importance. The former determines the final structure of the stiffness matrix and the latter its composition during assembly. Depending upon the type of algorithm used to solve the equations, these numbering systems can have a significant effect on the effort required to produce a solution.

The extent to which the final structure of the stiffness matrix is influenced by the degree of freedom numbering system, is illustrated in Fig. 5.1 and 5.2. Fig. 5.1 shows an assembly of bar elements with 16 unconstrained, degrees of freedom. These are numbered using alternative numbering systems (a) and (b). The elements are also numbered for future reference. The nonzero terms which occur in the stiffness matrices for each model are shown in Fig. 5.2(a) and (b). Their location, but not their magnitude, is indicated by a solid dot. These are placed by taking each element in turn, noting its degree of freedom numbers and inserting a marker in the appropriate rows and columns of an initially zero matrix. In comparing Fig. 5.2(a) and (b), we note immediately that the stiffness matrix obtained from numbering system (a) has a more 'compact' appearance than that for system (b). In particular, the nonzero terms are concentrated more tightly about the principal diagonal. This type of compactness can have a profound effect on the effort which is required to solve the equations. It is discussed further in section 5.3.

A second numbering system which can influence the computational effort required to solve the equations, is that of the elements, or to be more precise, the order in which they are assembled. Although this does not alter the final composition of the stiffness matrix, it does determine the location of the nonzero terms during assembly when some, but not all, of the elements have been processed. The instantaneous distribution of these terms within the matrix has a profound effect on the solution time when algorithms are used which simultaneously assemble *and* progressively solve the equations. This type of interaction is discussed in section 5.4.

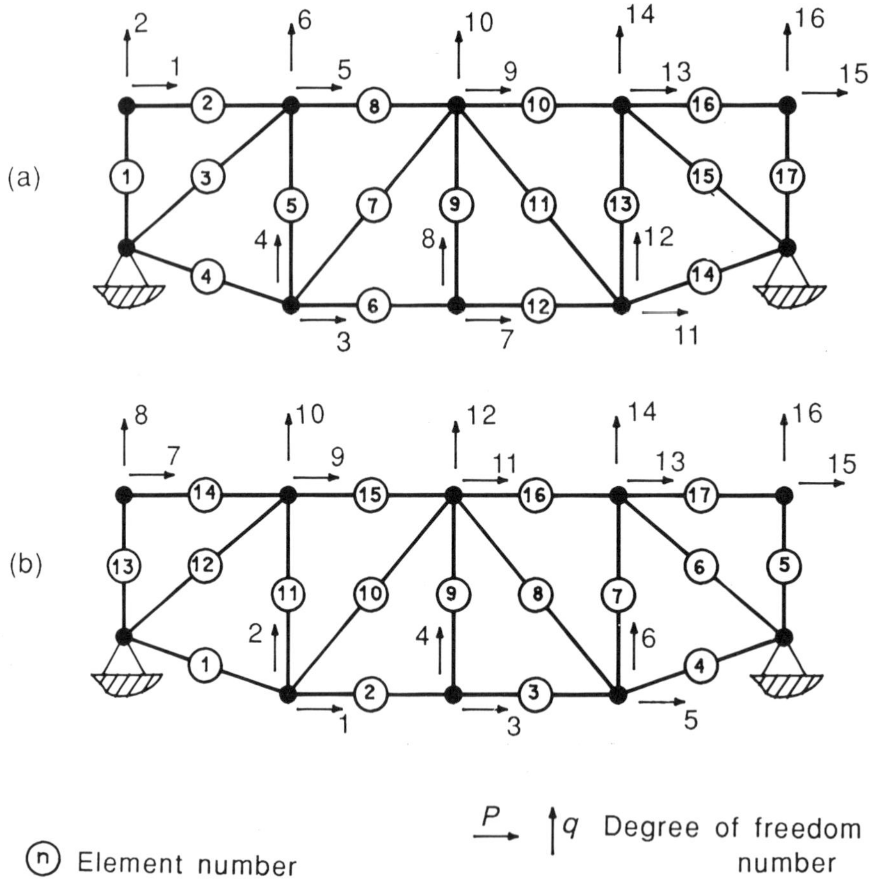

Fig. 5.1 *Element and degree of freedom numbering systems for an assembly of bars.*

Before looking in detail at either of these effects, it is helpful to review the methods commonly used to solve the finite element equations. A comprehensive treatment is not appropriate here, but a brief discussion of some of the more common methods forms a useful preliminary to further discussion of the numbering systems of the physical model.

5.2 SOLUTION ALGORITHMS

5.2.1 Direct and indirect methods

Consider first the general categories of algorithm which can be applied to the solution of the equation set:

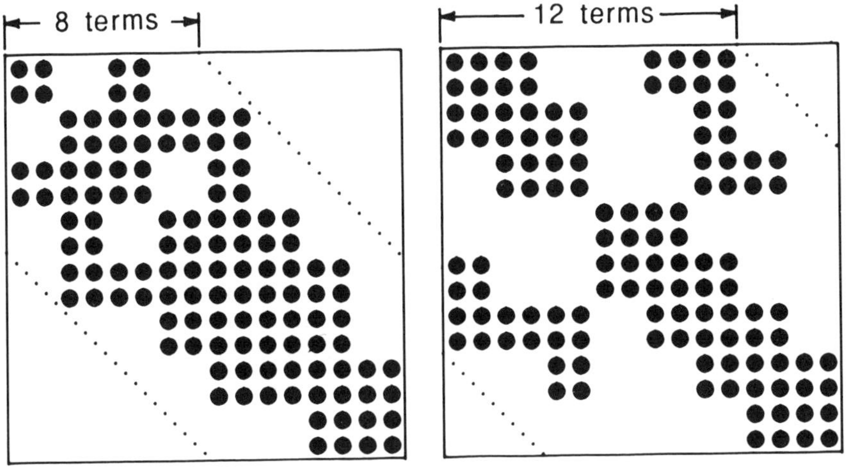

(a) (b)

Fig. 5.2 *Location of nonzero stiffness terms for the models of Fig. 5.1*

$$\boldsymbol{K}\,\boldsymbol{d}=\boldsymbol{f}. \tag{5.1}$$

Let us assume that the coefficient matrix $\boldsymbol{K}$ is symmetric and nonsingular. For the purposes of illustration, we will refer from time to time to the particular equations

$$\begin{bmatrix} 1 & -1 & 0 & 0 \\ -1 & 2 & -1 & 0 \\ 0 & -1 & 2 & -1 \\ 0 & 0 & -1 & 2 \end{bmatrix}\begin{bmatrix} \delta_1 \\ \delta_2 \\ \delta_3 \\ \delta_4 \end{bmatrix}=\begin{bmatrix} 1 \\ 0 \\ 0 \\ 0 \end{bmatrix}. \tag{5.2}$$

These have a solution; $\delta_1=4$, $\delta_2=3$, $\delta_3=2$ and $\delta_1=1$, and correspond to a suitably scaled four-element model of a uniform bar with a concentrated load at one end.

The first decision in selecting a solution algorithm, is whether to use a 'direct' or an 'indirect' method. A direct method gives the exact solution (within the limits of computational accuracy) after a finite number of steps. An indirect method gives a successively closer approximation until a prescribed accuracy is achieved. Common examples of direct methods are Gaussian elimination and a variety of LU ('lower-upper') decompositions. Gauss–Seidel iteration and its variants are the most widely used indirect methods, but gradient techniques are also increasing in popularity. The conventional wisdom is that direct methods are more effective in linear, finite element applications, indirect solvers being 'non- competitive except in . . . the solution of very large three-dimensional problems' [1]. This statement is being challenged to some extent by the introduction of parallel processing computers which can be used more effectively with iterative solvers, but most existing codes still use direct algorithms. We will confine our observations to direct methods in all that follows. Further information on indirect solvers can be found at an introductory level in [2].

5.2.2 Gaussian elimination

Gaussian elimination utilizes the property that a multiple of any row of the coefficient matrix may be added to, or subtracted from, any other row without altering the final solution, provided that the same operation is performed on the corresponding rows of the force vector. In algebraic terms, the process is equivalent to adding or subtracting multiples of the scalar equations which form the matrix relationship. The coefficient matrix and force vector of equation 5.1 can be systematically manipulated in this way to produce a new, but equivalent, system of equations

$$\boldsymbol{K}^* \boldsymbol{d} = \boldsymbol{f}^*, \tag{5.3a}$$

where the coefficient matrix $\boldsymbol{K}^*$ has zeros in all locations under the leading diagonal. It is termed an 'upper triangular' matrix. Equation 5.3 then looks like

$$\begin{bmatrix} k_{11}^* & k_{12}^* & k_{13}^* & \dots & k_{1n}^* \\ 0 & k_{22}^* & k_{23}^* & \dots & k_{2n}^* \\ 0 & 0 & k_{33}^* & \dots & k_{3n}^* \\ \vdots & \vdots & \vdots & & \vdots \\ 0 & 0 & 0 & & k_{nn}^* \end{bmatrix} \begin{bmatrix} \delta_1 \\ \delta_2 \\ \delta_3 \\ \vdots \\ \delta_n \end{bmatrix} = \begin{bmatrix} f_1^* \\ f_2^* \\ f_3^* \\ \vdots \\ f_n^* \end{bmatrix}. \tag{5.3b}$$

Its solution is straightforward since the last row immediately yields

$$\delta_n = f_n^* / k_{nn}^* .$$

This may be substituted into the row above to give

$$k_{n-1,n-1}^* \delta_{n-1} = f_{n-1}^* - k_{n-1,n}^* \delta_n$$

which, since δ_n is now known, yields a value for δ_{n-1}, and so on. When the top row is reached, all of the δ_i's have been evaluated and the solution is complete.

The above procedure involves two distinct phases: 'reduction', the manipulation of the initial matrix into upper triangular form, and 'back substitution', the consequent evaluation of the unknown displacements. The reduction algorithm may be summarized as follows (a more complete description and fortran listings can be found in [2] and [3])

1. select row i starting with $i = 1$ and continuing until $i = n - 1$,
2. select row j, starting with $j = i + 1$ and continuing until $j = n$,
3. replace row j of $\boldsymbol{K}^*$ and $\boldsymbol{f}^*$ by 'row $j - \alpha$. row i ' where $\alpha = k^*_{ji}/k^*_{ii}$ and where K^* and f^* are the *current* versions of coefficient matrix and force vector.

Consider for example the reduction of equations 5.2 using this method. Selecting row 1, and subtracting appropriate multiples of it from rows 2, 3 and 4, we obtain a new set of equations given by

$$\begin{bmatrix} 1 & -1 & 0 & 0 \\ 0 & 1 & -1 & 0 \\ 0 & -1 & 2 & -1 \\ 0 & 0 & -1 & 2 \end{bmatrix} \begin{bmatrix} \delta_1 \\ \delta_2 \\ \delta_3 \\ \delta_4 \end{bmatrix} = \begin{bmatrix} 1 \\ 1 \\ 0 \\ 0 \end{bmatrix}.$$

Next, selecting row 2 and subtracting multiples from rows 3 and 4, we obtain

$$\begin{bmatrix} 1 & -1 & 0 & 0 \\ 0 & 1 & -1 & 0 \\ 0 & 0 & 1 & -1 \\ 0 & 0 & -1 & 2 \end{bmatrix} \begin{bmatrix} \delta_1 \\ \delta_2 \\ \delta_3 \\ \delta_4 \end{bmatrix} = \begin{bmatrix} 1 \\ 1 \\ 1 \\ 0 \end{bmatrix}.$$

Finally, selecting row 3 and subtracting a multiple of it from row 4, gives

$$\begin{bmatrix} 1 & -1 & 0 & 0 \\ 0 & 1 & -1 & 0 \\ 0 & 0 & 1 & -1 \\ 0 & 0 & 0 & 1 \end{bmatrix} \begin{bmatrix} \delta_1 \\ \delta_2 \\ \delta_3 \\ \delta_4 \end{bmatrix} = \begin{bmatrix} 1 \\ 1 \\ 1 \\ 1 \end{bmatrix} \tag{5.4}$$

The reduction is now complete and the coefficient matrix is in upper triangular form. Back substitution then yields:

from row 4: $\delta_4 = 1$,
from row 3: $\delta_3 - \delta_4 = 1$, giving $\delta_3 = 2$,
from row 2: $\delta_2 - \delta_3 = 1$, giving $\delta_2 = 3$,
from row 1: $\delta_1 - \delta_2 = 1$, giving $\delta_1 = 4$,

which is indeed the correct solution.

5.2.3 LU methods, Choleski factorization

Although Gaussian elimination provides a simple and robust algorithm, it is seldom the most effective of the direct solution techniques in finite element applications. Solvers based on the LU (lower-upper) approach tend to perform more efficiently. These factor the stiffness matrix into the product of a lower triangular matrix $\boldsymbol{L}$, and an upper triangular matrix $\boldsymbol{U}$. In other words, $\boldsymbol{K}$ is written

$$\boldsymbol{K} = \boldsymbol{L}\,\boldsymbol{U} \tag{5.5}$$

where $\boldsymbol{L}$ and $\boldsymbol{U}$ have components

$$\boldsymbol{L} = \begin{bmatrix} l_{11} & 0 & 0 & \dots & 0 \\ l_{21} & l_{22} & 0 & \dots & 0 \\ l_{31} & l_{32} & l_{33} & \dots & 0 \\ \vdots & \vdots & \vdots & & \vdots \\ l_{n1} & l_{n2} & l_{n3} & \dots & l_{nn} \end{bmatrix}, \tag{5.6}$$

$$\boldsymbol{U} = \begin{bmatrix} u_{11} & u_{12} & u_{13} & \dots & u_{1n} \\ 0 & u_{22} & u_{23} & \dots & u_{2n} \\ 0 & 0 & u_{33} & \dots & u_{3n} \\ \vdots & \vdots & \vdots & & \vdots \\ 0 & 0 & 0 & \dots & u_{nn} \end{bmatrix}. \tag{5.7}$$

The original equation then becomes

$$[\boldsymbol{L}\,\boldsymbol{U}]\boldsymbol{d} = \boldsymbol{f},$$

and its solution is obtained by solving the two equations

$$\boldsymbol{L}\,\boldsymbol{a} = \boldsymbol{f} \text{ and } \boldsymbol{U}\,\boldsymbol{d} = \boldsymbol{a} \tag{5.8}$$

The second of these, $\boldsymbol{U}\,\boldsymbol{d} = \boldsymbol{a}$, is in the same upper triangular form as the reduced Gauss-elimination equation, $\boldsymbol{K}^*\boldsymbol{d} = \boldsymbol{f}^*$ (see equation 5.3b) and can be solved by an identical back-substitution process. The first, $\boldsymbol{L}\,\boldsymbol{a} = \boldsymbol{f}$, is in 'lower triangular' form — the coefficient matrix having zeros above the diagonal rather than below it — and can be solved using an analogous process, termed 'forward substitution' which starts at the top of the system of equations and works downwards. Once $\boldsymbol{K}$ has been factored, the solution of equations 5.8 is therefore relatively straightforward.

The LU factorization itself is nonunique and a number of different products may be used as the basis for a solution. One of the most effective for symmetric, positive definite matrices, and the only one discussed here, is the 'Choleski' or 'square root' method, in which $\boldsymbol{L}$ is obtained as the transpose of $\boldsymbol{U}$. The factorization is then

$$\boldsymbol{K} = \boldsymbol{U}^{\mathrm{T}}\,\boldsymbol{U}. \tag{5.9}$$

This is achieved in the following way. First, equation 5.5 is rewritten in full component form as,

$$\begin{bmatrix} k_{11} & k_{12} & k_{13} & \dots & k_{1n} \\ k_{21} & k_{22} & k_{23} & \dots & k_{2n} \\ k_{31} & k_{32} & k_{33} & \dots & k_{3n} \\ \vdots & \vdots & \vdots & & \vdots \\ k_{n1} & k_{n2} & k_{n3} & \dots & k_{nn} \end{bmatrix} = \begin{bmatrix} u_{11} & 0 & 0 & \dots & 0 \\ u_{12} & u_{22} & 0 & \dots & 0 \\ u_{13} & u_{23} & u_{33} & \dots & 0 \\ \vdots & \vdots & \vdots & & \vdots \\ u_{1n} & u_{2n} & u_{3n} & \dots & u_{nn} \end{bmatrix} \begin{bmatrix} u_{11} & u_{12} & u_{13} & \dots & u_{1n} \\ 0 & u_{22} & u_{23} & \dots & u_{2n} \\ 0 & 0 & u_{33} & \dots & u_{3n} \\ \vdots & \vdots & \vdots & & \vdots \\ 0 & 0 & 0 & \dots & u_{nn} \end{bmatrix} \tag{5.10}$$

The terms in $\boldsymbol{U}$ are then obtained by equating components on either side of this identity starting at the top left hand corner and moving from left to right across the top of the stiffness matrix equating components in each column down to and including the diagonal. This in effect assigns values to the corresponding locations of $\boldsymbol{U}$. A detailed description of the Choleski algorithm and a fortran listing are to be found in [2]. The basic idea is readily grasped by applying the method to the first few columns of equation 5.10. Equating 1–1 components on each side of the identity, we obtain

$$k_{11} = u_{11}^2, \text{ or } u_{11} = \sqrt{k_{11}}, \tag{5.11}$$

which determines the value of u_{11}. Equating the 1–2 and 2–2 terms gives

$$k_{12} = u_{11} \,.\, u_{12} \text{ or } u_{12} = k_{12}/u_{11},$$

and (5.12)

$$k_{22} = u_{12}^2 + u_{22}^2 \text{ or } u_{22} = \sqrt{(k_{22} - u_{12}^2)}.$$

This determines u_{12} and u_{22}. Equating the 1–3, 2–3 and 3–3 terms gives

$$\begin{aligned} u_{13} &= k_{13}/u_{11}, \\ u_{23} &= (k_{23} - u_{12}u_{13})/u_{22}, \\ u_{33} &= \sqrt{(k_{33} - u_{23}^2 - u_{13}^2)}, \end{aligned} \tag{5.13}$$

and so on. The process is continued until the whole of $\boldsymbol{U}$ has been evaluated. The presence of a square root within the expression for the diagonal terms causes no difficulty provided that the coefficient matrix, $\boldsymbol{K}$, is positive definite. This is always the case for correctly restrained, static, finite element models.

Applying Choleski decomposition to the 4×4 coefficient matrix of equation 5.2, the factorization is

$$\begin{bmatrix} 1 & -1 & 0 & 0 \\ -1 & 2 & -1 & 0 \\ 0 & -1 & 2 & -1 \\ 0 & 0 & -1 & 2 \end{bmatrix} = \begin{bmatrix} 1 & 0 & 0 & 0 \\ -1 & 1 & 0 & 0 \\ 0 & -1 & 1 & 0 \\ 0 & 0 & -1 & 1 \end{bmatrix} \begin{bmatrix} 1 & -1 & 0 & 0 \\ 0 & 1 & -1 & 0 \\ 0 & 0 & 1 & -1 \\ 0 & 0 & 0 & 1 \end{bmatrix}. \tag{5.14}$$

Verification that forward elimination followed by back substitution then yields the correct solution is left to the reader as an exercise.

5.2.4 Banded and profile ('skyline') solvers

A matrix is 'banded' when all elements are zero except those within a band on either side of the principal diagonal. The semibandwidth m of such a matrix is the maximum number of terms within the band to the right of (and including) the diagonal. The coefficient matrix of equation 5.2, for example, is banded with a semibandwidth of two. The stiffness matrices of fig. 5.2(a) and (b) have semibandwidths of eight and twelve, respectively. The semibandwidth is a useful quantitative measure of the compactness of a matrix about its diagonal.

It is not difficult to show that bandedness is preserved during Gaussian reduction or LU decomposition. In other words, the semibandwidth of the initial matrix $\boldsymbol{K}$ is the semibandwidth also of the reduced matrix $\boldsymbol{K}^*$ and of the $\boldsymbol{L}$ and $\boldsymbol{U}$ factors, as demonstrated, for example, in the reduction/decomposition of equation set 5.2, where the initial semibandwidth of two is preserved both in the reduced matrix (see expression 5.4) and also in the Choleski factors (expression 5.14). This means that we can store the initial stiffness matrix $\boldsymbol{K}$ in a compact, 'banded' form, secure in the knowledge that no new storage locations will be required during the solution of the equations. A banded storage format of this type is illustrated in Fig. 5.3(a). This shows the portion of the full stiffness matrix which must be stored in banded form for the bar assembly of Fig. 5.1(a). *All* of the terms, whether zero or non-zero, within the upper half band are stored in a regular array whose first column corresponds to the diagonal of the full matrix (since $\boldsymbol{K}$ is symmetric only the upper band need be stored). Some additional locations are added at the bottom of the banded storage area to complete each row of the compact array to its full complement of terms, eight in the current instance.

An even more compact form of storage is achieved in a profile or 'skyline' format. This type of storage is illustrated, once again for the model of Fig. 5.1(a), in Fig. 5.3(b). Here, only those terms above each diagonal up to and including the last nonzero term are stored. The tops of the columns then define a 'skyline' of variable height above the diagonal, but always less than or equal to the semibandwidth. The number of terms below the skyline forms the 'profile' of the matrix. As with a full banded format, it is not difficult to show that a skyline format is also preserved during Gaussian reduction or LU decomposition. The difference between the number of storage locations required in the 'full' banded format and in the skyline format is not great for the model of Fig. 5.1(a), but can be substantial in larger, less regular meshes.

When the coefficient matrix is manipulated within a computer program in either of these formats, not only are the storage requirements much less than

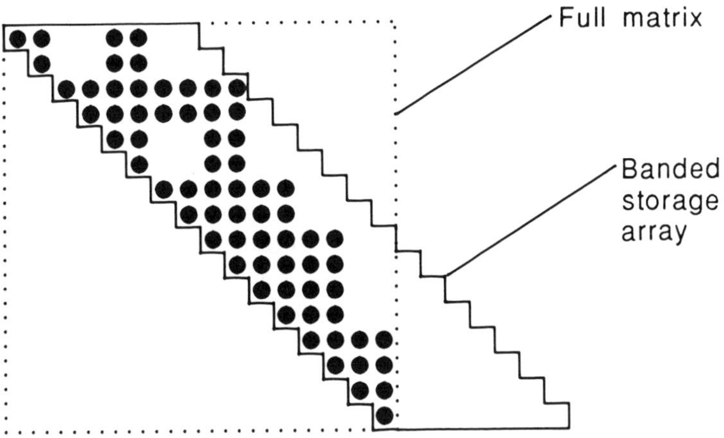

(a)

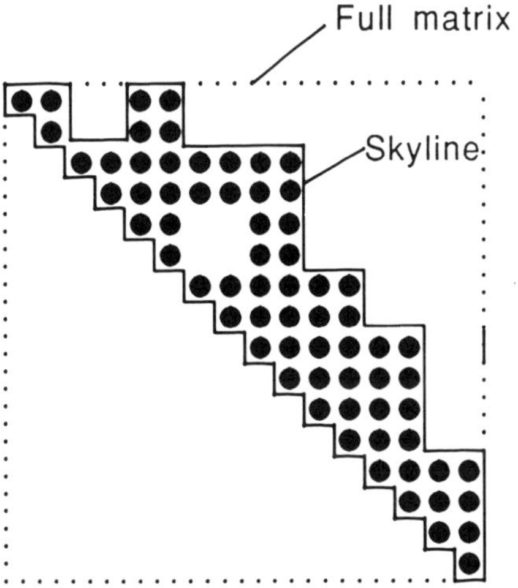

(b)

Fig. 5.3 *Storage formats (a) full banded, (b) skyline.*

would be required in a conventional representation, but the number of arithmetic operations which need to be performed is also greatly reduced. Consider, for example, the number of arithmetic operations required to reduce a banded matrix to upper triangular form using Gaussian reduction. If the semibandwidth m is known in advance, it is necessary when subtracting multiples of row i from all rows beneath it (step 3 in the reduction procedure, section 5.2.2) to perform this operation only for the $(m - 1)$ rows directly below the current row, all rows further down the matrix having zeros already in their ith columns. Moreover, it is only necessary to subtract multiples of the m nonzero terms to the right of the diagonal, all terms further along row i being zero. By making use of these economies, a 'banded' solver can be written in which the number of multiplications performed during the reduction phase is reduced by a factor of approximately $(m/n)^2$. Since the CPU time required for the solution of the equations is dominated by the large number of multiplications performed during the reduction phase, substantial savings in overall CPU time can be achieved by reducing the semibandwidth to as small a value as possible. The same economies of effort apply also to LU solvers although they are a little harder to demonstrate. In either event, applying the $(m/n)^2$ reduction factor to the stiffness matrices of Fig. 5.2(a) and (b), the solution time is reduced by a factor of two when numbering system (a) is used in place of numbering system (b). Reductions of an order of magnitude or more can easily be achieved in this way in large two and three-dimensional problems.

Somewhat greater savings can be obtained using solvers based on the skyline approach. The algorithm itself is a good deal more complex, however, since 'book keeping' must be included to map the physical address of each component of $\boldsymbol{K}$ onto an address within the stored profile. Some indication of the additional complexity is to be found in [2] which includes fortran listings of Choleski solvers for both banded and skyline formats. A more detailed discussion of such solvers is to be found also in [4].

5.3 EFFECTS OF DEGREE OF FREEDOM NUMBERING

It is clear from comments made in the preceding section that the solution of the stiffness equations using a direct solver, will be strongly influenced by the semibandwidth of the coefficient matrix or, in the case of a skyline solver, by the size of the profile under the skyline. In either case, it is desirable to concentrate the nonzero terms about the principal diagonal to as great a degree as possible. The degree of freedom numbering system is the most important factor in determining whether this can be achieved.

Consider first the effect on the bandwidth or skyline of the stiffness contribution from a single element. The comments which follow apply to *all* finite elements, and not just to the bar elements of chapter four. We will therefore assume a general element topology. Suppose that our general element has p degrees of freedom. If δ_i and δ_j are two of these, a contribution will certainly result from this element in the ith row and jth column of the assembled stiffness matrix. This logic works also in reverse. That is to say, a nonzero contribution will find its way into the i–jth place in the stiffness matrix, *only* if degrees of freedom δ_i and δ_j are common to at least one element. Suppose for arguments sake that i is smaller than j. The presence of a nonzero term in row i and column j then ensures that the skyline height above the diagonal and the semibandwidth are *at least* as great as $(j-i+1)$ (see Fig. 5.4). Moreover, if i is the smallest degree of freedom number of the element and j is the largest, *all* of the terms assembled from the element will then lie within this band and under this skyline. The largest difference in degree of freedom numbers, $(j_{max}-i_{min}+1)$, across each element is therefore the parameter which determines the semibandwidth of the matrix. Our objective, in numbering the degrees of freedom must be to minimize this quantity in an appropriate sense. In the case of a banded solver, we must minimize the extreme value. When a skyline solver is used, we must minimize the mean value.

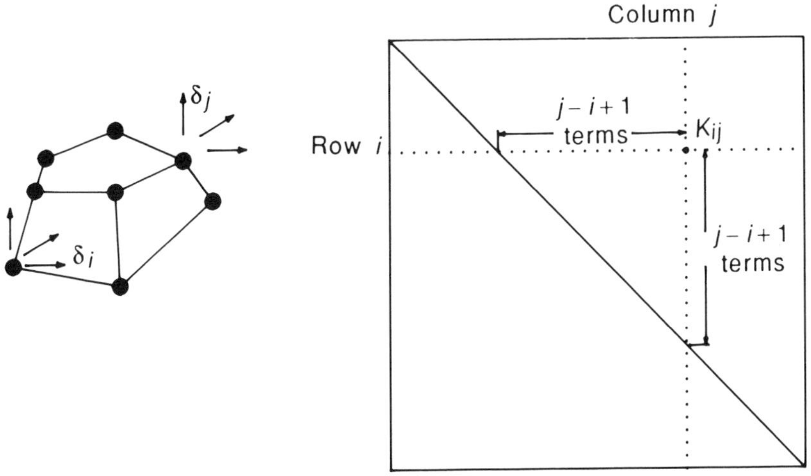

Fig. 5.4 *Location of element contributions in the assembled stiffness matrix.*

In many instances, an optimal numbering system can be achieved simply by numbering the degrees of freedom from left to right or top to bottom on as narrow a 'front' as possible. For example, the assembly of bar elements shown in Fig. 5.1, generates a semibandwidth of 8 for numbering system (a) and 12 for numbering system (b). In case (a), the degrees of freedom are numbered on a front, two nodes wide, which moves roughly from left to the right. In case (b), they are numbered on a broader front, five nodes wide, which moves from the bottom of the structure to the top.

Although this approach works quite well for regular meshes, it is inadequate in more complex models. Automatic algorithms which number the degrees of freedom in an optimal (or near optimal) way are available however for general meshes and are frequently incorporated in commercial finite element codes. In other words, user supplied node and degree of freedom numbers are adjusted — automatically or at the users request — to a near optimal state prior to analysis. The reverse Cuthill–McKee algorithm [5] is commonly used for this purpose although other approaches are claimed to be as effective and computationally less demanding [6]. It is important when using an existing code to find out whether or not the nodes *have* been renumbered in this way. If not, care must be exercised in preparing nodal data to ensure that the bandwidth or profile of the stiffness matrix is not excessive. CPU times, especially for large three-dimensional problems, can be orders of magnitude larger than necessary if nodes are numbered inappropriately.

An alternative to renumbering the degrees of freedom prior to the formation of the stiffness equations, is to renumber them as the elements are assembled. This procedure can be combined with the reduction phase of the solution to reduce still further the portion of the stiffness matrix that must be held in core storage at any one time. The order in which the elements are assembled then assumes critical importance.

5.4 FRONTAL SOLVERS, THE EFFECTS OF ELEMENT NUMBERING

It is often impossible to store the whole of the stiffness matrix in 'core' memory, even in banded or profile format. A model with 2000 degrees of freedom, for example (not unduly large for a static problem) and a semibandwidth of 200, requires storage in banded format for 400 000 stiffness components. This is beyond the 'in core' capacity of all but the largest computers. Fortunately, it is quite unnecessary to store the entire matrix in this way, given the relatively small number of stiffness components which are operated on at any particular point during the assembly or solution processes.

Consider, for example, the terms present in the stiffness matrix for the truss assembly of Fig. 5.1(a) when the first two elements have been assembled. The stiffness matrix at this stage in the assembly procedure is shown in Fig. 5.5. Its components are represented by solid, half solid and hollow circles. A solid circle represents a complete entry. It is 'complete' in the sense that it will receive no further contributions from the elements yet to be assembled. A half solid circle represents an 'incomplete' entry. That is, an entry which has already received at least one contribution, but will receive further contributions from the remaining elements. A hollow circle represents a component which has received no contribution as yet, but which *will* do so at some stage during the assembly phase. The entry in the 1–1 place, for example, is complete, since all the elements connected to degree of freedom δ_1, that is elements 1 and 2, have now been assembled. The 5–5 entry, on the other hand, is incomplete, since further contributions will be received from elements 3, 5 and 8 (Fig. 5.1(a)).

Suppose that it is our intention to solve the final stiffness equations — after assembling the remaining elements — using Gaussian elimination. The first

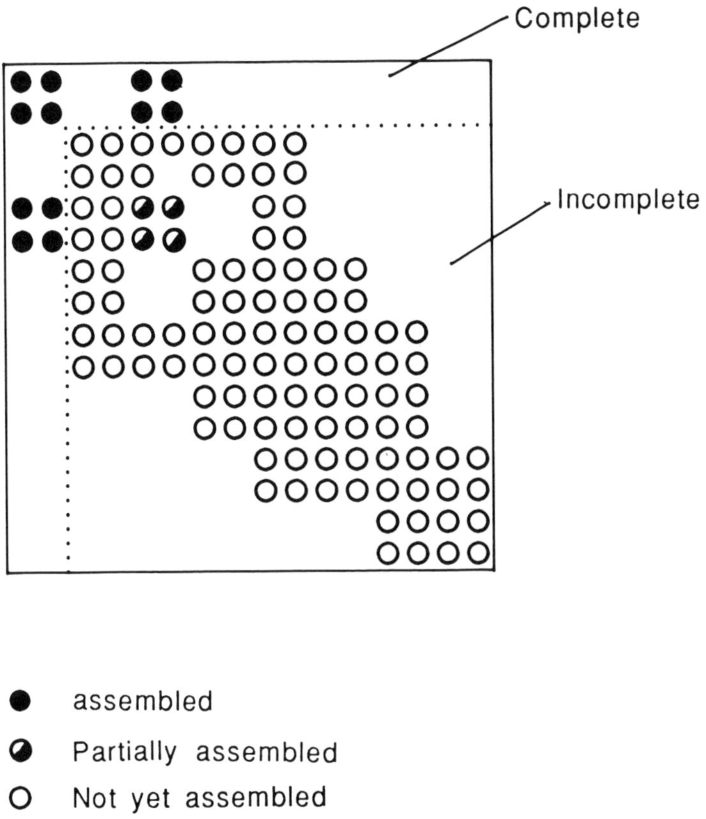

Fig. 5.5 *Stiffness components during assembly (model of Fig. 5.1(a)).*

step in the reduction process would be to subtract multiples of row one from all rows beneath it to produce zeros in column one under the diagonal. All of the terms in row one and column one are however complete at this stage of the assembly. There is therefore no need to assemble the remaining elements before performing the reduction of column one. The same argument applies to column two which is also complete. We therefore interrupt the assembly process at this point and perform the reduction of columns one and two before proceeding. When this has been done, the terms in rows one and two will not be required again during the rest of the reduction phase and can be relegated to a less accessible form of storage (in practice written to a disk file). We can then resume our assembly of the remaining elements until further rows become complete, at which point they are handled in a similar manner. In the present example, rows 3 and 4 are complete after assembly of elements 3–7, rows 5 and 6 are complete after assembly of element 8, and so on. As each row becomes complete the assembly process is interrupted and Gauss reduction is performed before copying the entire row to a disk file. By repeating this procedure every time that an element, or block of elements, is assembled, the number of 'active' rows and columns of the stiffness matrix (those which must actually be manipulated in the core at any given time) is greatly reduced. This philosophy forms the basis for the 'frontal' solver [7]. The only disadvantage of this approach is the complex 'book keeping' which is required to keep track of the degrees of freedom which are active at any instant. Programming details are to be found in [8] along with a fortran listing.

What is of importance to us here is the way in which the instantaneous size of the active matrix is affected by the order in which the elements are assembled. Obviously, the objective when using a frontal solver must be to keep this 'front size' to a minimum throughout the assembly. The degree of freedom numbers are unimportant in this regard since adjacent rows in the full matrix need not be adjacent in the active matrix, provided that a mapping is maintained from one to the other as part of the 'book keeping' activity already mentioned. What *is* important is the total number of active degrees of freedom at any instant, 'active' in the sense that some but not all of their stiffness contributions have been received. Their number at any instant defines the 'front size' of the matrix. Such degrees of freedom exist at nodes on the interface — or 'front' — between those elements which have been assembled and those which have not. The size and location of this interface is controlled by the order in which the elements are assembled and is reduced to a minimal value by ordering the elements to produce as narrow a front as possible, in much the same way that the semibandwidth is minimized by numbering the degrees of freedom in a similar way. This can sometimes be done by inspection. In Fig. 5.1, for example, model (a) which orders the elements on a relatively narrow front from left to right is much better than model (b), which orders them on a broader front, from bottom to top. Taking numbering system (a), for example, it is quite simple to estimate the instantaneous front size as each element is assembled. This is done by noting the active degrees of freedom — those for which some but not all stiffness contributions have been received — and tabulating them as shown in Table 1. The number of active degrees of freedom at each stage is then the instantaneous front size. For example, as element 1 is assembled, degrees of freedom δ_1 and δ_2 are active and the front size is 2. As element 2 is assembled, degrees of freedom δ_1, δ_2, δ_5 and δ_6 are active and the front size is 4, and so on. Note that each degree of freedom drops off the list once it has received all of its element contributions; δ_1 and δ_2, for example, are dropped from the list once element 2 has been assembled, δ_3 and δ_4 when element 7 has been assembled and so on. This process is represented graphically in Fig. 5.6 which shows the active nodes as the first element and as each successive second element is assembled. Since each node has two degrees of freedom, the front size takes a maximum value of 8 when four node are active, as elements 7 and 11 are assembled. Repeating the exercise for numbering system (b), we obtain a maximum front size of 12 as elements 8 and 12 are assembled (this calculation is left to the reader as an exercise).

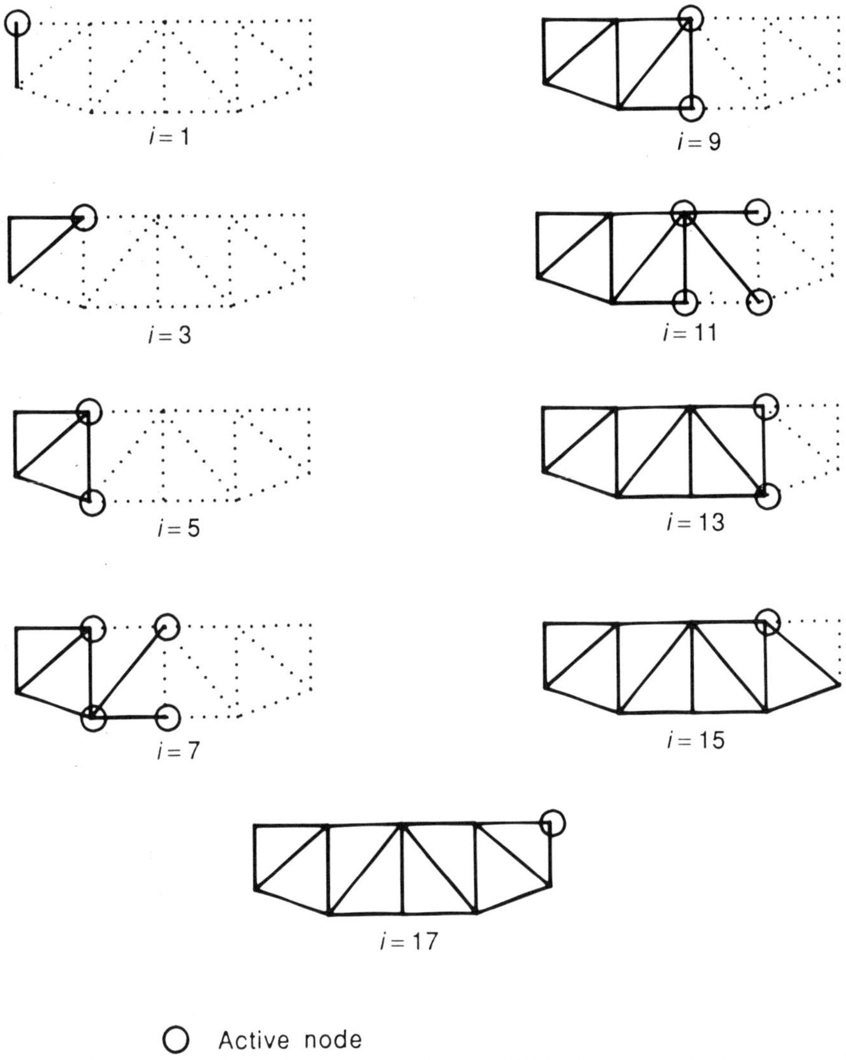

Fig. 5.6 *Active nodes during assembly (model of Fig. 5.1, element i in process of assembly).*

Although numbering the elements by inspection can work well for simple problems, a more general approach is required for complex models. Algorithms such as the reverse Cuthill–McKee are used in such instances and are applied to elements rather than nodes to produce an optimal or near optimal ordering. Once again, it is important when using an existing program with a frontal solver to find out whether such an option is implemented. If not, it may be necessary to order the element data by hand so that a reasonable front size is achieved.

TABLE 1 *Instantaneous front size as element i is assembled (Model of Fig. 5.1(a))*

Element number i	Active degrees of freedom as element i is assembled	Front size
1	δ_1, δ_2	2
2	$\delta_1, \delta_2, \delta_5, \delta_6$	4
3	δ_5, δ_6	2
4	$\delta_3, \delta_4, \delta_5, \delta_6$	4
5	$\delta_3, \delta_4, \delta_5, \delta_6$	4
6	$\delta_3, \delta_4, \delta_5, \delta_6, \delta_7, \delta_8$	6
7	$\delta_3, \delta_4, \delta_5, \delta_6, \delta_7, \delta_8, \delta_9, \delta_{10}$	8
8	$\delta_5, \delta_6, \delta_7, \delta_8, \delta_9, \delta_{10}$	6
9	$\delta_7, \delta_8, \delta_9, \delta_{10}$	4
10	$\delta_7, \delta_8, \delta_9, \delta_{10}, \delta_{13}, \delta_{14}$	6
11	$\delta_7, \delta_8, \delta_9, \delta_{10}, \delta_{11}, \delta_{12}, \delta_{13}, \delta_{14}$	8
12	$\delta_7, \delta_8, \delta_{11}, \delta_{12}, \delta_{13}, \delta_{14}$	6
13	$\delta_{11}, \delta_{12}, \delta_{13}, \delta_{14}$	4
14	δ_{13}, δ_{14}	2
15	$\delta_{13}, \delta_{14}, \delta_{15}, \delta_{16}$	4
16	δ_{15}, δ_{16}	2

5.5 CONCLUDING COMMENTS

The interaction of nodal (degree of freedom) and element numbering systems with matrix solvers, has been illustrated so far using examples drawn exclusively from assemblies of bar elements. The general philosophy extends however to all elements. The only difference in applying it to more complex, two and three-dimensional elements is that the number of degrees of freedom associated with each node is generally larger, as is their connectivity across each element. In the case of a plane bar element only four degrees of freedom are involved, whereas for the eight-noded block of Fig. 4.1, for example, twenty four nodal displacements are connected across each element (three at each of the eight nodes). The semibandwidth, skyline and 'front size' can be estimated in the same way for all assemblies of elements, however, and the general idea of numbering either the degrees of freedom or the elements (depending upon whether an in-core or frontal solver is used) on as narrow a front as possible still holds. Many commercial programs have 'built in' renumbering algorithms to assist the the user in this task and these should be invoked whenever possible.

REFERENCES

[1] Norrie D. and Kardestuncer H. (eds), (1987) *The Finite Element Handbook*, Part 4, McGraw-Hill, New York, section 1.1A.

[2] Griffiths D. V. and Smith, I. (1991) *Numerical Methods for Engineers*, Blackwell Scientific, Oxford, Chapter 2.
[3] Chapra, S. C. and Canale, R. P. (1989) *Numerical Methods for Engineers*, 2nd edn, McGraw-Hill, New York, Chapter 7.
[4] Zienkiewicz, O. C. and Taylor, R. L. (1990) *The Finite Element Method*, 4th edn, Vol. 1, McGraw-Hill, London, Chapter 15.
[5] Cuthill, E. H., (1972) *Several Strategies for Reducing the Bandwidth of Sparse Matrices, Sparse Matrices and their Applications* (edited by D. J. Roe and R. A. Willoughby), Plenum Press, New York.
[6] Everstine, G. C. (1979) A comparison of three resequencing algorithms for the reduction of matrix profile wavefront. *International Journal for Numerical Methods in Engineering*, **14** (6), 837–53.
[7] Irons, B. M. (1977) A frontal solution program. *International Journal for Numerical Methods in Engineering*, **2** (1), 5–32.
[8] Yeo, M. F. and Cheung, Y. K. (1979) *A Practical Guide to Finite Element Analysis*, Pitman, London.

PROBLEMS

1. Solve the matrix equation

$$\begin{bmatrix} 2 & 0 & -1 & 0 \\ 0 & 2 & -1 & -1 \\ -1 & -1 & 2 & 0 \\ 0 & -1 & 0 & 1 \end{bmatrix} \begin{bmatrix} \delta_1 \\ \delta_2 \\ \delta_3 \\ \delta_4 \end{bmatrix} = \begin{bmatrix} 1 \\ 0 \\ 0 \\ 0 \end{bmatrix}$$

   (a) by Gauss elimination,
   (b) by Choleski factorization.

   Confirm that both yield the correct solution, $\delta_1 = \delta_2 = \delta_3 = \delta_4 = 1$. Confirm also that the semibandwidth of the coefficient matrix and its 'skyline' are preserved during the reduction or decomposition. How many storage locations would be required to store this matrix (i) using a banded format, and (ii) using a profile or 'skyline' format?

2. An assembly of bar elements has 14 degrees of freedom numbered as in the figure overleaf. What is the semibandwidth of the stiffness matrix? Can it be decreased by renumbering? If so, how?

3. Sketch the 'skyline' of the stiffness matrix for problem two before and after renumbering. How many storage locations are required to store the profile of the matrix in each case? How many storage locations would be required if a banded format were used?

4. What is the maximum front size for the stiffness matrix of problem two when a frontal solver is used with the elements numbered as shown. Can this be reduced by renumbering? If so indicate how it may be done and

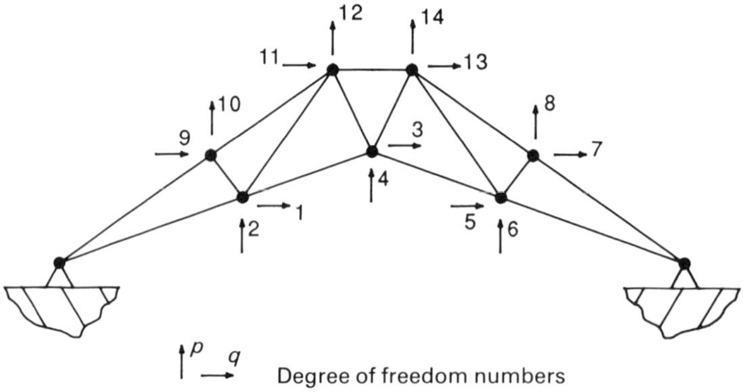

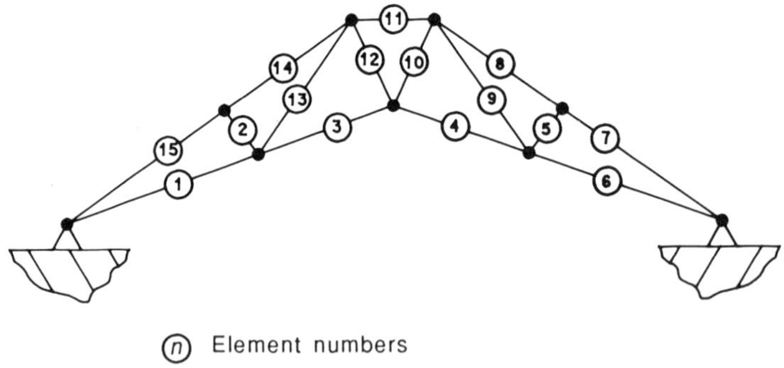

Problem 2.

construct a table similar to Table 1 showing the number of degrees of freedom active as each element is assembled. Do this for the original and modified numbering systems.

5. An elastic body is subdivided into a mesh of plane triangular elements as shown (see figure). Each node has two degrees of freedom and each element is defined by three nodes. If the unconstrained degrees of freedom are numbered as shown, what is the the semibandwidth of the stiffness matrix? Can this be reduced by renumbering? If so, how, and by how much?

6. The body of problem five is re-meshed using the same nodes and the same degrees of freedom but with the triangular elements replaced by four-noded, quadrilaterals as shown (see figure). Determine the semiband-

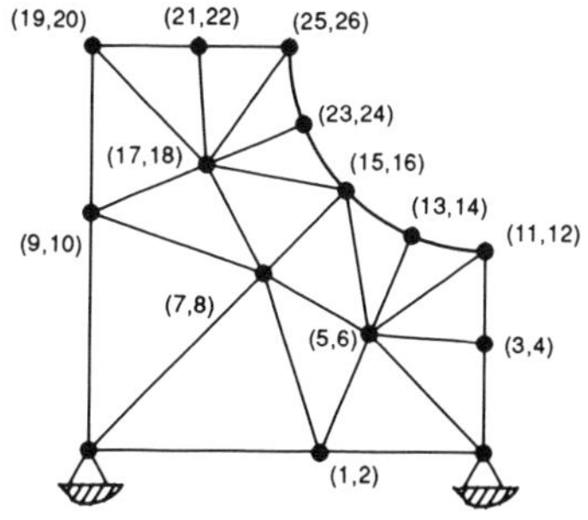

Problem 5.

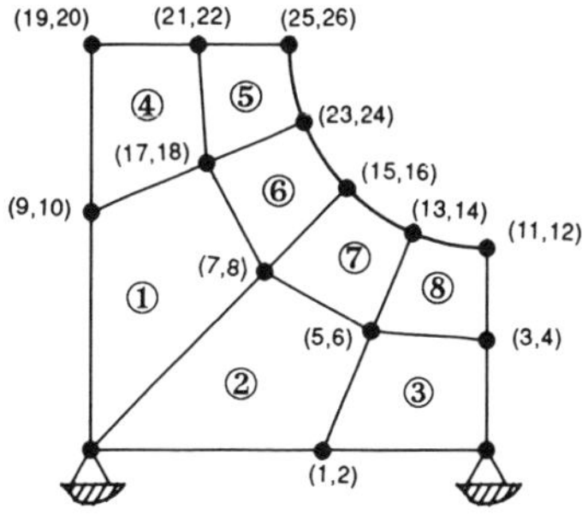

Problem 6.

width of the stiffness matrix and indicate whether it can be reduced by renumbering.

7. A frontal solver is used to solve the stiffness equations for problem six. The elements are assembled in the order shown. What is maximum front size of the active matrix? Indicate how it can be reduced by renumbering, and construct tables similar to Table 1 for the original and modified systems.

8. A plane framework with rigid joints is modelled by nine 'frame' elements as shown (see figure). They are defined by a node at each end. Each node has three degrees of freedom, two displacements and a rotation. What is the semibandwidth of the stiffness matrix? Can it be reduced by renumbering and if so how?

9. The framework of problem eight is to be solved using a frontal solver with the elements numbered as shown. Construct a table of active degrees of

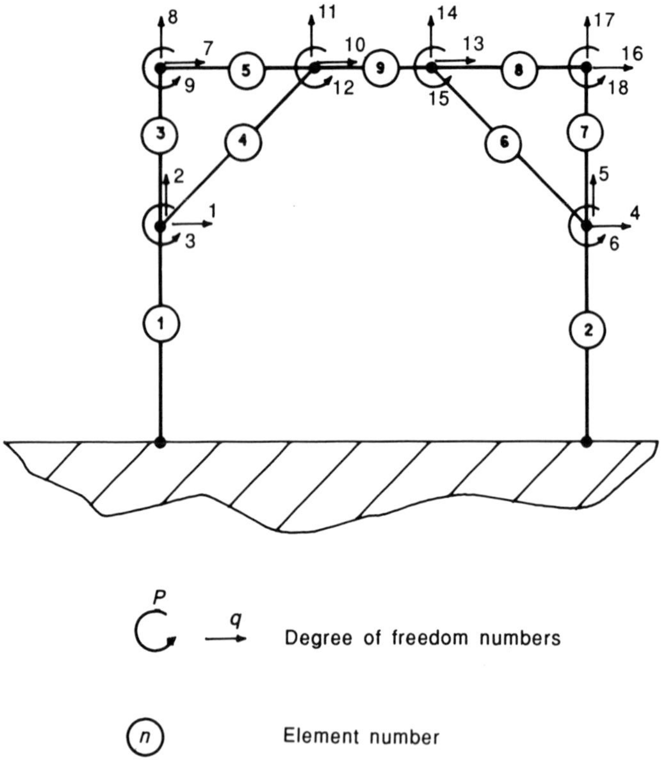

Problem 8.

freedom similar to Table 1, and calculate the maximum front size. Can this be reduced by renumbering? If so, how?

6

Linear elements for two-dimensional and three-dimensional analysis

6.1 INTRODUCTION

The bar elements of Chapter 4 were based on the notion of linear interpolation within an element. The same concept can readily extended to two and three-dimensions. The elements formed in this way are not generally as effective as those which use higher order interpolation (these are discussed in Chapter 8) but serve as a useful starting point for a discussion of two and three-dimensional problems. They were, as one might expect, the earliest elements of this type to be developed. In fact, the triangular plane-stress element, to be introduced shortly, formed the subject of Clough's seminal paper in 1960 [1]. Its three-dimensional equivalent, the linear tetrahedron, appeared in the literature a year or two later and is variously attributed to Gallagher [2] and Melosh [3].

In this chapter, linear elements are developed for two and three-dimensional problems. The term 'two-dimensional' encompasses not only objects which are 'thin' in the plane of analysis (plane stress) but also those which are 'long' in the out-of-plane direction (plane strain) and solids of revolution (axisymmetric stress). In all three cases, the problem is 'two-dimensional' in the sense that a plane area (rather than a solid volume) must be subdivided into elements. The simplest case is that of plane stress and this will be dealt with first.

In formulating elements for two-dimensional problems, there in no need to start absolutely from scratch. The general results of Chapter 4 — expressions 4.39 — apply to all elements provided that the stress–strain relationship has been cast in an appropriate multi-dimensional form. Most of the effort in formulating a new element is therefore expended in forming the shape functions and through them the shape matrix $\boldsymbol{N}^e$, and the strain–displacement matrix $\boldsymbol{B}^e$. Once these have been determined, the formulation of the element stiffness matrix and equivalent nodal loads is relatively straightforward.

6.2 ANALYSIS OF PLANE STRESS USING LINEAR TRIANGLES

Consider first a finite element model for a two-dimensional body in a state of plane stress. The body lies in the x–y plane, is of thickness h and is subject to in-plane loads which include a body force $\boldsymbol{g}(x, y)$ per unit volume. A temperature variation $T(x, y)$ is also permitted. This situation corresponds the state of stress illustrated in Fig. 2.4. The nonzero components of stress and strain in a cartesian system are σ_x, σ_y, τ_{xy}, and e_x, e_y, γ_{xy} respectively. They are taken to be independent of z, as are the in-plane displacements u and v.

6.2.1 Subdivision into elements, element topology

A finite element model is formed by subdividing the body into triangular subregions. Each of these forms a single element. The triangles have straight sides and are defined by nodal points at their vertices. A mesh of such elements is shown in Fig. 6.1. Also shown, in somewhat greater detail, is a typical element. Its vertices are defined by nodes 1, 2 and 3 with coordinates (x_1, y_1), (x_2, y_2) and (x_3, y_3). The nodal displacements are u_1, v_1, u_2, v_2, u_3 and v_3 as shown, and the displacement vector, $\boldsymbol{d}^e$, is defined as

$$\boldsymbol{d}^e = [u_1, v_1, u_2, v_2, u_3, v_3]^T. \tag{6.1}$$

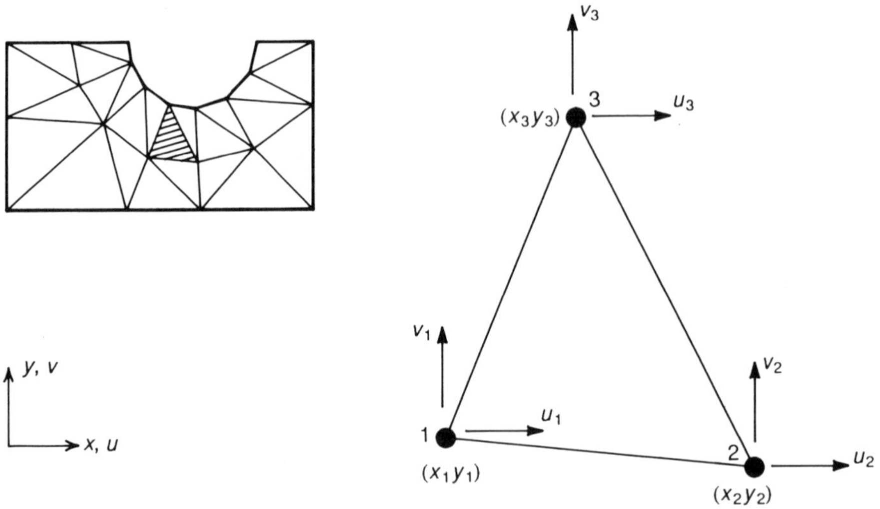

Fig. 6.1 *Subdivision of a thin plate into linear, plane stress triangles.*

6.2.2 Shape functions, the shape matrix and the strain–displacement matrix

The next step in the element formulation is the interpolation of displacement within the element. In the case of the bar element of Chapter 4, this was achieved by assuming that the axial displacement varied linearly with distance along the bar. In the case of the triangular element, the displacements are assumed to vary linearly with x and y. At any point (x, y) within the element, the displacements u and v are therefore given by,

$$u = \alpha_1 + \alpha_2 x + \alpha_3 y,$$
$$v = \beta_1 + \beta_2 x + \beta_3 y, \tag{6.2}$$

where α_i and $\beta_i (i = 1, 2, 3)$ are undetermined constants. This leads naturally to shape functions which are linear in x and y. The constants α_i and β_i are determined by imposing the requirement that u and v take nodal values u_1, v_1 etc at nodes (x_1, y_1), (x_2, y_2) and (x_3, y_3). In the case of u, this gives

$$\begin{aligned} \alpha_1 + \alpha_2 x_1 + \alpha_3 y_1 &= u_1, \\ \alpha_1 + \alpha_2 x_2 + \alpha_3 y_2 &= u_2, \\ \alpha_1 + \alpha_2 x_3 + \alpha_3 y_3 &= u_3. \end{aligned} \tag{6.3}$$

These are readily solved using Cramer's rule to give

$$\begin{aligned} \alpha_1 &= \frac{1}{2A}[a_1 u_1 + a_2 u_2 + a_3 u_3], \\ \alpha_2 &= \frac{1}{2A}[b_1 u_1 + b_2 u_2 + b_3 u_3], \\ \alpha_3 &= \frac{1}{2A}[c_1 u_1 + c_2 u_2 + c_3 u_3], \end{aligned} \tag{6.4}$$

where

$$\begin{aligned}
a_1 &= \det\begin{vmatrix} x_2 & y_2 \\ x_3 & y_3 \end{vmatrix}, & b_1 &= -\det\begin{vmatrix} 1 & y_2 \\ 1 & y_3 \end{vmatrix}, & c_1 &= \det\begin{vmatrix} 1 & x_2 \\ 1 & x_3 \end{vmatrix}, \\
a_2 &= -\det\begin{vmatrix} x_1 & y_1 \\ x_3 & y_3 \end{vmatrix}, & b_2 &= \det\begin{vmatrix} 1 & y_1 \\ 1 & y_3 \end{vmatrix}, & c_2 &= -\det\begin{vmatrix} 1 & x_1 \\ 1 & x_3 \end{vmatrix}, \\
a_3 &= \det\begin{vmatrix} x_1 & y_1 \\ x_2 & y_2 \end{vmatrix}, & b_3 &= -\det\begin{vmatrix} 1 & y_1 \\ 1 & y_2 \end{vmatrix}, & c_3 &= \det\begin{vmatrix} 1 & x_1 \\ 1 & x_2 \end{vmatrix},
\end{aligned} \tag{6.5a}$$

and

$$2A = \det\begin{vmatrix} 1 & x_1 & y_1 \\ 1 & x_2 & y_2 \\ 1 & x_3 & y_3 \end{vmatrix}. \tag{6.5b}$$

It is also quite simple (though somewhat tedious and not included here) to show that the constant A is equal to the area of the element provided that nodes 1, 2 and 3 are ordered anticlockwise around the perimeter of the element. If ordered clockwise, A is minus the area of the element.

Substitution of expressions 6.4 into the original interpolation for u gives, after some rearrangement of terms,

$$u = n_1(x, y)u_1 + n_2(x, y)u_2 + n_3(x, y)u_3, \tag{6.6}$$

where

$$n_i(x, y) = \frac{1}{2A}[a_i + b_i x + c_i y], \quad (i = 1,2,3). \tag{6.7}$$

The functions $n_i(x, y)$ are the shape functions of the element. As anticipated, they are linear in x and y.

Evaluation of the constants β_1, β_2, and β_3 yields an identical relationship for v,

$$v = n_1(x, y)v_1 + n_2(x, y)v_2 + n_3(x, y)v_3. \tag{6.8}$$

Taken together, equations 6.6 and 6.8 define an interpolation for the displacement field within the element in terms of its nodal variables. This is analogous to the one-dimensional interpolation (equation 4.5) which was obtained for the bar element of Chapter 4. Continuing the analogy, a shape matrix, $\boldsymbol{N}^{e}$, is formed by combining equations 6.6 and 6.8 to give a single matrix equality

$$\begin{bmatrix} u \\ v \end{bmatrix} = \begin{bmatrix} n_1(x, y) & 0 & n_2(x, y) & 0 & n_3(x, y) & 0 \\ 0 & n_1(x, y) & 0 & n_2(x, y) & 0 & n_3(x, y) \end{bmatrix} \begin{bmatrix} u_1 \\ v_1 \\ u_2 \\ v_2 \\ u_3 \\ v_3 \end{bmatrix}, \tag{6.9}$$

or, in the notation of Chapter 4 (*cf.* equation 4.6),

$$\boldsymbol{u} = \boldsymbol{N}^{e}\,\boldsymbol{d}^{e}.$$

The remainder of the formulation for the linear triangle then follows closely the general procedure outlined in Chapter 4. First, the strain components, in this case the in-plane strains e_x, e_y and γ_{xy}, are expressed in terms of the nodal displacements using the two-dimensional strain–displacement relationship. This gives

$$\begin{aligned} e_x &= \frac{\partial u}{\partial x} = \frac{\partial n_1}{\partial x}u_1 + \frac{\partial n_2}{\partial x}u_2 + \frac{\partial n_3}{\partial x}u_3, \\ e_y &= \frac{\partial v}{\partial y} = \frac{\partial n_1}{\partial y}v_1 + \frac{\partial n_2}{\partial y}v_2 + \frac{\partial n_3}{\partial y}v_3, \\ \gamma_{xy} &= \frac{\partial u}{\partial y} + \frac{\partial v}{\partial y} \\ &= \frac{\partial n_1}{\partial y}u_1 + \frac{\partial n_2}{\partial y}u_2 + \frac{\partial n_3}{\partial y}u_3 + \frac{\partial n_1}{\partial x}v_1 + \frac{\partial n_2}{\partial x}v_2 + \frac{\partial n_3}{\partial x}v_3. \end{aligned} \tag{6.10}$$

Rewriting these equations in matrix form, we obtain

$$\boldsymbol{e} = \boldsymbol{B}^{\mathrm{e}}\, \boldsymbol{d}^{\mathrm{e}}, \tag{6.11}$$

where

$$\boldsymbol{e} = \begin{bmatrix} e_x \\ e_y \\ \gamma_{xy} \end{bmatrix}, \tag{6.12a}$$

and

$$\boldsymbol{B}^{\mathrm{e}} = \begin{bmatrix} \partial n_1/\partial x & & \partial n_2/\partial x & & \partial n_3/\partial x & \\ 0 & \partial n_1/\partial y & 0 & \partial n_2/\partial y & 0 & \partial n_3/\partial y \\ \partial n_1/\partial y & \partial n_1/\partial x & \partial n_2/\partial y & \partial n_2/\partial x & \partial n_3/\partial y & \partial n_3/\partial x \end{bmatrix}. \tag{6.12b}$$

Once again, the equation 6.11 is consistent with the general form of the discrete strain–displacement relationship (equation 4.10) introduced in Chapter 4. The components of the strain–displacement matrix, $\boldsymbol{B}^{\mathrm{e}}$, are simplified in the current instance by using expressions 6.7 to evaluate the shape function derivatives. This gives

$$\boldsymbol{B}^{\mathrm{e}} = (1/2A) \begin{bmatrix} b_1 & 0 & b_2 & 0 & b_3 & 0 \\ 0 & c_1 & 0 & c_2 & 0 & c_3 \\ c_1 & b_1 & c_2 & b_2 & c_3 & b_3 \end{bmatrix}. \tag{6.13}$$

Since all of the components of $\boldsymbol{B}^{\mathrm{e}}$ are constant, the strains are constant throughout the element by virtue of equation 6.11. The element is for this reason referred to frequently as the 'constant strain triangle' (CST).

6.2.3 Stress–strain relationship

The $\boldsymbol{D}$ matrix for the element is derived by rewriting the general statement of Hooke's law (equations 2.22 and 2.25) for a cartesian state of stress with σ_z, τ_{xz} and τ_{yz} set equal to zero. This gives

$$\begin{aligned} e_x &= \frac{1}{E}(\sigma_x - \nu\sigma_y) + \alpha T, \\ e_y &= \frac{1}{E}(\sigma_y - \nu\sigma_x) + \alpha T, \\ \gamma_{xy} &= \frac{2(1+\nu)}{E}\tau_{xy}, \end{aligned} \tag{6.14}$$

and can be inverted to give

$$\sigma_x = \frac{E}{1-\nu^2}\,[(e_x - \alpha T) + \nu(e_y - \alpha T)]\,,$$

$$\sigma_y = \frac{E}{(1-\nu^2)}\,[(e_y - \alpha T) + \nu\,(e_x - \alpha T)], \tag{6.15}$$

$$\tau_{xy} = \frac{E}{2(1+\nu)}\,\gamma_{xy}.$$

By placing the stresses and strains in column vectors $\boldsymbol{s}$ and $\boldsymbol{e}$ (consistent with the definition of such vectors in Chapter 4) equations 6.15 can be written in matrix form as

$$\boldsymbol{s} = \boldsymbol{D}\,[\boldsymbol{e} - \boldsymbol{e}_{\mathrm{T}}], \tag{6.16}$$

where

$$\boldsymbol{s} = \begin{bmatrix} \sigma_x \\ \sigma_y \\ \tau_{xy} \end{bmatrix}, \quad \boldsymbol{e} = \begin{bmatrix} e_x \\ e_y \\ \gamma_{xy} \end{bmatrix}, \quad \boldsymbol{e}_{\mathrm{T}} = \begin{bmatrix} \alpha T \\ \alpha T \\ 0 \end{bmatrix}, \tag{6.17}$$

and

$$\boldsymbol{D} = \left(\frac{E}{1-\nu^2}\right) \begin{bmatrix} 1 & \nu & 0 \\ \nu & 1 & 0 \\ 0 & 0 & (1-\nu)/2 \end{bmatrix}. \tag{6.18}$$

The stress–strain relationship is now in the standard form required for the general formulation. The strain energy density, given by the scalar expression, $\frac{1}{2}\,[\sigma_x(e_x - \alpha T) + \sigma_y(e_y - \alpha T) + \tau_{xy}\gamma_{xy}]$, can also be written in the standard form as, $\frac{1}{2}\,\boldsymbol{s}^{\mathrm{T}}\,[\boldsymbol{e} - \boldsymbol{e}_{\mathrm{T}}]$.

6.2.4 Stiffness matrix

The stiffness matrix now follows directly from expression 4.39, that is

$$\boldsymbol{K}^{\mathrm{e}} = \int_V \boldsymbol{B}^{\mathrm{eT}}\boldsymbol{D}\boldsymbol{B}^{\mathrm{e}}\,\mathrm{d}V. \tag{6.19}$$

Since $\boldsymbol{B}^{\mathrm{e}}$, and $\boldsymbol{D}$ are constant matrices in the current instance (see equations 6.13 and 6.18), the entire integrand is a constant, and the integration can be performed trivially by multiplying the integrand by the volume of the element (Ah). This gives

$$\boldsymbol{K}^{\mathrm{e}} = (Ah)\,\boldsymbol{B}^{\mathrm{eT}}\boldsymbol{D}\boldsymbol{B}^{\mathrm{e}}. \tag{6.20}$$

Given that $\boldsymbol{B}^{\mathrm{eT}}$, $\boldsymbol{D}$ and $\boldsymbol{B}^{\mathrm{e}}$ are of order 6×3 and 3×3, and 3×6 equation 6.20 defines a 6×6 stiffness matrix for the element. The individual stiffness components are obtained by evaluating the triple product on the right hand side of equation 6.20. This can be done algebraically, in which case, general, closed-form expressions are obtained for each component (expressions 8.15 and 8.16 of ref. [4]) or numerically. Manual calculation of selected terms in $\boldsymbol{K}^{\mathrm{e}}$ for a particular element is included as an exercise at the end of this chapter (see problem 2).

6.2.5 Equivalent nodal forces, thermal loads and body forces

The nodal forces required to model temperature changes and body forces are given in the general case by (equations 3.39)

$$\boldsymbol{f}_{\mathrm{T}}^{\mathrm{e}} = \int_V (\boldsymbol{B}^{\mathrm{eT}}\, \boldsymbol{D}\, \boldsymbol{e}_{\mathrm{T}})\, \mathrm{d}V, \tag{6.21}$$

and

$$\boldsymbol{f}_{\mathrm{g}}^{\mathrm{e}} = \int_V (\boldsymbol{N}^{\mathrm{eT}} \boldsymbol{g})\, \mathrm{d}V. \tag{6.22}$$

In the current element, the thermal strain vector $\boldsymbol{e}_{\mathrm{T}}$ has components αT in its first two rows (see equation 6.17), and the distributed body force, $\boldsymbol{g}$, has in-plane components g_x and g_y .

Let us look first at the components of $\boldsymbol{f}_{\mathrm{T}}^{\mathrm{e}}$, that is, at the equivalent nodal forces due to thermal effects. If the temperature change is uniform over the element, all of the terms in the integrand of expression 6.21 are constant. The thermal load vector $\boldsymbol{f}_{\mathrm{T}}^{\mathrm{e}}$ is therefore obtained by multiplying the integrand, $\boldsymbol{B}^{\mathrm{eT}}\boldsymbol{D}\boldsymbol{e}_{\mathrm{T}}$, by the volume of the element, Ah. This gives

$$\boldsymbol{f}_{\mathrm{T}}^{\mathrm{e}} = (Ah)\boldsymbol{B}^{\mathrm{eT}}\boldsymbol{D}\boldsymbol{e}_{\mathrm{T}}. \tag{6.23}$$

In physical terms, $\boldsymbol{f}_{\mathrm{T}}^{\mathrm{e}}$ is a 6×1 vector whose components represent equivalent forces which must be applied to the nodes of the element in the directions of u_1, v_1, u_2, v_2, u_3 and v_3. Within the limitations of the finite element approximation, these model exactly the effects of a uniform temperature change over the element. The evaluation of the right hand side of equation 6.23 is left to the reader as an exercise (see problem 5 at the end of this chapter). In the event of a nonuniform temperature change, integral 6.21 must be evaluated with more care. It is not difficult, however, to show that the resulting nodal forces are identical to those given by equation 6.23 provided that the spatial average of the temperature is used to determine $\boldsymbol{e}_{\mathrm{T}}$.

The evaluation of the body force vector $\boldsymbol{f}_{\mathrm{g}}^{\mathrm{e}}$ is less straightforward. The integrand itself is no longer constant since the components of the shape matrix

$\boldsymbol{N}^{\mathrm{e}}$ are linear functions of x and y. Provided that the thickness of the element is constant, however, expression 6.22 can be rewritten

$$f_{\mathrm{g}}^{\mathrm{e}} = h \int_A (\boldsymbol{N}^{\mathrm{eT}}\boldsymbol{g})\,\mathrm{d}x\,\mathrm{d}y. \tag{6.24}$$

Inserting the components of $\boldsymbol{N}^{\mathrm{e}}$ and $\boldsymbol{g}$, we then obtain

$$f_{\mathrm{g}}^{\mathrm{e}} = h \int_A \begin{bmatrix} n_1 & 0 \\ 0 & n_1 \\ n_2 & 0 \\ 0 & n_2 \\ n_3 & 0 \\ 0 & n_3 \end{bmatrix} \begin{bmatrix} g_x \\ g_y \end{bmatrix} \mathrm{d}x\,\mathrm{d}y = h \int_A \begin{bmatrix} n_1 g_x \\ n_1 g_y \\ n_2 g_x \\ n_2 g_y \\ n_3 g_x \\ n_3 g_y \end{bmatrix} \mathrm{d}x\mathrm{d}y. \tag{6.24}$$

The individual components of $\boldsymbol{f}_{\mathrm{g}}^{\mathrm{e}}$ have the dimensions of force and correspond to equivalent, concentrated loads which must be applied at the nodes of the element. The first, third and fifth components act in the directions of u_1, u_2 and u_3, the second, fourth and sixth in the directions of v_1, v_2 and v_3. The equivalent nodal forces in the x and y directions at each node are therefore

$$h \int_A n_i(x, y) g_x \mathrm{d}x\,\mathrm{d}y, \quad (i = 1, 3), \text{ in the } x \text{ direction,} \tag{6.26}$$

and

$$h \int_A n_i(x, y) g_y \mathrm{d}x\,\mathrm{d}y, \quad (i = 1, 3), \text{ in the } y \text{ direction.} \tag{6.27}$$

If the body force is constant over the element, integrals 6.26 and 6.27 reduce to $(1/3)G_x$ and $(1/3)G_y$ respectively where G_x and G_y are resultant forces in the x and y directions. A simple physical interpretation may be placed upon this result; that a body force, distributed uniformly over the element, is exactly represented within the finite element model by one third of the resultant force acting at each vertex. This is analogous to the result obtained in Chapter 4, that a body force distributed uniformly over a bar element is exactly modelled by one half of the resultant load acting at each end. In both cases, the equipartition of the distributed load into two, or three, equal portions holds only if the distribution is uniform. If not, the body force components must be expressed explicitly as functions of position and placed within the appropriate integrals (equations 6.26 and 6.27 in the case of the linear triangle). This procedure results in a proportion, though not in general an *equal* proportion, of the total load being concentrated at each node.

6.2.6 Equivalent nodal forces, surface tractions

Tractions applied on the edges of a triangular element generate equivalent nodal forces in much the same way as a body force distributed throughout its volume. A loading of this type is illustrated in Fig. 6.2(a). There is no obvious parallel for this in the one-dimensional element of Chapter 4 and for this reason the nodal forces associated with such loads have not yet appeared in the general analysis. They can be added at this stage without difficulty.

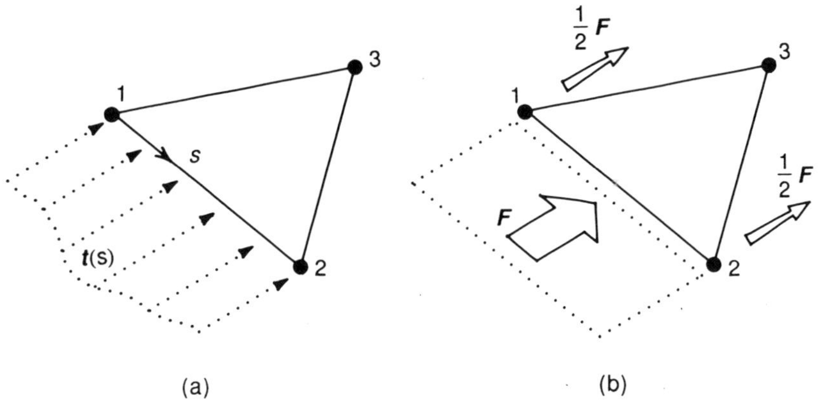

Fig. 6.2 *Surface traction applied to a linear triangle: (a) general distribution, (b) uniform distribution.*

In the general case, the potential energy associated with a traction $\boldsymbol{t}$ applied over a surface S is (see equation 3.37)

$$V = -\int_S (\boldsymbol{t} \cdot \boldsymbol{u})\, \mathrm{d}S. \tag{6.28}$$

Evaluation of this integral, element by element, yields a contribution to the potential energy of the system from any element whose bounding surface forms part of the external surface on which the traction is applied. Substitution of the interpolation relationship 6.9 into equation 6.28 then produces a potential energy contribution,

$$V^{\mathrm{e}} = -\boldsymbol{d}^{\mathrm{eT}} \boldsymbol{f}_{\mathrm{t}}^{\mathrm{e}},$$

where

$$\boldsymbol{f}_{\mathrm{t}}^{\mathrm{e}} = \int_S (\boldsymbol{N}^{\mathrm{eT}} \boldsymbol{t})\, \mathrm{d}S, \tag{6.29}$$

(S now denotes that portion of the loaded surface which coincides with the surface of the element). This defines a vector f_t^e of equivalent nodal forces due to the surface traction. In the case of the linear triangle of Fig. 6.2(a), a traction $\boldsymbol{t}$ with components t_x and t_y, is applied along the 1–2 side. Substitution of the components of $\boldsymbol{N}^e$ into equation 6.29, and replacement of the surface differential $\mathrm{d}S$ by $h\mathrm{d}s$ (s is measured along the side of the element) then gives

$$f_t^e = h\int_0^L \begin{bmatrix} n_1 t_x \\ n_1 t_y \\ n_2 t_x \\ n_2 t_y \\ n_3 t_x \\ n_3 t_y \end{bmatrix} \mathrm{d}S. \tag{6.30}$$

This is similar to expression 6.26 but involves integration along the side of the element rather than over its area. The cartesian expressions for n_1, n_2 and n_3 (equations 6.7) are not very helpful in evaluating this integral as it stands. Ideally we require expressions for the shape functions in terms of the coordinate s. Such expressions are not, in fact, difficult to obtain. Since u and v must vary linearly with s along the 1–2 side — by virtue of our original assumption of linear interpolation throughout the element — and since they must also take values u_1 (or v_1) and u_2 (or v_2) at nodes 1 and 2, the interpolation along the 1–2 side must take the form

$$u = \left(1 - \frac{s}{L}\right)u_1 + \left(\frac{s}{L}\right)u_2$$

$$v = \left(1 - \frac{s}{L}\right)v_1 + \left(\frac{s}{L}\right)v_2, \tag{6.31}$$

where L is the length of the 1–2 side and s takes the value 0 at node 1.

Comparing these expressions with equations 6.6 and 6.7, we conclude that the shape functions n_1, n_2 and n_3 are given along the 1–2 side by

$$n_1 = 1 - \frac{s}{L}, \quad n_2 = \frac{s}{L} \quad \text{and} \quad n_3 = 0. \tag{6.32}$$

These can be substituted into equation 6.30 and integrated to give the components of f_t^e. Since n_3 is identically zero, the equivalent nodal force at node 3 is zero in all cases. In the particular case of a uniformly distributed load (t_x and t_y constant with s) it is quite simple to show that the required nodal forces at nodes 1 and 2 are simply one half of the resultant load. This is illustrated in Fig. 6.3(b).

In the case on a nonuniform traction, the net load is again distributed between the end nodes but not generally in equal proportions. In such cases, the magnitudes of the equivalent nodal forces are obtained by substituting expressions for t_x and t_y as functions of s into equation 6.30 and evaluating the integral explicitly. Some exercises involving the integration and assembly of equivalent nodal forces for uniform and non-uniform edge loadings are included at the end of this chapter (see problems 6–10).

6.2.7 Effects of variable thickness

It has been assumed so far that the thickness h is constant over an element and can therefore be removed from element integrals when required. It is, however, free to vary from element to element. This permits modelling of thin sheets of variable thickness provided that the variation is small over a single element. If the variation is appreciable, the thickness must be included as a dependent variable within the various element integrals. The effect on the stiffness matrix and thermal load vector are relatively straightforward, the constant thickness h being replaced, in expressions 6.20 and 6.23, by a mean value, $\bar{h}$ say, where

$$\bar{h} = \frac{1}{A} \int_A h(x, y) \, dA .$$

The effect on the body-force and surface-traction load vectors is less obvious since an expression for $h(x, y)$ must then be inserted within integrals 6.25–6.27 and 6.30. Numerical integration (see Chapter 8) is often the most satisfactory approach, but analytic results can be obtained for simple cases (see problem 9).

6.3 LINEAR TRIANGLE FOR PLANE STRAIN AND AXISYMMETRIC STRESS

The two-dimensional region which was subdivided into triangular elements in Fig. 6.1, represented a body which was 'thin' in the z direction. A state of plane-stress was assumed to exist. The same region could also have been used to represent the cross-section of a prismatic body 'long' in the out-of-plane direction. An assumption of plane strain would then have been appropriate (Fig. 2.4 of Chapter 2). It might also have been interpreted as the generating section of an axisymmetric body, in which case a state of axisymmetric stress could have been assumed (Fig. 2.5). By the same argument, a subdivision of the region into triangular elements could have been interpreted in at least three different ways. The volume of material represented by the same triangular element in each of these interpretations is illustrated in Fig. 6.3. The plane stress element (Fig. 6.3(a)) is a thin triangular sheet, the plane strain element a long triangular bar (Fig. 6.3(b)), and the axisymmetric element an annular sector (Fig. 6.3(c)). The formulation of element matrices and vectors has been presented so far only for case (a), the 'thin' (plane stress) element. Modifications are now presented to accommodate interpretations (b) and (c).

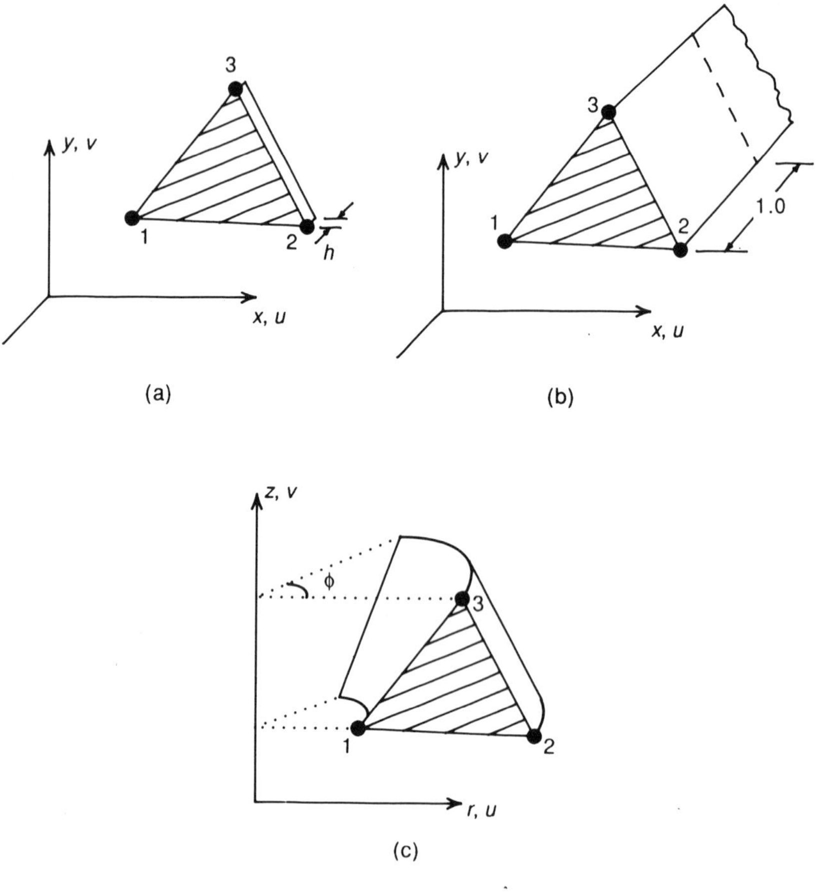

Fig. 6.3 *The linear triangle: (a) plane stress, (b) plane strain, (c) axisymmetric stress.*

6.3.1 Plane strain

The only modification required to accommodate an interpretation of plane strain, rather than plane stress, is a minor revision of the stress–strain relationship. In plane strain, the assumption of zero normal stress, σ_z, is replaced by an assumption that the out-of-plane strain e_z is zero. The shear stresses τ_{xz} and τ_{yz} are also taken to be zero. The stress–strain relationship is then (using equations 2.23 and 2.25)

$$e_x = \frac{1}{E}\left[\sigma_x - \nu\sigma_y - \nu\sigma_z\right] + \alpha T, \tag{6.33a}$$

$$e_y = \frac{1}{E}\left[\sigma_y - \nu\sigma_x - \nu\sigma_z\right] + \alpha T, \tag{6.33b}$$

$$0 = \frac{1}{E}\left[\sigma_z - \nu\sigma_x - \nu\sigma_y\right] + \alpha T, \tag{6.33c}$$

and

$$\gamma_{xy} = \tau_{xy}\,(1+\nu)2/E. \tag{6.33d}$$

The axial stress σ_z may be eliminated from equations 6.33(a) and (b) using equation 6.33(c) and the remaining equations inverted to give

$$\sigma_x = \frac{E\,(1-\nu)}{(1+\nu)\,(1-2\nu)}\left[e_x + \left(\frac{\nu}{1-\nu}\right)e_y - \left(\frac{1+\nu}{1-\nu}\right)\alpha T\right],$$

$$\sigma_y = \frac{E\,(1-\nu)}{(1+\nu)\,(1-2\nu)}\left[e_y + \left(\frac{\nu}{1-\nu}\right)e_x - \left(\frac{1+\nu}{1-\nu}\right)\alpha T\right], \tag{6.34}$$

$$\tau_{xy} = \frac{E}{2(1+\nu)}\,\tau_{xy}.$$

These in turn may be written in the 'standard' matrix form of Chapter 4 as

$$\boldsymbol{s} = \boldsymbol{D}\,[\boldsymbol{e} - \boldsymbol{e}_{\mathrm{T}}]$$

where $\boldsymbol{s} = \begin{bmatrix} \sigma_x \\ \sigma_y \\ \tau_{xy} \end{bmatrix}, \quad \boldsymbol{e} = \begin{bmatrix} e_x \\ e_y \\ \gamma_{xy} \end{bmatrix}, \quad \boldsymbol{e} = \begin{bmatrix} (1+\nu)\alpha T \\ (1+\nu)\alpha T \\ 0 \end{bmatrix},$ (6.35)

and where the stress–strain matrix, $\boldsymbol{D}$, is given by

$$\boldsymbol{D} = \frac{E(1-\nu)}{(1+\nu)\,(1-2\nu)}\begin{bmatrix} 1 & \nu/(1-\nu) & 0 \\ \nu/(1-\nu) & 1 & 0 \\ 0 & 0 & \frac{1}{2}(1-2\nu)/(1-\nu) \end{bmatrix}. \tag{6.36}$$

The formulation is then virtually identical to that of section 6.2, the only difference being that the matrices $\boldsymbol{D}$ and $\boldsymbol{e}_{\mathrm{T}}$ are now given by equations 6.35 and 6.36 instead of equations 6.17 and 6.18. Note that the thickness of the element in the z direction is not always clearly defined in the case of plane-strain, since the theory applies in principle to bodies which are indeterminately long. A definite value is required, however, in all element integrals, and is conveniently taken to be unity in the absence of other information (Fig. 6.3(b)).

6.3.2 Axisymmetric analysis

The modifications which must be made to the linear triangle for axisymmetric use are somewhat more extensive. Physically, the axisymmetric element represents a sector of a triangular annulus (Fig. 6.3(c)). The x and y coordinates of the original, plane stress representation are then replaced by radial and axial polar coordinates, r and z. The displacement at any point is again defined by components u and v, but these now represent displacements in the radial and axial directions. Nodal values of displacement are again denoted by u_1, v_1, u_2, v_2, u_3 and v_3. The shape functions for u and v are identical to those for plane-stress, except that the cartesian coordinates x and y are replaced by r and z. In other words, the interpolated displacements are governed by the matrix shape relationship

$$\begin{bmatrix} u \\ v \end{bmatrix} = \begin{bmatrix} n_1(r,z) & 0 & n_2(r,z) & 0 & n_3(r,z) & 0 \\ 0 & n_1(r,z) & 0 & n_2(r,z) & 0 & n_3(r,z) \end{bmatrix} \begin{bmatrix} u_1 \\ v_1 \\ \vdots \\ \vdots \\ v_3 \end{bmatrix}, \quad (6.37)$$

where, $n_i(r, z) = (1/2A)\,[a_i + b_i r + c_i z]$, $(i = 1, 2, 3)$, and where the coefficients a_i, b_i and c_i are given by expressions 6.5 with x_i and y_i replaced by r_i and z_i.

The strain–displacement relationship within the axisymmetric element is derived from the strain–displacement equations in cylindrical, rather than cartesian, coordinates. It is obtained from equations 2.16 by setting u_r and u_z equal to u and v (from symmetry, $u_\theta = 0$). In addition to the in-plane, strain components, e_r, e_z and γ_{rz}, this produces a nonzero 'hoop' strain e_θ. In terms of the displacement components u and v, the strains are given by

$$\begin{aligned}
e_r &= \frac{\partial u}{\partial r} = \frac{\partial n_1}{\partial r} u_1 + \frac{\partial n_2}{\partial r} u_2 + \frac{\partial n_3}{\partial r} u_3, \\
e_z &= \frac{\partial v}{\partial z} = \frac{\partial n_1}{\partial z} v_1 + \frac{\partial n_2}{\partial z} v_2 + \frac{\partial n_3}{\partial z} v_3, \\
e_\theta &= \frac{u}{r} = \frac{n_1}{r} u_1 + \frac{n_2}{r} u_2 + \frac{n_3}{r} u_3, \\
\gamma_{rz} &= \frac{\partial v}{\partial r} + \frac{\partial u}{\partial z} = \frac{\partial n_1}{\partial r} v_1 + \frac{\partial n_2}{\partial r} v_2 + \frac{\partial n_3}{\partial r} v_3 + \frac{\partial n_1}{\partial z} u_1 + \frac{\partial n_2}{\partial z} u_2 + \frac{\partial n_3}{\partial z} u_3,
\end{aligned} \quad (6.38)$$

and can be written in matrix form as

$$\boldsymbol{e} = \begin{bmatrix} e_r \\ e_z \\ e_\theta \\ \gamma_{rz} \end{bmatrix} = \begin{bmatrix} \partial n_1/\partial r & 0 & \partial n_2/\partial r & 0 & \partial n_3/\partial r & 0 \\ 0 & \partial n_1/\partial z & 0 & \partial n_2/\partial z & 0 & \partial n_3/\partial z \\ n_1/r & 0 & n_2/r & 0 & n_3/r & 0 \\ \partial n_1/\partial z & \partial n_1/\partial r & \partial n_2/\partial z & \partial n_2/\partial r & \partial n_3/\partial z & \partial n_3/\partial r \end{bmatrix} \begin{bmatrix} u_1 \\ v_1 \\ \vdots \\ v_3 \end{bmatrix}. \qquad (6.39)$$

The 4×6 matrix on the right hand side of this equation then defines the $\boldsymbol{B}^{\mathrm{e}}$ matrix for the element, replacing the analogous 3×6 matrix for the case of plane stress. When the expressions for $n_i(r, z)$ are substituted into $\boldsymbol{B}^{\mathrm{e}}$, we obtain

$$\boldsymbol{B}^{\mathrm{e}} = \left(\frac{1}{2A}\right) \begin{bmatrix} b_1 & 0 & b_2 & 0 & b_3 & 0 \\ 0 & c_1 & 0 & c_2 & 0 & c_3 \\ (a_1 + b_1 r + c_1 z)/r & 0 & (a_2 + b_2 r + c_2 z)/r & 0 & (a_3 + b_3 r + c_3 z)/r & 0 \\ c_1 & b_1 & c_2 & b_2 & c_3 & b_3 \end{bmatrix}. \qquad (6.40)$$

Three rows of the above matrix, the first, third and fourth, are identical to those of the analogous, plane stress matrix (*cf.* equation 6.13). The third row, associated with the hoop strain, has no equivalent in plane stress and contains terms which are rational functions of r and z.

The stress–strain relationship is defined once again by equations 2.22 and 2.25, but these must now be written in terms of cylindrical polar components and with the shear components $\tau_{r\theta}$ and $\tau_{z\theta}$ set to zero. This gives

$$e_r = \frac{1}{E}[\sigma_r - \nu\sigma_z - \nu\sigma_\theta] + \alpha T,$$

$$e_z = \frac{1}{E}[\sigma_z - \nu\sigma_r - \nu\sigma_\theta] + \alpha T,$$

$$e_\theta = \frac{1}{E}[\sigma_\theta - \nu\sigma_z - \nu\sigma_r] + \alpha T,$$

$$\gamma_{rz} = \frac{2(1+\nu)}{E}\tau_{rz},$$

which can be inverted to give the 'standard' matrix relationship

$$\boldsymbol{s} = \boldsymbol{D}\,[\boldsymbol{e} - \boldsymbol{e}_{\mathrm{T}}],$$

where

$$\boldsymbol{s} = \begin{bmatrix} \sigma_r \\ \sigma_z \\ \sigma_\theta \\ \tau_{rz} \end{bmatrix}, \quad \boldsymbol{e} = \begin{bmatrix} e_r \\ e_z \\ e_\theta \\ \gamma_{rz} \end{bmatrix}, \quad \boldsymbol{e}_{\mathrm{T}} = \begin{bmatrix} \alpha T \\ \alpha T \\ \alpha T \\ 0 \end{bmatrix}, \qquad (6.41)$$

and

$$\boldsymbol{D} = \frac{E(1-\nu)}{(1+\nu)(1-2\nu)} \begin{bmatrix} 1 & \nu/(1-\nu) & \nu/(1-\nu) & 0 \\ \nu/(1-\nu) & 1 & \nu/(1-\nu) & 0 \\ \nu/(1-\nu) & \nu/(1-\nu) & 1 & 0 \\ 0 & 0 & 0 & (1-2\nu)/2(1-\nu) \end{bmatrix}. \quad (6.42)$$

The stiffness matrix and equivalent nodal force vectors for the element are now obtained from the standard integrals of equations 4.39 with the volume of the element taken to be the annular region shown in Fig. 6.3(c). If the angle φ is taken to be one radian, the volume differential $\mathrm{d}V$ becomes '$r\mathrm{d}r\mathrm{d}z$' and the stiffness matrix is given by

$$\boldsymbol{K}^{e} = \int_{A} \boldsymbol{B}^{eT}\boldsymbol{D}\,\boldsymbol{B}^{e} r\,\mathrm{d}r\,\mathrm{d}z. \quad (6.42)$$

The expressions for the load vectors $\mathbf{f}_g^e$ and $\mathbf{f}_T^e$ are modified in a similar manner.

The integration of these quantities is no longer trivial. Each integrand contains the factor r, and the matrix $\boldsymbol{B}^e$ has a row of nonconstant terms. Analytic integration is difficult but not impossible (see [5]). As an alternative, approximate procedure, we can evaluate the integrand at the centroid of the element and simply multiply it by the area of integration. This is a crude 'one point' form of numerical integration, but gives satisfactory results provided that the element is sufficiently distant from the axis of symmetry. Numerical schemes which give a better estimate of such integrals will be discussed in Chapter 8.

6.4 THREE-DIMENSIONAL ANALYSIS, THE LINEAR TETRAHEDRON

The linear triangle has a three-dimensional counterpart in the four-noded linear tetrahedron shown in Fig. 6.4. The element is defined by nodes 1, 2, 3 and 4, which displace in three directions giving a total of twelve degrees of freedom. These are labelled u_i, v_i, w_i ($i = 1, \ldots, 4$) and form the components of a nodal displacement vector

$$\boldsymbol{d}^{e} = [u_1, v_1, w_1, u_2, v_2, w_2, u_3, v_3, w_3, u_4, v_4, w_4]^{T}. \quad (6.43)$$

Interpolation within the element is based on the assumption that the displacements vary linearly with x, y and z. The displacement in the x direction, for example, is taken to be

$$u = \alpha_1 + \alpha_2 x + \alpha_3 y + \alpha_4 z, \quad (6.44)$$

where α_1, α_2, α_3 and α_4 are constants which are determined by equating u to its nodal values. The procedure is analogous to that applied to the triangular element except that there are now four constants to be determined instead of three. These

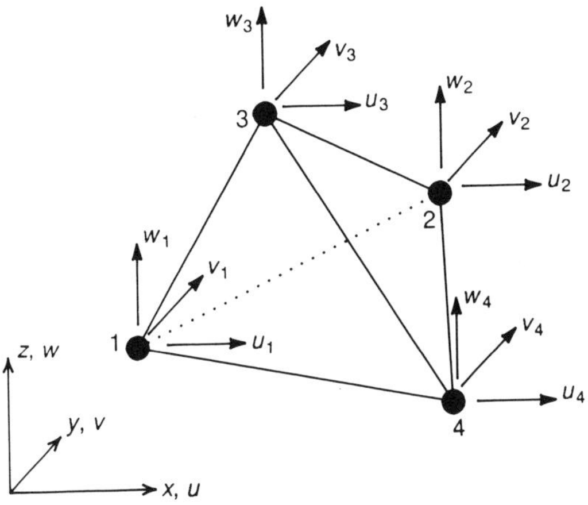

Fig. 6.4 *The linear tetrahedron, geometry and topology.*

are obtained by solving a set of *four* simultaneous equations analogous to equations 6.3. They are then substituted back into equation 6.44 to give

$$u = n_1(x, y, z)u_1 + n_2(x, y, z)u_2 + n_3(x, y, z)u_3 + n_4(x, y, z)u_4, \qquad (6.45)$$

where the shape functions $n_i(i = 1, \ldots, 4)$ are given by

$$n_i = \frac{1}{6V}(a_i + b_i x + c_i y + d_i z), \quad (i = 1, \ldots, 4). \qquad (6.46)$$

Here V is the volume of the tetrahedron and the geometric constants a_i, b_i, c_i, d_i $(i = 1, \ldots, 4)$ are given by

$$a_i = \det \begin{vmatrix} x_j & y_j & z_j \\ x_k & y_k & z_k \\ x_l & y_l & z_l \end{vmatrix}, \quad b_i = -\det \begin{vmatrix} 1 & y_j & z_j \\ 1 & y_k & z_k \\ 1 & y_l & z_l \end{vmatrix},$$

$$c_i = -\det \begin{vmatrix} x_j & 1 & z_j \\ x_k & 1 & z_k \\ x_l & 1 & z_l \end{vmatrix}, \quad d_i = -\det \begin{vmatrix} x_j & y_j & 1 \\ x_k & y_k & 1 \\ x_l & y_l & 1 \end{vmatrix},$$

$$6V = \det \begin{vmatrix} 1 & x_i & y_i & z_i \\ 1 & x_j & y_j & z_j \\ 1 & x_k & y_k & z_k \\ 1 & x_l & y_l & z_l \end{vmatrix},$$

where i, j, k and l are cyclic rotations of the integers 1, 2, 3 and 4. (Note: the above expressions are valid only if nodes 1, 2 and 3 are taken in a counter-clockwise direction when viewed from node 4, see Fig. 6.4).

The same shape functions apply to v and w and yield a matrix shape relationship of the form

$$\begin{bmatrix} u \\ v \\ w \end{bmatrix} = \begin{bmatrix} n_1 & 0 & 0 & n_2 & 0 & 0 & n_3 & 0 & 0 & n_4 & 0 & 0 \\ 0 & n_1 & 0 & 0 & n_2 & 0 & 0 & n_3 & 0 & 0 & n_4 & 0 \\ 0 & 0 & n_1 & 0 & 0 & n_2 & 0 & 0 & n_3 & 0 & 0 & n_4 \end{bmatrix} \begin{bmatrix} u_1 \\ v_1 \\ w_1 \\ \vdots \\ w_4 \end{bmatrix}. \tag{6.47}$$

This is clearly analogous to the general shape relationship of Chapter 4 (equation 4.9), the 3×12 matrix on the right hand side forming the shape matrix, $\boldsymbol{N}^{\mathrm{e}}$, for the element.

The strains are obtained from the three-dimensional strain–displacement equations. The strain e_x, for example, is given by (see equation 2.13)

$$e_x = \frac{\partial u}{\partial x} = \frac{\partial n_1}{\partial x} u_1 + \frac{\partial n_2}{\partial x} u_2 + \frac{\partial n_3}{\partial x} u_3 + \frac{\partial n_4}{\partial x} u_4$$

$$= \frac{b_1}{6V} u_1 + \frac{b_2}{6V} u_2 + \frac{b_3}{6V} u_3 + \frac{b_4}{6V} u_4.$$

Similar treatment of the remaining strain components produces the discrete strain–displacement relationship

$$\begin{bmatrix} e_x \\ e_y \\ e_z \\ \gamma_{xy} \\ \gamma_{yz} \\ \gamma_{xz} \end{bmatrix} = \frac{1}{6V} \left[\begin{array}{ccc|ccc|ccc|ccc} b_1 & 0 & 0 & b_2 & 0 & 0 & b_3 & 0 & 0 & b_4 & 0 & 0 \\ 0 & c_1 & 0 & 0 & c_2 & 0 & 0 & c_3 & 0 & 0 & c_4 & 0 \\ 0 & 0 & d_1 & 0 & 0 & d_2 & 0 & 0 & d_3 & 0 & 0 & d_4 \\ c_1 & b_1 & 0 & c_2 & b_2 & 0 & c_3 & b_3 & 0 & c_4 & b_4 & 0 \\ 0 & d_1 & c_1 & 0 & d_2 & c_2 & 0 & d_3 & c_3 & 0 & d_4 & c_4 \\ d_1 & 0 & b_1 & d_2 & 0 & b_2 & d_3 & 0 & b_3 & d_4 & 0 & b_4 \end{array} \right] \begin{bmatrix} u_1 \\ v_1 \\ w_1 \\ u_2 \\ \vdots \\ w_4 \end{bmatrix}, \tag{6.48}$$

which conforms to the 'standard', matrix form of this relationship (equation 4.10) and defines in the process a 6×12 strain–displacement matrix $\boldsymbol{B}^{\mathrm{e}}$ which has constant components.

The stress–strain relationship within the element is obtained from the full statement of Hooke's law (equations 1.19a and 1.21) in a cartesian system. It can be inverted to give an equivalent set of equations expressing the stresses in terms of the strains. These are

$$\sigma_x = \frac{E(1-\nu)}{(1+\nu)(1-2\nu)} \left[(e_x - \alpha T) + \frac{\nu}{1-\nu}(e_y - \alpha T) + \frac{\nu}{1-\nu}(e_z - \alpha T) \right],$$

$$\sigma_y = \frac{E(1-\nu)}{(1+\nu)(1-2\nu)} \left[(e_y - \alpha T) + \frac{\nu}{1-\nu}(e_x - \alpha T) + \frac{\nu}{1-\nu}(e_z - \alpha T) \right],$$

$$\sigma_z = \frac{E(1-\nu)}{(1+\nu)(1-2\nu)}\left[(e_z - \alpha T) + \frac{\nu}{1-\nu}(e_y - \alpha T) + \frac{\nu}{1-\nu}(e_x - \alpha T)\right],$$

$$\tau_{xy} = \frac{E}{2(1+\nu)}\gamma_{xy},$$

$$\tau_{xz} = \frac{E}{2(1+\nu)}\gamma_{xz},$$

$$\tau_{yz} = \frac{E}{2(1+\nu)}\gamma_{yz},$$

and can be written in the standard notation of Chapter 4 as

$$\boldsymbol{s} = \boldsymbol{D}[\boldsymbol{e} - \boldsymbol{e}_{\mathrm{T}}].$$

Here the stress vector $\boldsymbol{s}$ contains six components (ordered in the same way as the components of $\boldsymbol{e}$), and the stress–strain matrix $\boldsymbol{D}$, and thermal strain vector $\boldsymbol{e}_{\mathrm{T}}$ are given by

$$\boldsymbol{D} = \frac{E(1-\nu)}{(1+\nu)(1-2\nu)}\begin{bmatrix} 1 & \frac{\nu}{1-\nu} & \frac{\nu}{1-\nu} & 0 & 0 & 0 \\ \frac{\nu}{1-\nu} & 1 & \frac{\nu}{1-\nu} & 0 & 0 & 0 \\ \frac{\nu}{1-\nu} & \frac{\nu}{1-\nu} & 1 & 0 & 0 & 0 \\ 0 & 0 & 0 & \frac{(1-2\nu)}{2(1-\nu)} & 0 & 0 \\ 0 & 0 & 0 & 0 & \frac{(1-2\nu)}{2(1-\nu)} & 0 \\ 0 & 0 & 0 & 0 & 0 & \frac{(1-2\nu)}{2(1-\nu)} \end{bmatrix} \tag{6.49}$$

and

$$\boldsymbol{e}_{\mathrm{T}} = [\alpha T, \alpha T, \alpha T, 0, 0, 0]^{\mathrm{T}}. \tag{6.50}$$

The stiffness matrix and load vectors are then given by the general integrals of expressions 4.39. As with the linear triangle, the components of $\boldsymbol{B}^{\mathrm{e}}$ and $\boldsymbol{D}$ are constants. The stiffness integral is therefore trivial giving a stiffness matrix

$$\boldsymbol{K}^{\mathrm{e}} = (\boldsymbol{B}^{\mathrm{eT}}\boldsymbol{D}\boldsymbol{B}^{\mathrm{e}})\,V,$$

where V is the volume of the element. The nodal forces associated with a uniform temperature rise are similarly

$$\boldsymbol{f}_{\mathrm{T}}^{\mathrm{e}} = (\boldsymbol{B}^{\mathrm{eT}}\boldsymbol{D}\boldsymbol{e}_{\mathrm{T}})\,V.$$

The nodal forces associated with distributed body and surface loads are a little harder to evaluate since they involve non-trivial integrals of the shape functions n_i over the volume or surface of the tetrahedron. It can be shown, after some manipulation, that a uniformly distributed body force is modelled by four equal loads applied at the vertices of the element (Fig. 6.5(a)), and that a uniform pressure acting on one face is modelled by three equal loads acting at the corners (Fig. 6.5(b)).

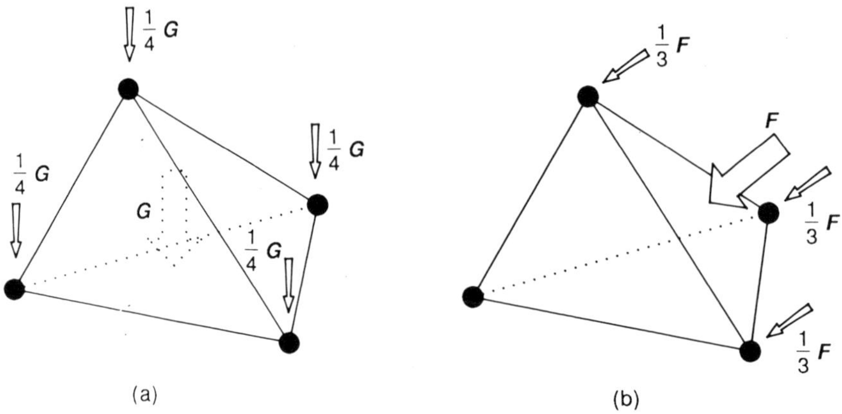

Fig. 6.5 *Equivalent nodal forces on a linear tetrahedron, (a) uniformly distributed body force, (b) uniformly distributed surface load.*

6.5 CONCLUDING COMMENTS

In theory, the triangular and tetrahedral elements described in this chapter can be used to solve any two-dimensional or three-dimensional problem in linear elasticity. In practice, they are less than ideal in most situations, a large number of elements being required to represent quite simple stress states. This is inevitable given that a linear interpolation of displacement implies constant strain within an element. Quite simple distributions of stress must therefore be approximated by a series of constant 'steps', one per element. Elements which are capable of approximating the stress field in a much smoother fashion and of achieving the same accuracy with far fewer nodes can be developed quite easily, however, once the assumption of linear variation has been abandoned. The formulation of such elements is discussed in Chapter 8.

First, however, we must investigate the important matter of 'convergence'. In other words, do we have an assurance, even using the current elements, that the finite element solution will approach the exact solution as the mesh is refined? This is essential if we are to place any reliance upon such results. Such an assurance can, in fact, be given for most continuum elements without difficulty.

REFERENCES

[1] Clough, R. W. (1960) *The Finite Element in Plane Stress Analysis*. Proceedings second conference on electronic computation, American Society of Civil Engineers, pp. 345–77.

[2] Gallagher, R. H., Padlog, J. and Bijlaard, P. P. (1962) Stress analysis of heated complex shapes. *Journal of Aerospace Sciences* **32**, 700–7.

[3] Melosh, R. J. (1963) Structural analysis of solids. *Proceedings ASCE, Journal of the Structural Division*, **89** (st4), 205–23.

[4] Dawe, D. J. (1984) *Matrix and Finite Element Displacement Analysis of Structures*, Clarendon Press, Oxford.

[5] Utku, S. (1968) Explicit expressions for triangular torus element stiffness matrix. *AIAA Journal* **6**, 1174–5.

PROBLEMS

1. Obtain explicit expressions for the shape functions $n_i(x, y)$ $(i = 1, \ldots, 3)$ for triangles (a) and (b) in the illustration below. Calculate the terms in the strain–displacement matrix $\boldsymbol{B}^e$ for each element and confirm in each case that a rigid body translation gives zero strain at all points (a rigid body translation is defined by nodal displacements $u_i = u_0$ and $v_i = v_0$, $i = 1, \ldots, 3$ u_0 and v_0 constants).

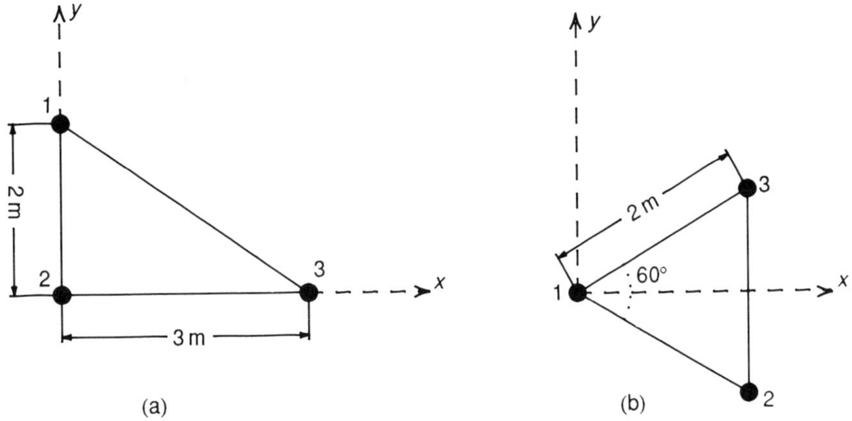

Problem 1.

2. A thin plate of the dimensions shown in problem figure 2 is modelled using a coarse mesh of three linear triangles. The plate is of uniform thickness 0.15 m, Young's modulus 210 GPa and Poisson's ratio 0.3. The unconstrained degrees of freedom are numbered $\delta_1, \delta_2, \ldots, \delta_7$ as shown. Calculate the 3–3 term in the assembled stiffness matrix. (Note: element 1 has the same geometry as element (a) of problem 1).

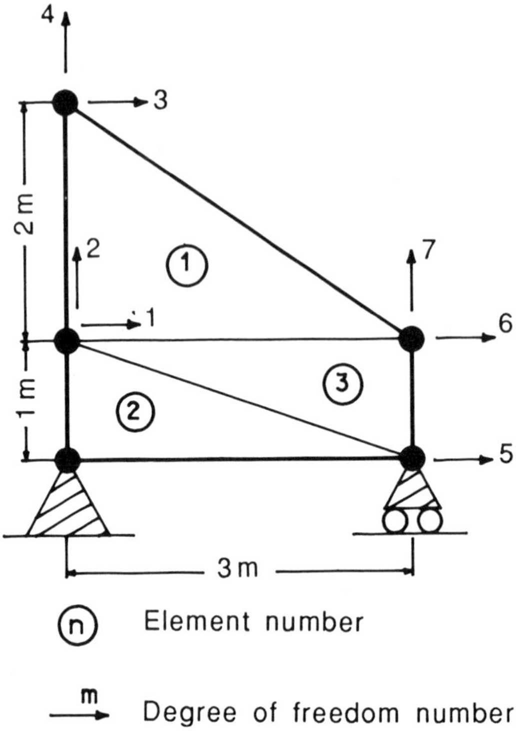

Problem 2.

3. The self-weight of the body of problem two is 60 kN and is distributed uniformly. What nodal forces must be applied to the model to duplicate this loading? (Calculate the equivalent nodal forces associated with each element and assemble them in the usual way.)

4. A horizontal load of 60 kN is distributed uniformly on the vertical face of the plate of problem 2 as shown. What nodal forces must be applied at nodes B and C to model this loading?

5. Show that the equivalent nodal force vector which must be applied to the triangular plane-stress element of Fig. 6.1 to model the effect of a uniform temperature rise T has components

$$f_T^e = \frac{Eh\alpha T}{2(1-\nu)} \begin{bmatrix} b_1 \\ c_1 \\ b_2 \\ c_2 \\ b_3 \\ c_3 \end{bmatrix}.$$

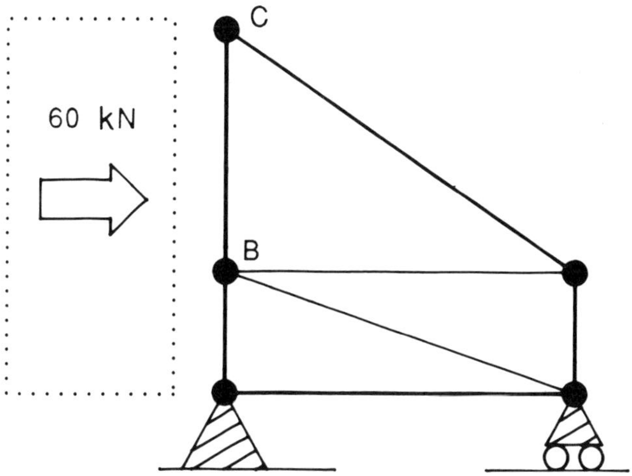

Problem 4.

(The geometric constants a_i, b_i and c_i are given by expressions 6.5). Confirm that the forces applied at each node act at right angles to the opposite face of the triangle.

6. A surface traction $\boldsymbol{t}(s)$ is distributed along the 1–2 side of a linear triangle of constant thickness. Consider the two cases

$$\text{(a)}\quad \boldsymbol{t}(s) = \boldsymbol{t}_0(s/L), \qquad \text{(b)}\quad \boldsymbol{t}(s) = \boldsymbol{t}_0,\ 0 < s < L/2$$

$$= 0,\ L/2 < s < L,$$

where L is the length of side 1–2, and s is measured from node 1. Show that the proportions of the resultant load which must be applied at the nodes are; 1/3 and 2/3 in case (a), and 3/4 and 1/4 in case (b), as shown.

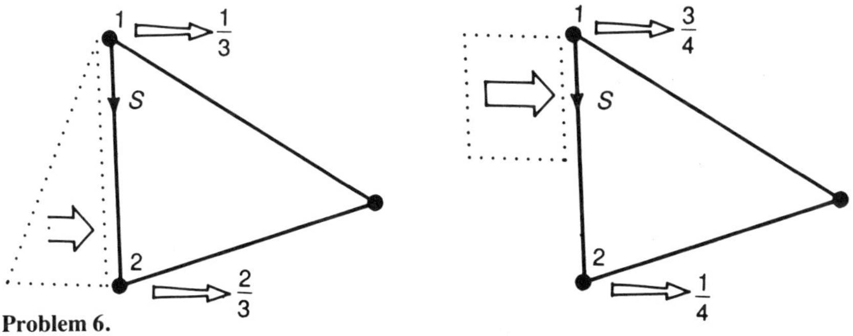

Problem 6.

7. A plain-strain model for a retaining structure is formed using a regular mesh of identical triangular elements as shown. The vertical face AE carries

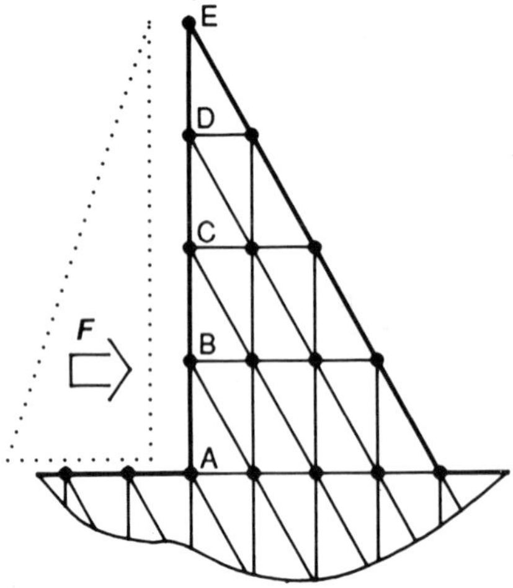

Problem 7.

a distributed load F whose intensity varies linearly as shown. Show that the proportions of F which must be applied as equivalent nodal forces at nodes A, B, C, D and E, are 11/48, 3/8, 1/4, 1/8 and 1/48, respectively. [Note: the surface load on each element can be treated as a combination of a uniformly distributed load and a linearly varying load].

8. A surface load, F, is distributed uniformly along a portion of the upper edge of a thin plate which is modelled using two meshes of linear triangles (see (a) and (b) in the diagram below). In each case, the nodes on the upper surface are spaced regularly a distance a apart. Show that in the case of mesh (a), the resultant load F should be distributed between nodes A, B, and C in the proportion; 1/4:1/2:1/4. In what proportion should it be distributed between nodes D, E, F and G, in the case of mesh (b)?

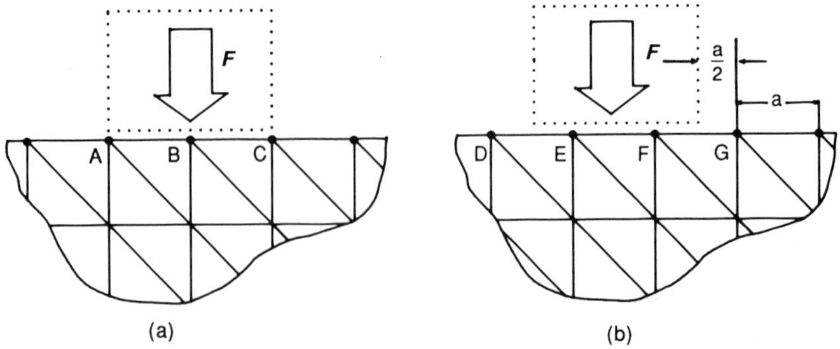

Problem 8.

9. A surface load F acts uniformly on the 1–2 edge of a plane-stress triangle whose thickness varies linearly from h_0 at node 1 to $2h_0$ at node 2 (as

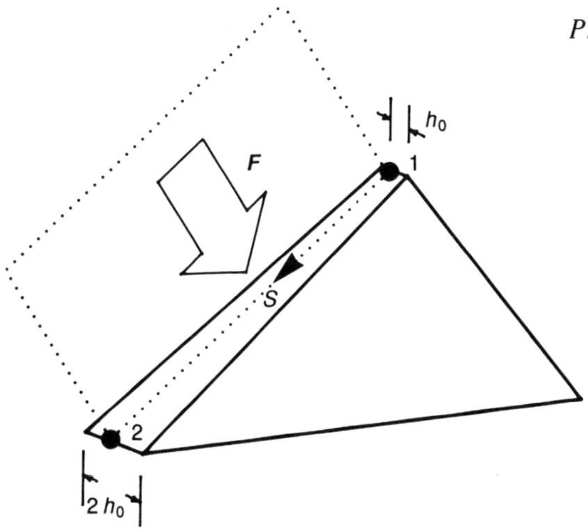

Problem 9.

shown). Show that the resultant load is distributed between nodes 1 and 2 in the proportion 4/9:5/9. (Suggestion: write $h(s) = h_0(1 + s/L)$ and place this within the integral in equation 6.30).

10. A uniform pressure p_0 acts on the upper surface of an axisymmetric triangular element which subtends an angle of one radian at the axis of symmetry (as shown). Show that the equivalent nodal forces which must be applied at nodes A and B are ring loads of magnitude $p_0(b^2 - 2a^2 + ab)/6$ and $p_0(2b^2 - a^2 - ab)/6$, respectively. (Suggestion: write the surface differential dS of equation 6.29 as (rdr) and integrate from $r = a$ to $r = b$).

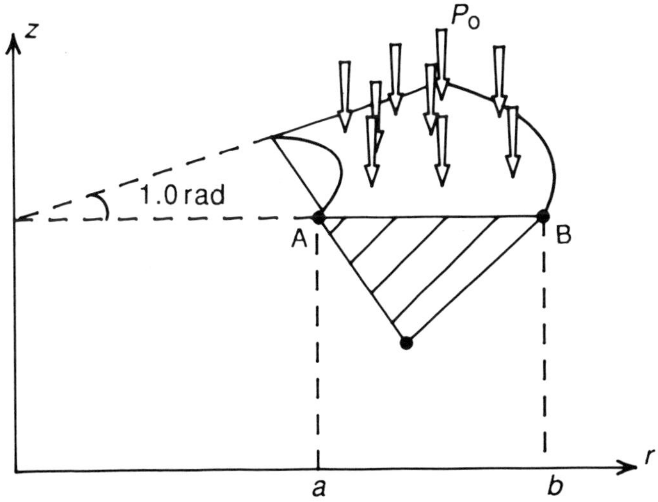

Problem 10.

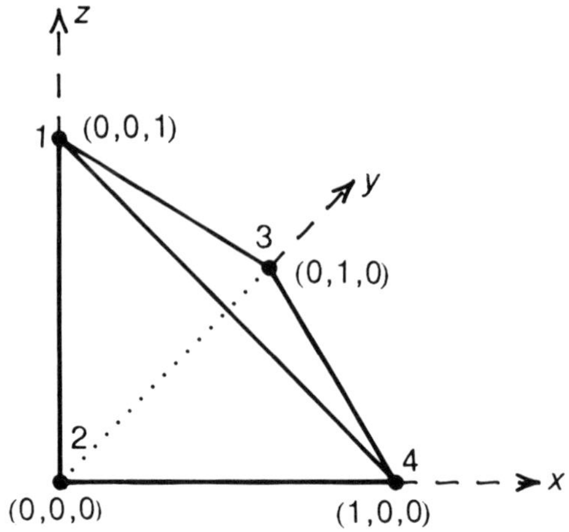

Problem 11.

11. A linear tetrahedron is defined by four nodes located and numbered as shown. Obtain an explicit expression for the shape function $n_1(x, y, z)$. Integrate n_1 over the volume of the tetrahedron, and confirm the validity of the statement, in section 6.4, that an equivalent nodal force of **G**/4 must be applied to node 1 to correctly model a net body force ***G*** distributed uniformly over the element.

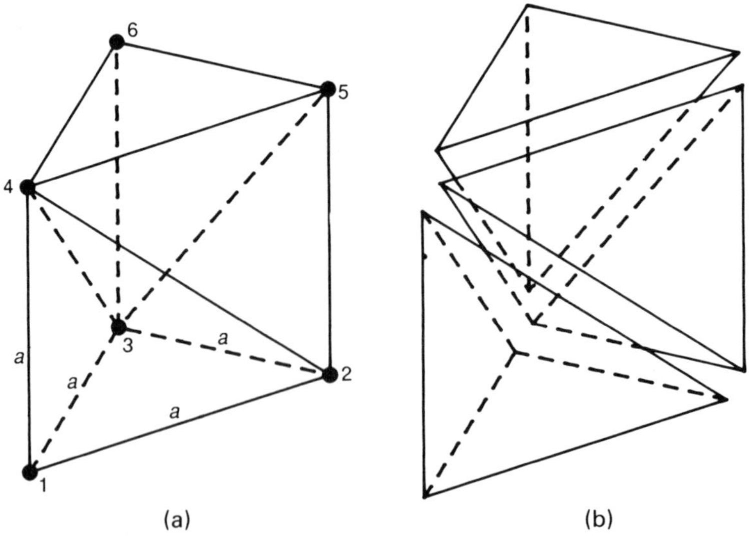

Problem 12.

12. A solid triangular prism of height a, whose base is an equilateral triangle of side a, is assembled from three linear tetrahedra (as shown in (a)). The individual elements are shown slightly offset in (b). A body force is distributed uniformly over the prism. Confirm that the resultant force should be distributed between nodes 1, . . ., 6 in the ratio; 1/12:1/6:1/4:1/4:1/6:1/12.
(Suggestion: calculate the equivalent nodal forces on each tetrahedral element and assemble them to obtain the nodal forces on the prism)

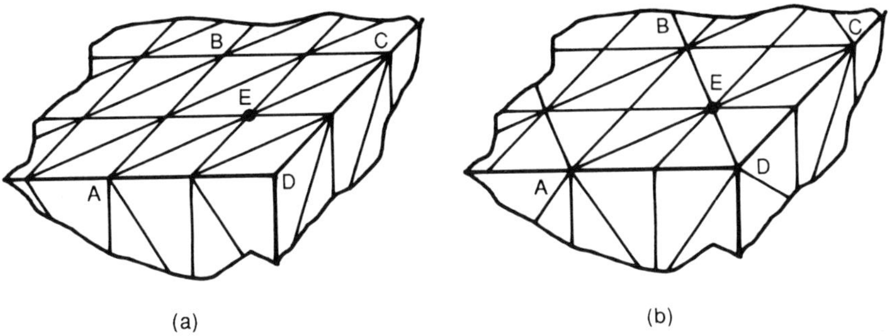

Problem 13.

13. A uniformly distributed load acts over the region ABCD on the surface of a three-dimensional body as shown. Two meshes, (a) and (b), of linear tetrahedra are used to model the body. In both cases, the surface nodes in the vicinity of ABCD are located on a regular rectangular grid. Show that the proportion of the net load which must be applied as an equivalent equivalent nodal force at E is 1/4 and 1/3 for meshes (a) and (b), respectively.

7

Convergence, compatibility and completeness

7.1 INTRODUCTION

A finite element model is expected to yield more accurate results as the number of elements in the mesh is increased and their size reduced. In terms of the underlying Rayleigh–Ritz statement, there is an expectation that the approximate solution will converge to the continuum solution as the number of basis functions in the trial expansion is increased. In some instances, this expectation is not realized. The discrete solution then diverges or, worse still, converges to an incorrect result. Convergence is assured only if the trial displacement field satisfies certain criteria. These have in fact been met by the linear elements introduced so far, but it is important to define them more precisely before attempting to formulate elements of greater complexity.

Convergence of a finite element model is associated with the notions of 'compatibility' and 'completeness'. Of these, compatibility is the most straightforward. The idea of compatibility has already been discussed for the full continuum problem in Chapter 2 (section 2.1). Moreover, it is explicitly stated in the principle of stationary total potential that the trial solution should belong to a class of 'compatible' displacement fields. This excludes, for example, discontinuous displacements of the type illustrated in Fig. 2.1(b). A natural question which then arises is, 'Do the same requirements apply to the Rayleigh–Ritz solution?'. In other words, given that we have approximated the continuum by a discrete system, can we also 'approximate' the exact compatibility condition? The short answer is 'yes'. Some relaxation of the full compatibility requirement *is* permitted. This is seldom necessary in continuum applications but is relatively common in plate and shell formulations.

'Completeness' is a somewhat more abstract notion. A sequence of functions $f_i(\boldsymbol{x})$ $(i = 1, 2, \ldots)$ is 'complete' with respect to a particular class of functions, when any member of that class can be approximated by a truncated series, $\Sigma a_i f_i(\boldsymbol{x})$, to a prescribed degree of accuracy. For example, the sequence of functions 1, $\sin(x)$, $\cos(x)$, $\sin(2x)$, $\cos(2x)$, . . . , is complete for continuous functions on the interval $(0, 2\pi)$, since any continuous function can be matched as closely as we like to a truncated fourier series, provided that a sufficiently large number of terms is taken. On the other hand, the same set of functions with the first term missing, that is sin (x), cos (x), sin ($2x$), cos ($2x$), . . . , is *not* complete, since a fourier expansion minus the first (constant) term can never adequately represent any function whose mean value is not zero. Similarly, a series of polynomial terms is 'complete' provided that all integer powers are present in ascending order. If one term is omitted, however, the sequence is no longer complete, irrespective of the number of terms which are subsequently included. In both instances, it is clearly important to ensure that 'lower order' terms are consistently included ahead of 'higher order' ones within any sequence of complete functions.

How this relates to the basis functions of a finite element model is not immediately obvious. The idea of a 'complete' displacement field is somewhat easier to visualize if we think in terms of the derived strains. It then seems a reasonable assumption that an arbitrary strain field can be represented in the last resort by a large number of constant 'steps', and that any desired accuracy can be achieved provided that the steps are sufficiently small. Using this somewhat heuristic approach, we might anticipate that any finite element representation which possesses the intrinsic ability to model constant stress and strain within an element will be 'complete' in some sense. This is, in fact, the case.

7.2 SUFFICIENT CRITERIA FOR CONVERGENCE

Convergence of a finite element model to the continuum solution is assured when the following conditions are satisfied.

1. The assumed displacement field is continuous at all points within the model.
2. Each element is capable of representing exactly a state of constant strain.

A formal proof of this statement is by no means straightforward and lies well beyond the scope of the current theoretical treatment. Readers wish to satisfy themselves regarding the validity of the statement will find a fairly rigorous justification in [1] and varying levels of informal argument in [2]–[4]. The origins of the two conditions are not difficult to discern. Criterion 1 is simply a restatement of the full compatibility condition for the original continuum problem. Criterion 2 is a completeness condition expressed in terms of the strain field, as foreshadowed in the last paragraph of the preceding section. The two conditions are referred to in all that follows simply as the **compatibility criterion** and the **completeness criterion**.

Note that, although convergence is assured if both criteria are satisfied, it is not precluded if they are *not*. They constitute, in this sense, sufficient, rather than a necessary, conditions for convergence and leave open the question of what happens if one or both is violated. This is further explored in section 7.3.

7.2.1 Compatibility criterion

Let us look first at the implications of criterion 1. It is conveniently restated as separate subcriteria:

(i) that displacements are continuous within each element; and
(ii) that displacements are continuous across element boundaries.

These are simply the 'full' compatibility requirements of the original continuum problem transposed onto the finite element model. The conditions under which they are satisfied are easily described.

Criterion (i) is satisfied provided that the shape functions are continuous within an element. This is certainly the case for elements based on polynomial interpolation. These include the linear elements of Chapter 6 and all higher order elements to be presented in Chapter 8. In practice, criterion (i) is trivially satisfied in all but the most bizarre elements.

Criterion (ii), the imposition of continuity at element boundaries, is not so straightforward. Consider, for example, the boundary between linear triangles (a) and (b), which have nodes 1 and 2 in common (Fig. 7.1). Let s be a local coordinate measured along the common edge from node 1. The question we must ask with regard to compatibility is: 'Does this mesh automatically satisfy criterion (ii) for all values of the displacements at nodes 1 and 2?'. In other words, does the mesh permit or exclude discontinuities of displacement across this boundary?

Since the displacements vary linearly within each element, they vary linearly with s along 1–2. They do this moreover in *both* elements, and take com-

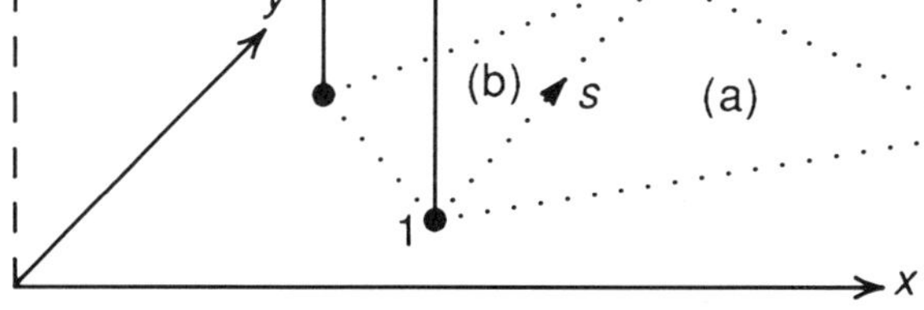

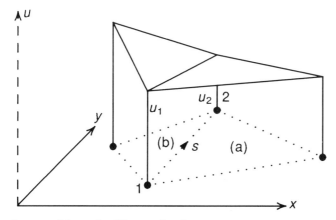

Fig. 7.1 *A compatible mesh of linear triangles.*

mon, nodal values at each end. This is sufficient to ensure the *same* definition on either side of the interface, any linear function being *uniquely* defined by two values. In other words, u (or v) is defined in the same way on both sides of the interface. Condition (ii) is therefore satisfied irrespective of the nodal values of u_1, u_2, v_1 and v_2.

Compatibility across the 1–2 interface can be visualized by plotting the interpolated displacement, u (or v) as a surface above the element. If the surface is continuous for all nodal values of u (or v) compatibility is achieved. In the case of Fig. 7.1, this construction gives two triangular facets, one above each element. These clearly join along the 1–2 interface to give a continous surface over both elements.

The same construction can be used to check the compatibility of a mesh involving three elements joined as shown in Fig. 7.2. Here, three linear triangles, (a), (b) and (c), are joined along a common interface 1–2–3. Nodes 1 and 3 define the common side of element (a), whereas nodes 1 and 2, and nodes 2 and 3, define the common sides of elements (b) and (c). Fig. 7.2 shows the displacement u plotted as a surface above each element. Once again we obtain a series of triangular facets, but they no longer form a continuous surface for *all* values of u_2, u_2 and u_3. The resulting discontinuity in the surface in the vicinity of node 2 then represents a potential physical discontinuity of displacement across the interface 1–2–3 and hence a violation of criterion (ii).

Figure 7.2 serves to illustrate that inter-element compatibility depends not only on the interpolation within each element (in this case the linear variation of displacement within each triangle) but also on the manner in which the elements are joined together. For a mesh of linear triangles, the rules are very simple. Compatibility is assured provided that the elements are joined so that adjacent triangles match exactly, node for node and side for side. The same logic applies to three-dimensional elements such as the linear tetrahedra of

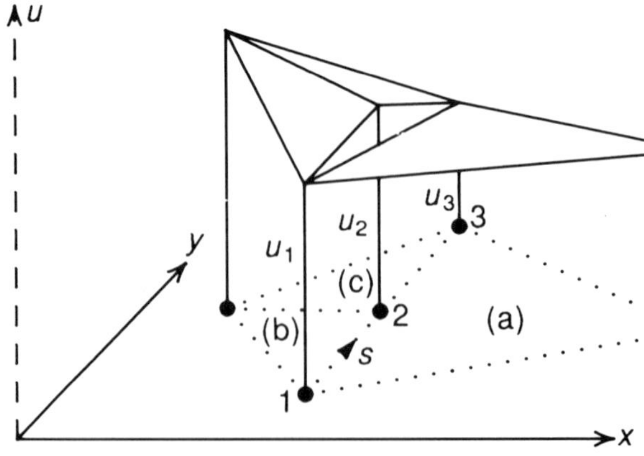

Fig. 7.2 *An incompatible mesh of linear triangles.*

Chapter 6. These must be meshed so that each triangular face matches exactly the triangular face of an adjacent element.

The requirements imposed by compatibility on the formulation of higher order continuum elements are relatively straightforward. They will be discussed further when these elements are introduced in Chapter 8. The implications for 'structural' elements (beams, plates and shells) are more far reaching. In such cases, compatibility of displacement at *all* points, both on and off the the neutral axis or plane, requires that both the lateral displacement of the neutral plane *and* its slope(s) are continuous. Criterion (ii) then becomes much more difficult to satisfy, a point which will be discussed at much greater length in Chapter 10.

7.2.2 Completeness criterion

The second of the two sufficiency criteria requires that an element model exactly a state of constant strain. The condition is satisfied when the imposition of nodal displacements consistent with an arbitrary state of constant strain reproduces the exact strain at all points within an element. In the case of zero strain, it ensures that rigid body translations and rotations give zero strain within an element (and hence zero strain energy) an intuitively sensible provision.

The completeness criterion is satisfied by any continuum element whose trial interpolation includes a complete first-order polynomial in the global cartesian coordinates. The proof of the above statement is quite straightforward and proceeds as follows for the two-dimensional case. Consider an arbitrary, two-dimensional state of constant strain,

$$e_x = \varepsilon_a,\ e_y = \varepsilon_b \text{ and } \gamma_{xy} = \gamma_c\ (\varepsilon_a,\ \varepsilon_b \text{ and } \gamma_c \text{ constants}).$$

The corresponding displacement field is obtained by integrating the strain–displacement equations;

$$\frac{\partial u}{\partial x} = \varepsilon_a,\quad \frac{\partial u}{\partial y} = \varepsilon_b \text{ and } \frac{\partial u}{\partial y} + \frac{\partial u}{\partial x} = \gamma_c.$$

The general solution is

$$\begin{aligned} u &= \varepsilon_a x + \tfrac{1}{2}\gamma_c y + Ay + B, \\ v &= \tfrac{1}{2}\gamma_c x + \varepsilon_b y - Ax + C, \end{aligned} \qquad (7.1)$$

where A, B and C are constants of integration which represent a translation (B and C) and a rotation (A) in the x–y plane. If nodal displacements for a two-dimensional element are calculated using expressions 7.1 and substituted into the strain–displacement relationship 4.10, the completeness criterion is satisfied if the original state of strain is recovered. This procedure can be bypassed in most instances by returning to the original form of the interpolation within an element and demonstrating that it is capable of exactly duplicating equations 7.1. If this is the case, the imposition of nodal displacements evaluated using equations 7.1 will simply reproduce the exact displacement field at all points within the element and hence the exact strains. In the case of the linear triangle, for example, the interpolation within the element is of the form

$$u = \alpha_1 + \alpha_2 x + \alpha_3 y,$$

$$v = \beta_1 + \beta_2 x + \beta_3 y$$

This represents exactly the 'constant strain' displacement field (equation 7.1) when the constants α_i and β_i are chosen as

$$\alpha_1 = B,\ \alpha_2 = \varepsilon_a,\ \alpha_3 = \tfrac{1}{2}\gamma_c + A,\ \beta_1 = C,\ \beta_2 = \tfrac{1}{2}\gamma_c - A \text{ and } \beta_3 = \varepsilon_b. \qquad (7.2)$$

The interpolation within the element (and at the nodes) is then identical to expression 7.1 and the completeness criterion is satisfied. By the same logic, *any* two-dimensional element with an interpolation of the form,

$$u = \alpha_1 + \alpha_2 x + \alpha_3 y + \text{higher order polynomial terms in } x \text{ and } y$$

$$v = \beta_1 + \beta_2 x + \beta_3 y + \text{higher order polynomial terms in } x \text{ and } y$$

also satisfies the completeness criterion, since the higher order polynomial coefficients can be equated to zero and the coefficients α_i and $\beta_i (i = 1, \ldots, 3)$ given values (equation 7.2) above.

The same argument holds for three-dimensional elements. In other words, the completeness criterion is automatically satisfied by any element whose interpolated displacement field contains a complete set of first order terms in a cartesian coordinate system. The linear tetrahedron of section 6.4 clearly satisfies this requirement as do all of the more complex elements which will be derived in Chapter 8.

The completeness criterion applies also to 'structural' elements (beams, plates and shells), but imposes more stringent restrictions on their shape functions. Since the strain in such elements is determined by the curvature of the neutral plane, a state of 'constant strain' must be re-interpreted as a state of constant curvature. The completeness criterion then requires that a beam or plate is capable of exactly representing an arbitrary state of constant curvature. The implications of this are discussed further in Chapters 9 and 10.

7.3 NON-CONFORMING ELEMENTS, THE PATCH TEST

Although the criteria of section 7.2 are 'sufficient' to ensure convergence, they are not essential and elements which do not satisfy them should not necessarily be discarded as unsatisfactory. In particular, inter-element compatibility can be violated quite easily, without invalidating an element. Elements and meshes which do this are termed 'non-conforming'.

Non-conforming elements are seldom used for continuum problems for the simple reason that full inter-element compatibility is easily enforced. In the case of plates and shells, however, full compatibility is much more difficult to achieve and elements which violate criterion (ii) are quite common. Many of these are perfectly serviceable and give convergent solutions. This is not infallibly the case, however, and such elements must be tested in some way to ensure that they do indeed converge.

The **patch test** provides a simple procedure for doing this. It is a straightforward numerical test which determines whether *any* element — conforming or non-conforming — will produce convergent results. Even when an element openly violates the compatibility or completeness criteria of section 7.2, it will still converge to the continuum solution provided that it passes the patch test.

The patch test is presented here in its simplest form. A number of more sophisticated variants have developed in the two and a half decades since it was first proposed [5]. What started as a pragmatic test has acquired not only a degree of mathematical respectability [1] but also a scope and generality far beyond that originally envisaged for it. A full discussion of the test in all its forms is to be found in [2] which devotes an entire chapter to the topic.

The patch test in its most straightforward form involves the following procedure. First a number of elements are assembled to form a 'patch'. This is a mesh of elements which contains at least one interior node. A two-dimensional patch of quadrilateral elements is shown in Fig. 7.3. The arrangement of the elements is arbitrary in the sense that the patch must pass the test for any arrangement that we select. This is most important since some elements will pass the test for particular geometries but fail for others. In applying the test, it is therefore important to make the shape of the elements as idiosyncratic as possible, and to avoid orthogonal geometries and regular spacings.

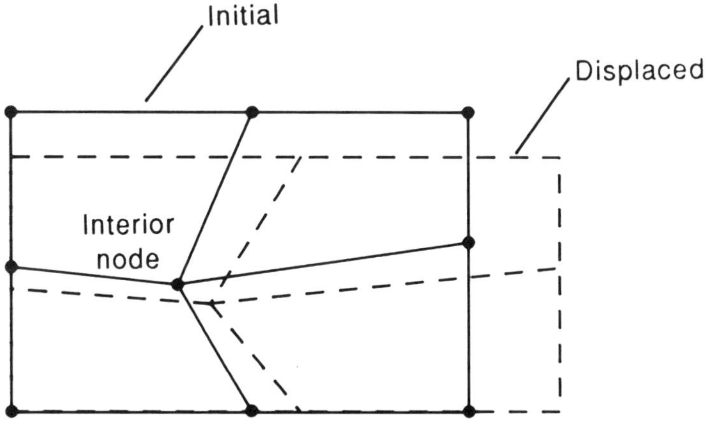

Fig. 7.3 *A 'patch' of quadrilateral elements.*

The test itself involves the application *either* of nodal displacements *or* of equivalent nodal forces, consistent with an exact state of constant strain. Both are applied at the boundary of the patch. The test is passed if the computed solution reproduces the exact strains (or stresses) at all points within the patch, within the limits of computational accuracy. This is the 'strong' form of the test. A weaker version, which also guarantees convergence, is passed when, as the patch is repeatedly subdivided, the stresses or strains within it converge to the expected, constant values. All elements which satisfy the criteria of section 7.2 will by definition pass both forms of the patch test, but the reverse is not necessarily true.

7.4 CONCLUDING REMARKS

To summarize, convergence of the finite element model to a continuum solution is assured if the completeness and compatibility criteria of section 7.2 are satisfied, *or* if the elements pass the patch test of section 7.3. In practice, convergence is seldom an issue with continuum elements. Almost without exception, they satisfy the compatibility and completeness criteria in full, and convergence is therefore assured. In some instances — and these lie well beyond the scope of the present discussion — a measure of incompatibility is deliberately incorporated in continuum elements to remedy other deficiencies. A relatively common example is the introduction of 'incompatible' displacement modes into lower order elements to improve their notoriously poor performance in bending [6]. Continuum elements of this type along with the more numerous nonconforming, structural elements cannot be assumed to be convergent however unless they pass the patch test in one of its forms. If an element fails the patch test both in the simple and 'weak' versions, it should be not be used.

REFERENCES

[1] Strang, G. and Fix, G. J. (1973) *An Analysis of the Finite Element Method*, Prentice-Hall, Englewood Cliffs, Chapters 1–4.

[2] Zienkiewicz, O. C. and Taylor, R. L. (1990) *The Finite Element Method*, 4th edn, Vol. 1, McGraw-Hill, London, Chapters 2 and 11.

[3] Norrie, D. H. and De Vries, G. (1978) *An Introduction to Finite Element Analysis*, Academic Press, New York, Chapter 8.

[4] Cook, R. D., Malkus, D. S. and Plesha, M. E. (1989) *Concepts and Applications of Finite Element Analysis*, 3rd edn sections 4.5–4.6. John Wiley & Sons, New York, sections 4.5 and 4.6.

[5] Bazeley, G. P., Cheung, Y. K., Irons, B. M. and Zienkiewicz, O. C. (1965) *Triangular Elements in Bending — Conforming and Nonconforming* solutions. Proceedings of the Conference on Matrix Methods in Structural Mechanics, Air Force Institute of Technology, Wright–Patterson Air Force Base, Dayton, Ohio.

[6] Taylor, R. L., Beresford, P. J. and Wilson, E. L. (1976) A nonconforming element for stress analysis. *International Journal for Numerical Methods in Engineering*, **10**, 1211–20.

PROBLEMS

1. A one-dimensional bar element of length L, Young's modulus E and cross-sectional area A has the topology of the linear element shown in Fig. 4.2. An interpolation for the axial displacement within the element is proposed of the form

$$u(x') = \alpha_1 + \alpha_2 x'^2,$$

where α_1 and α_2 are constants, and x' is a local axial coordinate. Show that the shape matrix $\boldsymbol{N}^e$ and strain–displacement matrix $\boldsymbol{B}^e$ of the element are

$$\boldsymbol{N}^e = \left[1 - \left(\frac{x'}{L}\right)^2, \left(\frac{x'}{L}\right)^2 \right], \text{ and } \boldsymbol{B}^e = \left[-\frac{2x'}{L^2}, \frac{2x'}{L^2} \right].$$

Confirm that the element does *not* satisfy the completeness criterion of section 7.2. (Suggestion: apply nodal displacements consistent with a state of uniform axial strain and calculate the strain within the element).

2. Apply nodal displacements consistent with a state of constant axial strain to the three-noded quadratic bar element of Chapter 4, problem 2. Demonstrate that the element satisfies the completeness criterion of section 7.2.

3. A four-noded rectangular element is proposed for plane-stress analysis. It has nodes at each corner and sides which are parallel to the x and y-axes of a cartesian coordinate system. Three formulations are proposed for the interpolation within the element. These are

(a) $u(x, y) = \alpha_1 + \alpha_2 x + \alpha_3 y + \alpha_4 xy,$
$v(x, y) = \beta_1 + \beta_2 x + \beta_3 y + \beta_4 xy,$

(b) $u(x, y) = \alpha_1 + \alpha_2 x + \alpha_3 y + \alpha_4 (x^2 + y^2),$
$v(x, y) = \beta_1 + \beta_2 x + \beta_3 y + \beta_4 (x^2 + y^2),$

(c) $u(x, y) = \alpha_1 + \alpha_2 x^2 + \alpha_3 y^2 + \alpha_4 xy,$
$v(x, y) = \beta_1 + \beta_2 x^2 + \beta_3 y^2 + \beta_4 xy.$

The element is joined along one side to a linear triangle of the type described in section 6.2. Explain why formulation (a) is the only one which does not violate inter-element compatibility. Which, if any, of the three formulations satisfies the completeness criterion?

4. A plane stress element has nodes 1, 2, 3 and 4 at (0, 0), (2, 0), (1, 1) and (2, 1) as shown. It is to be formulated using a trial interpolation

$$u(x, y) = \alpha_1 + \alpha_2 x + \alpha_3 y + \alpha_4 xy,$$

$$v(x, y) = \beta_1 + \beta_2 x + \beta_3 y + \beta_4 xy.$$

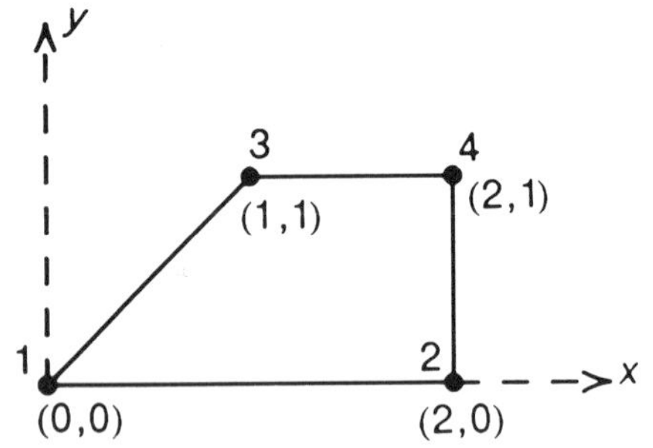

Problem 4.

Evaluate the constants $\alpha_1, \ldots, \alpha_4, \beta_1, \ldots, \beta_4$ in terms of the nodal displacements and show that the shape functions $n_i(x, y)$, $(i = 1, \ldots, 4)$ are given by

$$n_1(x, y) = \left(1 - \frac{x}{2} - y + \frac{xy}{2}\right), \quad n_2(x, y) = \left(\frac{x - xy}{2}\right),$$

$$n_3(x, y) = (2y - xy), \quad \text{and} \qquad n_4(x, y) = (xy - y).$$

Obtain an explicit expression for the variation of u (or v) along the 1–2 and 1–3 sides. Hence demonstrate that the element can be compatibly joined to a linear triangle along the 1–2 side but not along the 1–3 side.

5. Confirm that the two-dimensional displacement field, $u(x, y) = \varepsilon x$, $v(x, y) = 0$, corresponds to a state of constant strain, $e_x = \varepsilon$, $e_y = \gamma_{xy} = 0$. When evaluated at the nodes of the element of problem 4, confirm that the above displacement field gives nodal displacements;

$$u_1 = 0,\ u_2 = 2\varepsilon,\ u_3 = \varepsilon,\ u_4 = 2\varepsilon,\ v_1 = v_2 = v_3 = v_4 = 0.$$

Calculate the $\boldsymbol{B}^e$ matrix for the element and confirm that the exact value of strain is recovered when these nodal values are substituted into the discrete strain–displacement relationship. Repeat the exercise for the nodal displacements

(a) $u_1 = u_2 = u_3 = u_4 = 0,\ v_1 = 0,\ v_2 = 0,\ v_3 = v_4 = \varepsilon,$

and

(b) $u_1 = u_2 = 0,\ u_3 = u_4 = \gamma,\ v_1 = v_2 = v_3 = v_4 = 0.$

Hence show that the element is capable of representing exactly an arbitrary state of of constant strain and demonstrate that the completeness criterion of section 7.2 is satisfied. Could this have been deduced directly from the original form of the interpolation?

6. A two-dimensional element with nodes at (0,0), (1,0), (1,1) and (2,1) as shown is formulated using a trial interpolation

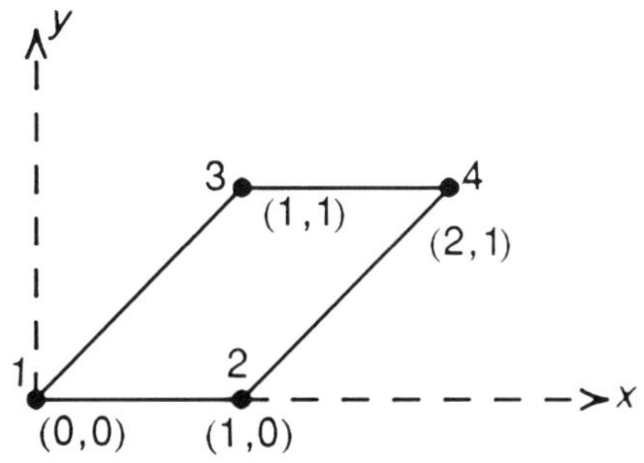

Problem 6.

$$u(x, y) = \alpha_1 + \alpha_2(x - y) + \alpha_3 y + \alpha_4(x - y),$$

$$v(x, y) = \beta_1 + \beta_2(x - y) + \beta_3 y + \beta_4(x - y).$$

Evaluate the constants $\alpha_1, \ldots, \alpha_4, \beta_1, \ldots, \beta_4$ and show that the shape functions $n_i(x, y), (i = 1, \ldots, 4)$ are given by

$$n_1(x, y) = (xy - y^2 - x + 1), \quad n_2(x, y) = (x - y - yx + y^2),$$

$$n_3(x, y) = (y - xy + y^2), \qquad n_4(x, y) = (xy - y^2).$$

Confirm that the element can be compatibly joined to a linear triangle along the 1–2 side and the 1–3 side.

7. Show that the stiffness matrix for the two-noded element of problem one has components

$$\boldsymbol{K}^{\mathrm{e}} = \left(\frac{4}{3}\right)\begin{bmatrix} k & -k \\ -k & k \end{bmatrix}, \text{ where } k = \frac{EA}{L}.$$

Two such elements are joined together to form a bar of length $2L$. Impose known displacements at each end consistent with an exact state of constant axial strain and solve for the displacement of the central node. By treating this model as a one-dimensional 'patch' and the central node as an interior node, show that the element fails the simple patch test (suggestion: plot the axial strain in each element as a function of distance along the bar).

8. Repeat problem seven using three equal elements, once again applying known displacements at the ends of the bar consistent with a state of constant axial strain. Show that the element again fails the patch test. Explain carefully why this will always be the case irrespective of the number of elements used and comment on the failure or success of the element to pass the 'weak' form of the patch test.

8

Higher order elements for two-dimensional and three-dimensional analysis

8.1 INTRODUCTION

The linear elements introduced in Chapter 6 allow us to perform finite element analysis on arbitrarily shaped elastic bodies. They are simple to derive and implement, but have been largely superseded in current practice by higher-order elements with more complex topologies. These tend to be more efficient in the sense that they produce more accurate results for a given number of nodes. The reason for this improvement can be illustrated using the one-dimensional formulations of Chapter 4. Consider, for example, the relative merits of representing a short section of a prismatic bar by:

(i) a single three-noded element using quadratic interpolation (see problem two of Chapter 4 for details of this formulation); or
(ii) an assembly of two linear elements.

The two models are illustrated in Fig. 8.1. Both have the same number of nodes and the same number of degrees of freedom. In the case of the three-noded element, however, the shape functions are quadratic, whereas in the two-element model they are linear. The axial displacement is therefore represented as a smooth quadratic curve in case (i), but as two, piecewise-continuous linear segments in case (ii). Similarly, the axial stress is approximated by a single linear function in case (i), but by two constant 'steps' in case (ii). Clearly, the quadratic model is intrinsically more suited to the representation of continuous, smoothly-varying stress and displacement fields than is the linear one. The same arguments apply in two and three dimensions.

Higher-order elements provide remedies also for other deficiencies in the linear formulation; its inability, for example, to model within a single element the deformation of a straight fibre into a circular arc. This causes 'parasitic shear' [1], a characteristic of linear elements which renders them notoriously inaccurate in situations where substantial bending strains are present. The problem can be remedied by a number of *ad hoc* adjustments [2] but is most easily cured by the use of higher-order interpolation.

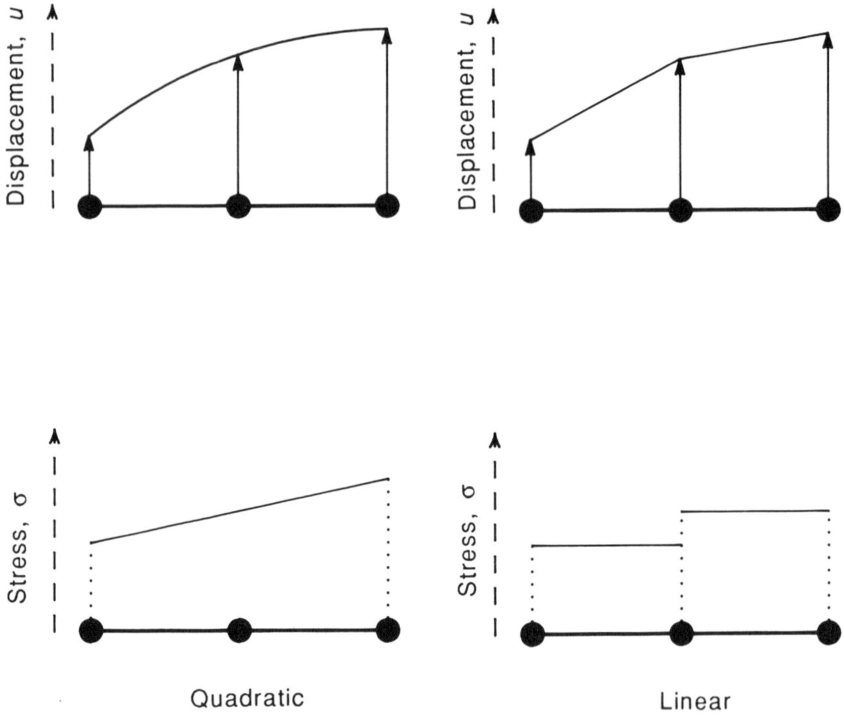

Fig. 8.1 *Quadratic and linear models of stress and displacement.*

The use of higher-order elements is not without its disadvantages. The stiffness matrix for a system modelled using such elements is generally fuller than for an equivalent mesh of linear elements, a consequence of the increased connectivity over each element (higher-order elements connect a larger number of nodes). Although the computational effort required to solve the stiffness equations is thereby increased, this is offset in most instances by improved accuracy in the resulting solution. Greater effort is required also in the integration of element stiffnesses. Since stress and strain are no longer constant within an element, the integration of stiffness terms and equivalent nodal forces is therefore no longer trivial. In practice, however, the benefits to be derived from the use of higher-order interpolation generally outweigh the inconveniences and higher-order continuum elements, particularly those which incorporate the 'isoparametric' mapping to be introduced later in this chapter, feature prominently in the libraries of most commercial programs.

8.2 DIRECT FORMULATION OF SHAPE FUNCTIONS

In Chapters 4 and 6, shape functions were obtained for linear elements in one, two and three-dimensions by adopting the following procedure:

(i) A polynomial interpolation was proposed within the element, the number of unknown polynomial coefficients being equal to the number of nodes defining the topology of the element.
(ii) The interpolation function was evaluated at each node and equated to the nodal displacement. This gave a set of simultaneous linear equations which were solved to yield the unknown polynomial coefficients.
(iii) The resulting expressions for the coefficients were substituted into the original interpolation and the terms rearranged to yield an expression of the form

$$u(\boldsymbol{x}) = n_1(\boldsymbol{x})u_1 + n_2(\boldsymbol{x})u_2 + \ \ldots . \tag{8.1}$$

This defined the shape functions $n_i(\boldsymbol{x})$ of the element.

In the formulations considered so far, polynomials with two, three and four unknown coefficients have been used to obtain shape functions for the two-noded bar, the three-noded triangle and the four-noded tetrahedron, respectively. Even for relatively primitive elements such as these, the evaluation of the polynomial coefficients becomes quite cumbersome. If the same procedure is applied to elements with more extensive topologies, this type of formulation becomes quite unmanageable.

In developing alternative procedures, we will confine our attention initially to interpolating functions which are polynomials of the global, cartesian coordinates x, y and z. These will later be mapped to curvilinear coordinates (section 8.5) to give a more general class of element.

8.2.1 Higher-order interpolation, the implications for inter-element compatibility

The powers which are included in the interpolating polynomial of a two-dimensional or three-dimensional element determine whether compatibility can be achieved on its boundary. For a two-dimensional element, an interpolating polynomial takes the form

$$u = \alpha_1 + \alpha_2 x + \alpha_3 y + \alpha_4 x^2 + \alpha_5 xy + \alpha_6 y^2 + \ \ldots . \tag{8.2}$$

The number of terms in this expansion must be equal to the number of nodes in the topology of the element if the unknown coefficients α_i are to be uniquely determined. Within this overall constraint, the individual polynomial terms can be selected as required. A useful way of representing the potential candidates for inclusion in such a series is to place the homogeneous terms of a particular order on separate (descending) levels of Pascal's triangle. This produces the arrangement shown in Fig. 8.2. The terms down to and including those on a particular level form a 'complete' polynomial of a given order. For example, the constant and linear terms form a complete linear polynomial, the constant, linear and quadratic terms form a complete linear polynomial, the constant, linear and quadratic terms form a complete quadratic polynomial, and so on. Complete polynomials of this type are intrinsically attractive as a basis for element interpolation since they have the same form in any rectangular coordinate system irrespective of its orientation. They are used as the basis for a family of triangular elements to be introduced in section 8.4. Incomplete polynomials — those which mix terms on different, lower levels of the triangle — are also used for element interpolation, and form the basis for several families of rectangular and block elements to be introduced shortly.

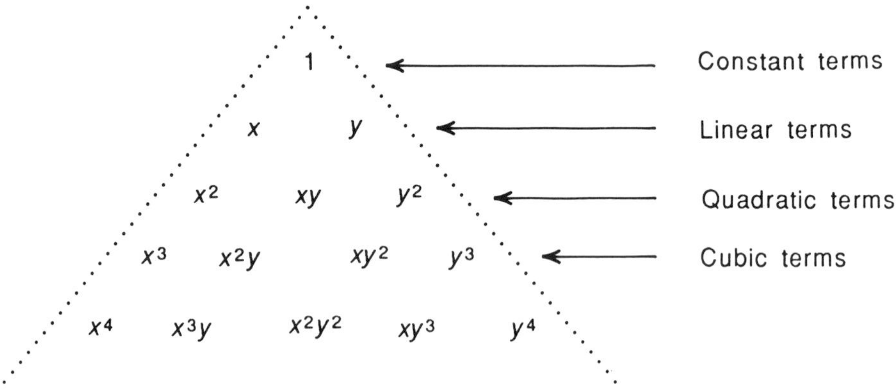

Fig. 8.2 *Pascal's triangle of polynomial terms.*

Before looking at specific forms of expression 8.2, let us look in more general terms at the variation of displacement along one side of an element whose displacement field is given by a polynomial of this type. Assume at this stage that the side is straight and lies parallel to the x-axis. The variation of displacement is then defined by setting y equal to a constant in expression 8.2. This gives an ordinary polynomial in x. Suppose that this is complete up to and including the nth power. The implications for inter-element compatibility are then as follows. For the displacement to be uniquely defined along the boundary, its value must be specified at $n+1$ points, these being necessary to define the $n+1$ unknown coefficients in an nth order polynomial in x. In other words, $n+1$ nodal values must be specified along the side. If these are common to adjacent elements of the same type, the displacement will then be uniquely defined in the same way on both sides of the boundary and the compatibility criterion of section 7.2 will be satisfied. Linear variation ($n=1$) along a side therefore requires *two* nodes for compatibility (Fig. 8.3(a)), quadratic variation ($n=1$) requires three nodes (Fig. 8.3(b)), cubic variation ($n=3$) requires four nodes (Fig. 8.3(c)), and so on.

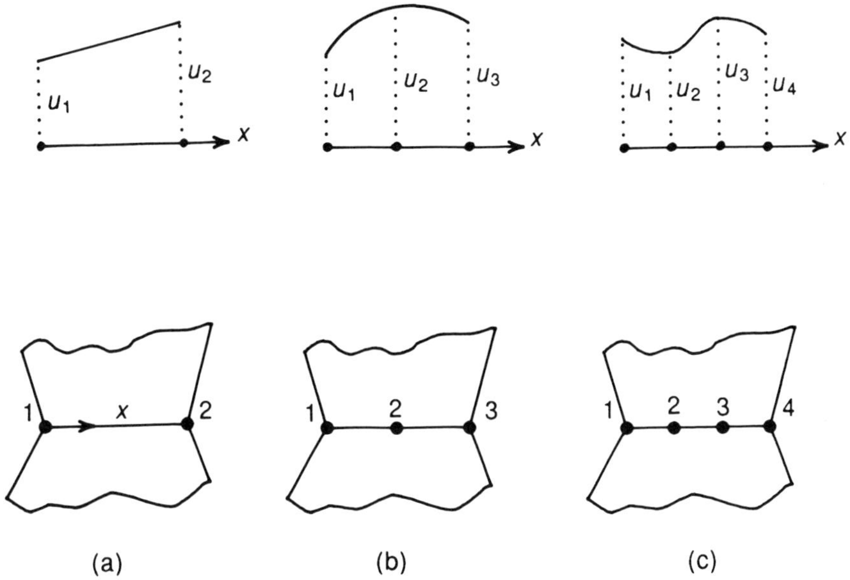

Fig. 8.3 *Compatible matching of (a) linear, (b) quadratic and (c) cubic elements.*

Arguments of this type are extremely useful in selecting appropriate polynomials for higher-order elements. If the shape functions are required to preserve inter-element compatibility, as is usually the case, the number of nodes which can be placed along any side is entirely determined by the separable powers present in the interpolating polynomial. In addition, the *total* number of terms in the polynomial must be equal to the number of nodes in the element topology. By balancing such requirements, shape functions can be formulated for quite complex elements without explicitly solving for the unknown coefficients of expression 8.2.

8.2.2 Some general properties of element shape functions

Universal conditions which must be satisfied by the shape functions of all elements, irrespective of element type, are established here for a general element which has p nodes. Suppose that suitable shape functions have been determined for this element so that the interpolation of displacement takes the form,

$$u(\boldsymbol{x}) = n_1(\boldsymbol{x})u_1 + n_2(\boldsymbol{x})u_2 + \ \dots \ + n_p(\boldsymbol{x})u_p, \tag{8.3}$$

where $n_i(\boldsymbol{x})$ is the ith shape function and u_i the ith nodal value of $u(\boldsymbol{x})$. Evaluation of equation 8.3 at the ith node yields the identity

$$u_i = n_1(\boldsymbol{x}_i)u_1 + n_2(\boldsymbol{x}_i)u_2 + \ldots + n_i(\boldsymbol{x}_i)u_i + \ldots + n_p(\boldsymbol{x}_i)u_p. \qquad (8.4)$$

This holds for *all* values of $u_1, u_2, \ldots, u_p$. In particular, it holds when all of the u_j's are zero except for u_i which is equal to 1.0. Substituting these values into identity 8.4 we obtain

$$1 = n_i(\boldsymbol{x}_i).$$

Similarly, by selecting all of the u_j's equal to zero except for $u_k (k \neq i)$ and setting this equal to 1.0, we obtain

$$0 = n_k(\boldsymbol{x}_i)\ (k \neq i).$$

These relationships hold for all values of i and k and can be summarized as

$$\begin{aligned} n_k(\boldsymbol{x}_i) &= 1 \quad (i = k), \\ &= 0 \quad (i \neq k). \end{aligned} \qquad (8.5)$$

All shape functions therefore have the property that they take the value of unity at their own node and zero at any other.

It is quite simple to demonstrate that the linear, shape functions which have already been derived in Chapters 4 and 6 do indeed possess this characteristic (see problems 1–4 at the end of this chapter). In the case of more complex elements, the same property, when taken in conjunction with the compatibility requirements discussed in section 8.2, often provides sufficient information to completely determine the individual shape functions of the element without requiring the coefficients α_i of expression 8.2 to be evaluated explicitly. This direct approach is now used to obtain shape function for some common families of higher-order elements.

8.3 SERENDIPITY ELEMENTS IN TWO AND THREE DIMENSIONS

The 'serendipity' family of rectangular and block elements is illustrated in Fig. 8.4. The first three elements of each type are shown. Although their orthogonal geometry is somewhat restrictive, they form a useful starting point for a general discussion of the 'direct' approach to element formulation, and are used later as the basis for a popular family of 'isoparametric' elements.

Consider first the two-dimensional elements shown in the left-hand column of Fig. 8.4. The simplest of these is the four-noded rectangle shown as element (a). The sides of the element are defined by $x' = \pm a$ and $y' = \pm b$, where x' and y' are 'local', cartesian coordinates whose origin is at the centre of the element. Since the element has four nodes, the interpolating polynomial must have four unknown coefficients. A suitable expression of type 8.2 is

$$\alpha_1 + \alpha_2 x' + \alpha_3 y' + \alpha_4 x'y'. \qquad (8.6)$$

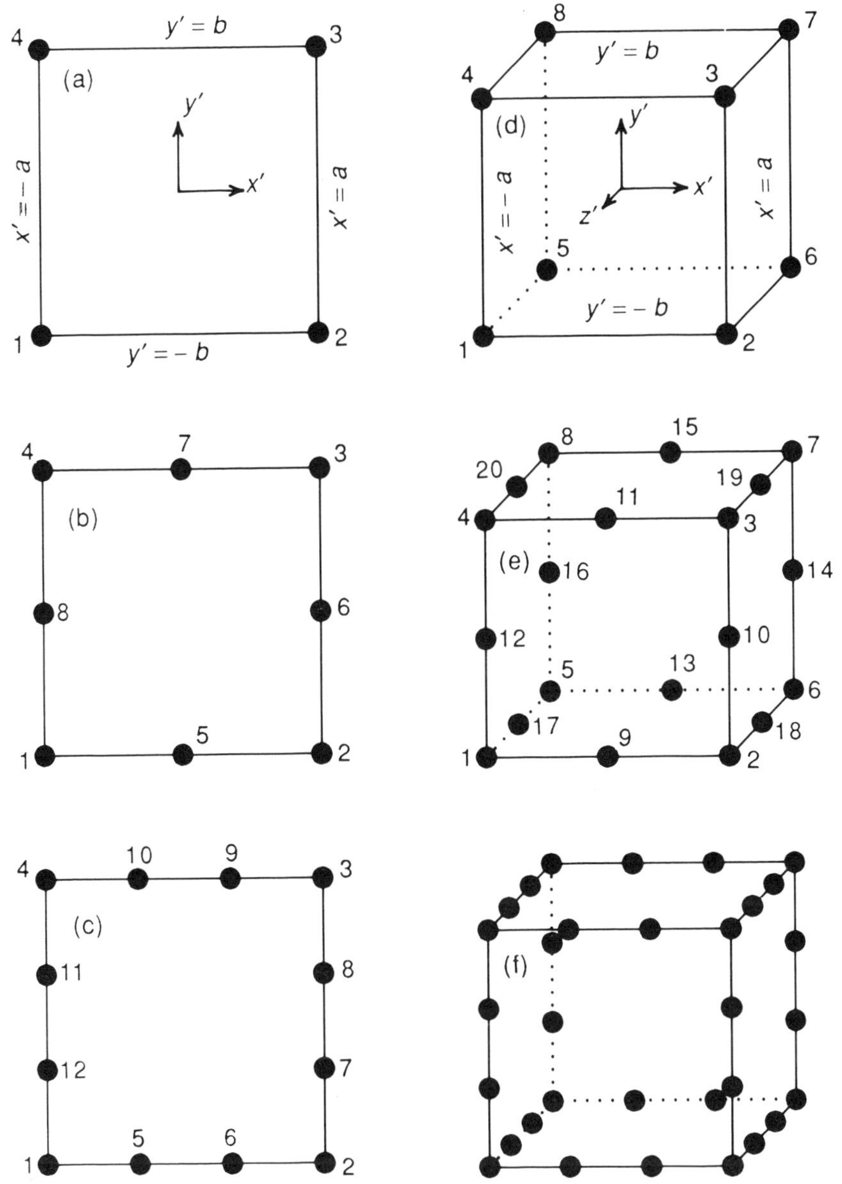

Fig. 8.4 *Serendipity elements in two- and three-dimensions.*

This is the natural selection from several possible candidates. 'Natural' because it contains a complete first order polynomial (as required by the completeness criterion) and also because it is 'isotropic' in the sense that it favours neither x' nor y'. This would not be the case, for example, if the last term were $\alpha_4 x'^2$ or $\alpha_4 y'^2$ instead of $\alpha_4 x'y'$. Equally important, expression 8.6 exhibits linear variation with x' or y' along each side of the element. This ensures inter-element compatibility when the element is joined to another of the same type, or indeed to any element with linear variation along the common side (a linear triangle, for example). The location of the terms in expression 8.6 within Pascal's triangle are indicated in Fig. 8.5. They fall somewhere between a complete linear and a complete quadratic polynomial but are closer to the former than the latter. The element will be referred to in all that follows as a **linear serendipity rectangle**.

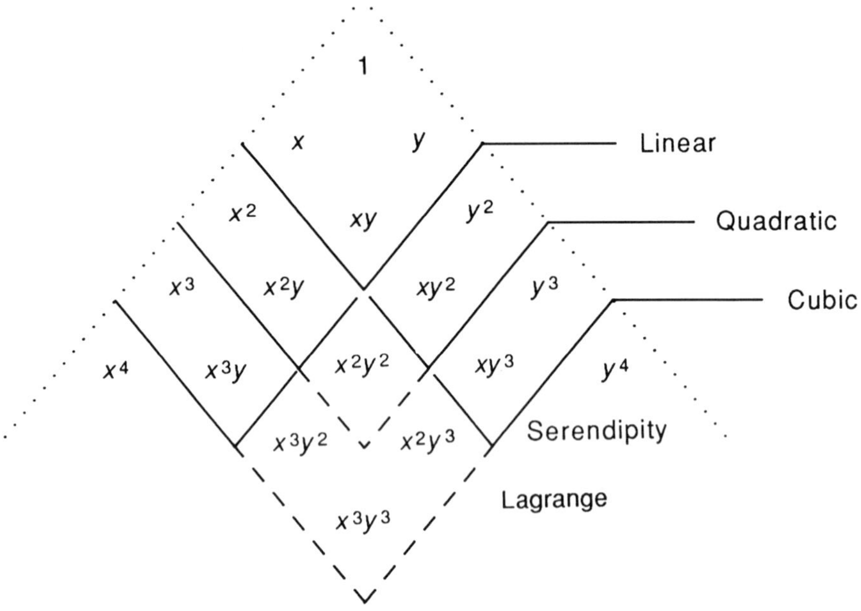

Fig. 8.5 *Polynomial terms for serendipity and Lagrange rectangles.*

The shape functions must now satisfy the following requirements:

(i) they must be of the same polynomial form as expression 8.6; and
(ii) they must equate to unity and zero at their own and other nodes, respectively.

This provides enough information to determine them uniquely. In the case of node 1, for example, a function of the form *constant* $(1 - x'/a)(1 - y'/b)$, satisfies condition (i) and satisfies condition (ii) to the extent that it takes the value zero at nodes 2, 3 and 4. If the constant is chosen to be $\frac{1}{4}$, it also satisfies the requirement at node 1 and condition (ii) is then satisfied in full. The shape function for node 1 is, therefore,

$$n_1 = \frac{1}{4}\left(1 - \frac{x'}{a}\right)\left(1 - \frac{y'}{b}\right), \tag{8.7a}$$

Similarly, the shape functions for nodes 2, 3 and 4 are found to be

$$\begin{aligned} n_2 &= \frac{1}{4}\left(1 + \frac{x'}{a}\right)\left(1 - \frac{y'}{b}\right), \\ n_3 &= \frac{1}{4}\left(1 + \frac{x'}{a}\right)\left(1 + \frac{y'}{b}\right), \\ n_4 &= \frac{1}{4}\left(1 - \frac{x'}{a}\right)\left(1 + \frac{y'}{b}\right). \end{aligned} \tag{8.7b}$$

The same general approach can be applied to the higher-order members of the family. The eight-noded rectangle, for example (element (b) of Fig. 8.4) requires an interpolation with eight constants. Since each side of the element is defined by three nodes, one at each end and one at the centre, inter-element compatibility requires that the displacements vary quadratically with x' or y' along each side (Fig. 8.2(b)). This suggests an interpolation of the form

$$\alpha_1 + \alpha_2 x' + \alpha_3 y' + \alpha_4 x'^2 + \alpha_5 x' y' + \alpha_6 y'^2 + \alpha_7 x'^2 y' + \alpha_8 x' y'^2. \tag{8.8}$$

Once again, the individual shape functions must be of the same polynomial form as expression 8.8 and must take values of unity and zero at their own and other nodes, respectively. In the case of node 1, a function of the form *constant* $\times (1 - x'/a)\,(1 - y'/b)\,(x'/a + y'/b + 1)$, takes the value zero at nodes 2, 3, . . . , 8. If the constant is chosen to be $-1/4$, it also takes the value of unity at node 1. The shape function n_1 is, therefore,

$$n_1(x', y') = \frac{1}{4}\left(1 - \frac{x'}{a}\right)\left(1 - \frac{y'}{a}\right)\left(-\frac{x'}{a} - \frac{y'}{b} - 1\right). \tag{8.9a}$$

Similarly, the shape functions at nodes 2, 3 and 4 are given by

$$\begin{aligned} n_2(x', y') &= \frac{1}{4}\left(1 + \frac{x'}{a}\right)\left(1 - \frac{y'}{b}\right)\left(\frac{x'}{a} - \frac{y'}{b} - 1\right), \\ n_3(x', y') &= \frac{1}{4}\left(1 + \frac{x'}{a}\right)\left(1 + \frac{y'}{b}\right)\left(\frac{x'}{a} + \frac{y'}{b} - 1\right), \\ n_4(x', y') &= \frac{1}{4}\left(1 - \frac{x'}{a}\right)\left(1 + \frac{y'}{b}\right)\left(-\frac{x'}{a} + \frac{y'}{b} - 1\right). \end{aligned} \tag{8.9b}$$

The shape functions for the midside nodes are obtained in a similar way: In the case of node 5, a function of the form *constant*$\times(1+x'/a)(1-x'/a)$ $(1-y'/b)$, satisfies all the requirements except for the normalization condition at node 5. The required constant in this case is $\frac{1}{2}$, and n_5 is then given by,

$$n_5(x', y') = \frac{1}{2}\left(1+\frac{x'}{a}\right)\left(1-\frac{x'}{a}\right)\left(1-\frac{y'}{b}\right). \tag{8.9c}$$

Similarly,

$$\begin{aligned} n_6(x', y') &= \frac{1}{2}\left(1+\frac{y'}{b}\right)\left(1-\frac{y'}{b}\right)\left(1+\frac{x'}{a}\right), \\ n_7(x', y') &= \frac{1}{2}\left(1+\frac{y'}{b}\right)\left(1-\frac{x'}{a}\right)\left(1+\frac{x'}{a}\right), \\ n_8(x', y') &= \frac{1}{2}\left(1+\frac{y'}{b}\right)\left(1-\frac{y'}{b}\right)\left(1-\frac{x'}{a}\right). \end{aligned} \tag{8.9d}$$

Derivation of the shape functions for the next element in this sequence, the twelve-noded, cubic rectangle, is left to the reader as an exercise (see problem 7 at the end of this chapter). A methodical approach for determining the shape functions for higher-order elements of this type is described in section 7.6 of [4].

The same general philosophy extends to the three-dimensional, elements shown on the right hand side of Fig. 8.4. The simplest of these, element (d), is an eight-noded 'linear' block whose faces are defined by the planes $x' = \pm\ a$, $y' = \pm\, b$ and $z' = \pm\, c$. The interpolating polynomial requires eight constants (one per node) and must be chosen so that linear variation exists along each edge. The natural choice is

$$\alpha_1 + \alpha_2 x' + \alpha_3 y' + \alpha_4 z' + \alpha_5 x'y' + \alpha_6 y'z' + \alpha_7 x'z' + \alpha_8 x'y'z'. \tag{8.10}$$

Shape functions of this form which take values of unity and zero at their own and other nodes, respectively, are,

$$\begin{aligned} n_1(x', y', z') &= \frac{1}{8}\left(1-\frac{x'}{a}\right)\left(1-\frac{y'}{b}\right)\left(1+\frac{z'}{c}\right), \\ n_2(x', y', z') &= \frac{1}{8}\left(1+\frac{x'}{a}\right)\left(1-\frac{y'}{b}\right)\left(1+\frac{z'}{c}\right), \\ n_3(x', y', z') &= \frac{1}{8}\left(1+\frac{x'}{a}\right)\left(1+\frac{y'}{b}\right)\left(1+\frac{z'}{c}\right), \\ n_4(x', y', z') &= \frac{1}{8}\left(1-\frac{x'}{a}\right)\left(1+\frac{y'}{b}\right)\left(1+\frac{z'}{c}\right), \end{aligned} \tag{8.11}$$

$$n_5(x', y', z') = \frac{1}{8}\left(1 - \frac{x'}{a}\right)\left(1 - \frac{y'}{b}\right)\left(1 - \frac{z'}{c}\right),$$

$$n_6(x', y', z') = \frac{1}{8}\left(1 + \frac{x'}{a}\right)\left(1 - \frac{y'}{b}\right)\left(1 - \frac{z'}{c}\right),$$

$$n_7(x', y', z') = \frac{1}{8}\left(1 + \frac{x'}{a}\right)\left(1 + \frac{y'}{b}\right)\left(1 - \frac{z'}{c}\right),$$

$$n_8(x', y', z') = \frac{1}{8}\left(1 - \frac{x'}{a}\right)\left(1 + \frac{y'}{b}\right)\left(1 - \frac{z'}{c}\right).$$

The derivation of shape functions for higher-order blocks such as elements (e) and (f) proceeds along similar lines. Full details are to be found in [4].

A second family of rectangular elements can be formulated using interpolating polynomials which span more levels of Pascal's triangle. Those which are used for the quadratic and cubic members of this 'Lagrangian family' are indicated Fig. 8.5. These elements are similar in appearance to those of the serendipity family but include additional interior nodes. The 'quadratic', Lagrangian rectangle, for example, has eight boundary nodes plus an additional ninth node at the element centroid. The cubic element has the same 16 boundary nodes as the cubic serendipity element, but includes four additional interior nodes. No further discussion of these elements is included here but full details are to be found in [3] and [4]. The derivation of shape functions for the nine-noded rectangle is included as an exercise at the end of this chapter (problem 8).

The application of serendipity (or Lagrangian) elements to real engineering problems is severely limited by their orthogonal geometry, the very characteristic which makes their shape functions relatively simple to calculate. The subdivision of an arbitrarily shaped body into elements with orthogonal sides not only imposes an inflexible mesh spacing in each direction over the whole body, but necessitates the use of crude 'stepped' approximations for curved boundaries. Both limitations are removed by the use of mappings, to be introduced shortly. An alternative solution to the problem of generating computationally efficient meshes of variable spatial resolution is to use higher-order triangles or tetrahedra. Higher-order triangles, in particular, are common in many finite element codes, either as elements in their own right or as the basis for isoparametric formulations. A brief discussion of these elements follows.

8.4 HIGHER-ORDER TRIANGULAR ELEMENTS

The formulation of shape functions for orthogonal rectangles and blocks is facilitated by the fact that the sides or faces of the element are defined by constant values of x', y' and z'. The boundaries of a triangular region are much harder to define, certainly in terms of cartesian coordinates. The formulation of shape functions for higher-order triangles is therefore extremely cumbersome if a cartesian representation is retained. It becomes relatively straightforward however when 'natural' coordinates (also termed 'area' coordinates) are used. These are defined in the following way.

In a two-dimensional, cartesian representation, a point is specified by two orthogonal coordinates x and y. If the same point lies inside a triangle with vertices 1, 2 and 3 (Fig. 8.6), its location can be specified equally well by the values of the three triangular sub-areas A_1, A_2, and A_3. In terms of x and y, these sub-areas are given by

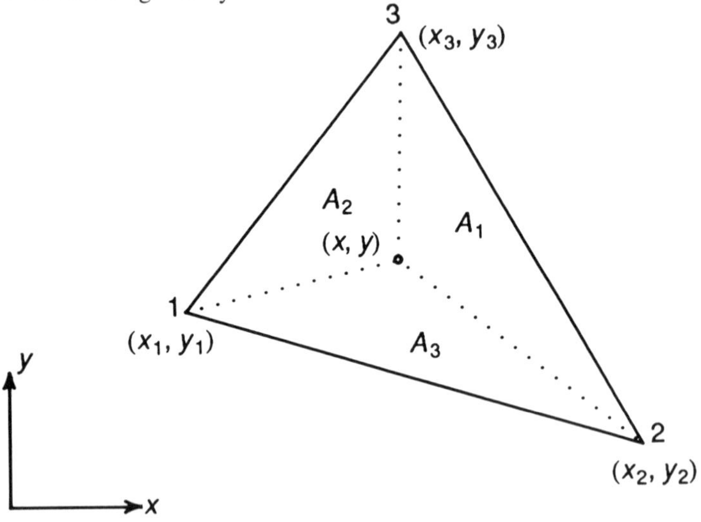

Fig. 8.6 *Sub-areas of a triangular element.*

$$A_i = \tfrac{1}{2}(a_i + b_i x + c_i y), \qquad (i = 1, 2, 3), \tag{8.12}$$

where a_i, b_i and c_i are the geometric constants derived in Chapter 6 (equation 6.5(a)). The reader is referred to problem 11 at the end of this chapter for a simple derivation of this relationship. The 'natural' coordinates (L_1, L_2, L_3) of the point are then obtained by normalizing A_1, A_2 and A_3 with respect to the area of the triangle. In other words, we define

$$L_i = A_i/A \qquad (i = 1, 2, 3). \tag{8.13}$$

By combining equations 8.12 and 8.13, L_1, L_2 and L_3 can also be expressed directly in terms of the cartesian coordinates, if required. This gives

$$L_i = \frac{1}{2A}(a_i + b_i x + c_i y), \qquad (i = 1, 2, 3). \tag{8.14}$$

It is important to note that the natural coordinates are not independent quantities. Given that the sum of the subareas A_i is always equal to the area of the triangle, the natural coordinates must satisfy the identity

$$L_1 + L_2 + L_3 \equiv 1. \qquad (8.15)$$

A major advantage of using natural coordinates in the formulation of shape functions for triangular elements is the ease with which the sides and nodes of an element can now be defined. The vertices, for example, are defined in the natural coordinate system by the 'points', (1, 0, 0), (0, 1, 0) and (0, 0, 1). The sides are given by the equations, $L_1 = 0$ (side 2–3), $L_2 = 0$ (side 1–3) and $L_3 = 0$ (side 1–2). These expressions hold irrespective of the size, shape or location of the triangle.

Given that the natural coordinates are themselves complete linear functions of x and y (equation 8.14) it is simple to show that a complete polynomial in x and y translates into a homogeneous polynomial of the same order in L_1, L_2 and L_3. In other words, a complete linear polynomial in cartesian coordinates x and y is equivalent, in natural coordinates, to a homogeneous expression of the form

$$\alpha_1 L_1 + \alpha_2 L_2 + \alpha_3 L_3. \qquad (8.16)$$

A complete quadratic polynomial is equivalent to an expression

$$\alpha_1 L_1^2 + \alpha_2 L_2^2 + \alpha_3 L_3^2 + \alpha_4 L_1 L_2 + \alpha_5 L_2 L_3 + \alpha_6 L_3 L_1, \qquad (8.17)$$

and a complete cubic polynomial is equivalent to

$$\alpha_1 L_1^3 + \alpha_2 L_2^3 + \alpha_3 L_3^3 + \alpha_4 L_1^2 L_2 + \alpha_5 L_2^2 L_1 + \alpha_6 L_3^2 L_2 + \alpha_7 L_2^2 L_3 + \alpha_8 L_3^2 L_1 + \alpha_9 L_1^2 L_3 + \alpha_{10} L_1 L_2 L_3. \qquad (8.18)$$

These may be used as the basis for linear, quadratic and cubic triangular elements. The first of these, the linear triangle, has already been formulated in Chapter 6. Its shape functions were derived at that time in terms of x and y, but can be rederived now — with considerably less effort — as functions of L_1, L_2 and L_3. The shape function n_1, for example, is sought as a linear function of L_1, L_2 and L_3, which takes the value of unity at node 1 and zero at nodes 2 and 3. A homogeneous expression of type 8.16 which satisfies all these requirements is simply

$$n_1 = L_1.$$

Similarly, n_2 and n_3 are given by $n_2 = L_2$ and $n_2 = L_3$. Note that if equation 8.14 is substituted into any of these expressions we obtain

$$n_i = \frac{1}{2A}\,[a_i + b_i x + c_i y], \qquad (i = 1, 2, 3),$$

which is in exact agreement with the same expression obtained, after much algebraic manipulation, in Chapter 6 (*cf.* equation 6.7).

The linear triangle is the first in a hierarchy of triangular elements. The next, in order of increasing complexity, is the six-noded triangle shown as element (b) in Fig. 8.7. The sides of this element are defined by three nodes, one at each end and one at the midpoint. Inter-element compatibility is therefore satisfied if the displacement varies quadratically along each side. Also, since the element has a total of six nodes, six constants are required in the interpolating function. A quadratic polynomial in L_1, L_2 and L_3 (equation 8.17) is the obvious choice. Given that this represents a complete quadratic polynomial in terms of cartesian coordinates x and y, inter-element compatibility is assured notwithstanding the oblique orientation of the sides. Given also that the natural coordinates of the midside nodes are $(0, \frac{1}{2}, \frac{1}{2})$, $(\frac{1}{2}, 0, \frac{1}{2})$ and $(\frac{1}{2}, \frac{1}{2}, 0)$, and those of the corners $(1, 0, 0)$, $(0, 1, 0)$ and $(0, 0, 1)$, it is relatively simple to show that functions of type 8.17 which take values of unity and zero at their own and other nodes, respectively, are

$$\begin{aligned} n_1 &= L_1(L_1 - L_2 - L_3), & n_4 &= 4L_1L_2, \\ n_2 &= L_2(L_2 - L_1 - L_3), & n_5 &= 4L_2L_3, \\ n_3 &= L_3(L_3 - L_1 - L_2), & n_6 &= 4L_3L_1. \end{aligned} \tag{8.19}$$

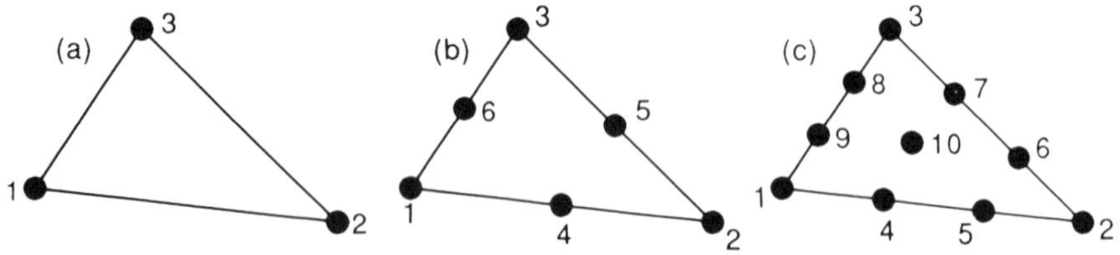

Fig. 8.7 *Triangular elements. (a) Linear, (b) quadratic, (c) cubic.*

Further elements may be developed along similar lines. The next element in the sequence is a ten-noded, cubic element shown as element (c). Its shape functions are homogeneous, cubic expressions of type 8.18. Their formulation is left to the reader as an exercise (problem 13 at the end of this chapter). Compact expressions for the shape functions of all elements of this type are to be found in [4].

Although the shape functions for triangular elements are relatively simple to formulate in terms of their natural coordinates, some practical difficulties might reasonably be anticipated when it comes to differentiating these expressions with respect to x and y to obtain the components of $\boldsymbol{B}^e$. Problems might also be anticipated in integrating the resulting stiffness terms over the area of the element. Such problems are, in fact, minimal. Differentiation of the shape functions with respect to x and y is easily accomplished using the 'chain rule' of partial differentiation. That is to say, the differential operator $\partial/\partial x$ can be written

$$\frac{\partial}{\partial x}=\frac{\partial L_1}{\partial x}\frac{\partial}{\partial L_1}+\frac{\partial L_2}{\partial x}\frac{\partial}{\partial L_2}+\frac{\partial L_3}{\partial x}\frac{\partial}{\partial L_3},$$

or, using expression 8.15 and writing $\partial L_1/\partial x$, $\partial L_2/\partial x$ and $\partial L_3/\partial x$ in terms of the constants b_i and A,

$$\frac{\partial}{\partial x}=\frac{1}{2A}\left(b_1\frac{\partial}{\partial L_1}+b_2\frac{\partial}{\partial L_2}+b_3\frac{\partial}{\partial L_3}\right). \tag{8.20}$$

Similarly, '$\partial/\partial y$' becomes

$$\frac{\partial}{\partial y}=\frac{1}{2A}\left(c_1\frac{\partial}{\partial L_1}+c_2\frac{\partial}{\partial L_2}+c_3\frac{\partial}{\partial L_3}\right). \tag{8.21}$$

Terms such as $\partial n_i/\partial x$ and $\partial n_i/\mathrm{d}y$ (the components of $\boldsymbol{B}^{\mathrm{e}}$) can therefore be evaluated quite simply as polynomial expressions in L_1, L_2 and L_3.

The integration of element stiffnesses and equivalent nodal forces over the area of the element, then reduces to the integration of products of integer powers of L_1, L_2 and L_3 over the area of a triangle. Such integrals can be evaluated relatively easily using the identity

$$\int_A L_1^a\, L_2^b\, L_3^c\, \mathrm{d}x\, \mathrm{d}y \equiv \frac{2A\, a!b!c!}{(a+b+c+2)!}. \tag{8.22}$$

8.5 ISOPARAMETRIC FORMULATION

The elements introduced so far, although they permit the use of higher-order interpolation, are still limited in their ability to model objects of arbitrary shape. Rectangular and block elements, in particular, are severely restricted by the orthogonal nature of their geometries. Triangular elements are more flexible in this regard, but curved boundaries can still be modelled only by a series of straight lines or plane facets. Such limitations are removed when mapped elements are used. A discussion of the most common formulation of this type, that based on the 'isoparametric' mapping, now follows.

8.5.1 Isoparametric formulation, a two-dimensional example

Mapped elements are obtained by transforming a 'parent' element, drawn from those described already, onto a 'physical' element which can be used to model a real body. The most common transformation of this type is defined by an 'isoparametric' mapping. Some typical two-dimensional isoparametric elements and their 'parents' are shown in Fig. 8.8. Rectangular and triangular elements can both be transformed in this way.

To fix ideas, let us look first at the simplest of the two-dimensional, isoparametric elements, the distorted four-noded rectangle shown as element (a) in Fig. 8.8. The parent element and mapped element are shown in somewhat greater detail in Fig. 8.9. The parent is a four-noded serendipity rectangle which exists in an imaginary ξ–η plane and has corners at $(\pm 1, \pm 1)$. The mapped element exists in the physical x–y plane and has nodes at $(x_1, y_1), (x_2, y_2), \ldots, (x_4, y_4)$. The mapping between them is defined by the equations

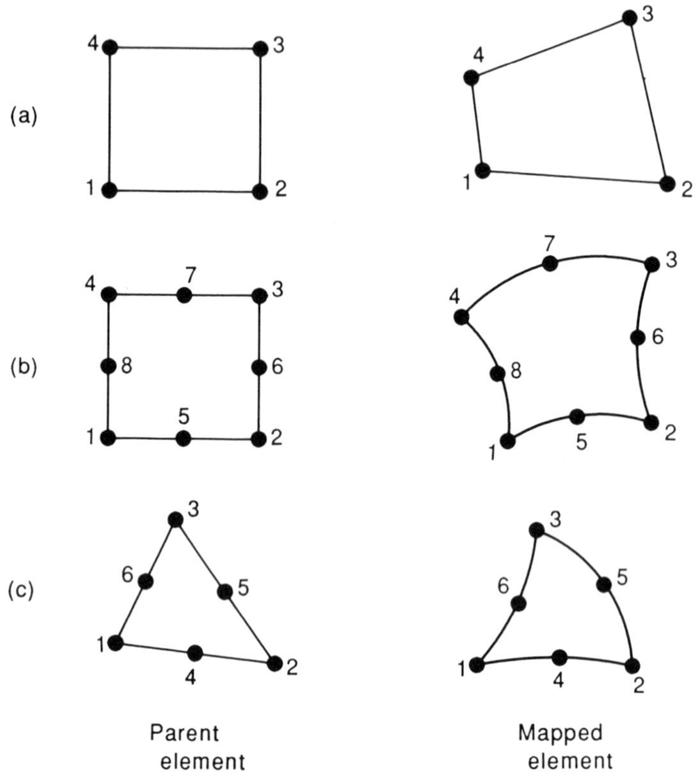

Fig. 8.8 *Two-dimensional isoparametric elements.*

$$x = n_1(\xi, \eta)x_1 + n_2(\xi, \eta)x_2 + n_3(\xi, \eta)x_3 + n_4(\xi, \eta)x_4, \tag{8.23a}$$

and

$$y = n_1(\xi, \eta)y_1 + n_2(\xi, \eta)y_2 + n_3(\xi, \eta)y_3 + n_4(\xi, \eta)y_4, \tag{8.23b}$$

where $n_i(\xi, \eta)$ $(i = 1, \ldots, 4)$ are the shape functions of the parent element in the parent space. These are the standard, serendipity shape functions for the element and are obtained from equations 8.7 by putting $x' = \xi$, $y' = \eta$ and $a = b = 1$. In other words,

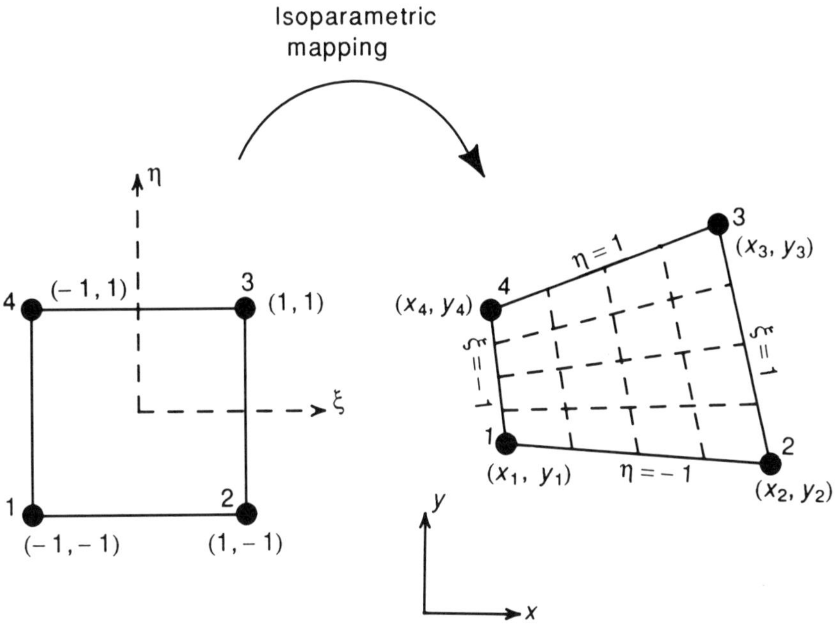

Fig. 8.9 *The isoparametric mapping for a four-noded rectangle.*

$$n_1(\xi, \eta) = \tfrac{1}{4}(1-\xi)(1-\eta), \qquad n_2(\xi, \eta) = \tfrac{1}{4}(1+\xi)(1-\eta),$$
$$n_3(\xi, \eta) = \tfrac{1}{4}(1+\xi)(1+\eta), \qquad n_4(\xi, \eta) = \tfrac{1}{4}(1-\xi)(1+\eta) \tag{8.24}$$

Equations 8.23 can be regarded as a transformation of each point (ξ, η) within the parent element onto a unique point (x, y) in the mapped element. The transformation is such that each node of the parent element maps onto the corresponding node of the mapped element, $(-1, -1)$ to (x_1, y_1) and so on. Alternatively, the mapping can be regarded as defining a set of curvilinear coordinates ξ and η *within* the physical element. The resulting lines of constant ξ and η are shown as broken lines in Fig. 8.9. In particular, the lines $\xi = \pm 1$ and $\eta = \pm 1$ form the element boundaries.

The shape functions for the physical element have not yet been determined, but must satisfy the usual requirements. That is, they must give compatible behaviour along each side of the element and take values of unity and zero at their own and other nodes, respectively. The required functions are by no means obvious, particularly in terms of the global coordinates x and y. In terms of the parent coordinates ξ and η, however, their formulation is relatively straightforward. In fact, the shape functions of the parent element, which have already been calculated for use in the mapping, possess exactly the properties that we require. In other words, they take values of unity and zero at their own and other nodes, respectively — this is true both in the parent element and the mapped element since the parent nodes map to the physical nodes — and vary linearly with ξ or η along the sides of the element, this latter property giving compatibility with adjacent elements of the same type.

The realization that the shape functions of the parent element can be used both as mapping functions and as shape functions within the mapped element, lies at the heart of the isoparametric formulation. It is, of course, feasible to use shape functions in the physical element which are of higher (or lower) order than those used in the mapping. This gives a 'superparametric' or 'subparametric' formulation. The *iso*parametric approach is by far the most common however and possesses an intrinsic economy of computational effort which sets it apart from other mappings.

8.5.2 Evaluation of element matrices and vectors

The stiffness matrix and equivalent nodal forces for an isoparametric element are obtained from the standard expressions of Chapter 4 (equations 4.39). In the case of a two-dimensional element, such as the four-noded quadrilateral of Fig. 8.9, the stiffness matrix is given by

$$\boldsymbol{K}^{\mathrm{e}} = \int_A h\, \boldsymbol{B}^{\mathrm{eT}} \boldsymbol{D}^{\mathrm{e}} \boldsymbol{B}^{\mathrm{e}}\, \mathrm{d}x\mathrm{d}y, \tag{8.25}$$

where A is the area of the mapped element and h is its thickness. Two practical problems now present themselves. First, we must evaluate the components of the strain–displacement matrix, $\boldsymbol{B}^{\mathrm{e}}$, and second, we must perform the integration of expression 8.25 over the area of the mapped element.

The strain–displacement matrix $\boldsymbol{B}^{\mathrm{e}}$ has components which are global derivatives of the element shape functions, that is to say, terms such as $\partial n_i/\partial x$ and $\partial n_i/\partial y$. The shape functions, n_i, however, have been defined in terms of the parent coordinates ξ and η. Their derivatives with respect to x and y are not therefore immediately obvious. However, they are related to the parent derivatives, $\partial n_i/\partial \xi$ and $\partial n_i/\partial \eta$, by the chain rule of differentiation;

$$\begin{bmatrix} \partial n_i/\partial \xi \\ \partial n_i/\partial \eta \end{bmatrix} = \boldsymbol{J} \begin{bmatrix} \partial n_i/\partial x \\ \partial n_i/\partial y \end{bmatrix} \quad \text{or} \quad \begin{bmatrix} \partial n_i/\partial x \\ \partial n_i/\partial y \end{bmatrix} = \boldsymbol{J}^{-1} \begin{bmatrix} \partial n_i/\partial \xi \\ \partial n_i/\partial \eta \end{bmatrix}, \tag{8.26a}$$

where $\boldsymbol{J}$, the Jacobian matrix, has components

$$\boldsymbol{J} = \begin{bmatrix} \partial x/\partial \xi, & \partial y/\partial \xi, \\ \partial x/\partial \eta, & \partial y/\partial \eta, \end{bmatrix}. \tag{8.26b}$$

These are obtained quite easily by differentiating equations 8.23. This gives

$$\frac{\partial x}{\partial \xi} = \frac{\partial n_1}{\partial \xi} x_1 + \frac{\partial n_2}{\partial \xi} x_2 + \frac{\partial n_3}{\partial \xi} x_3 + \frac{\partial n_4}{\partial \xi} x_4,$$

$$\frac{\partial y}{\partial \xi} = \frac{\partial n_1}{\partial \xi} y_1 + \frac{\partial n_2}{\partial \xi} y_2 + \frac{\partial n_3}{\partial \xi} y_3 + \frac{\partial n_4}{\partial \xi} y_4, \tag{8.27}$$

$$\frac{\partial x}{\partial \eta} = \frac{\partial n_1}{\partial \eta} x_1 + \frac{\partial n_2}{\partial \eta} x_2 + \frac{\partial n_3}{\partial \eta} x_3 + \frac{\partial n_4}{\partial \eta} x_4,$$

$$\frac{\partial y}{\partial \eta} = \frac{\partial n_1}{\partial \eta} y_1 + \frac{\partial n_2}{\partial \eta} y_2 + \frac{\partial n_3}{\partial \eta} y_3 + \frac{\partial n_4}{\partial \eta} y_4.$$

The local derivatives $\partial n_i/\partial \xi$ and $\partial n_i/\partial \eta$, which occur on the right hand side of the above expressions, are quite simple to calculate, given that the shape functions are themselves polynomials in ξ and η. The Jacobian components can therefore be determined for given values of ξ and η, and, by combining equations 8.26 and 8.27, the global derivatives of the shape functions can be computed with relative ease, even though they cannot be written in closed algebraic form. Let us emphasize the 'computability' of these terms for given values of ξ and η by writing the strain displacement matrix $\boldsymbol{B}^{\mathrm{e}}$ as $\boldsymbol{B}(\xi,\eta)^{\mathrm{e}}$. Equation 8.24 then becomes

$$\boldsymbol{K}^{\mathrm{e}} = \int_A (h\boldsymbol{B}\,(\xi,\eta)^{\mathrm{eT}}\,\boldsymbol{D}\,\boldsymbol{B}\,(\xi,\eta)^{\mathrm{e}})\,\mathrm{d}x\mathrm{d}y = \int_A \boldsymbol{k}(\xi,\eta)\,\mathrm{d}x\mathrm{d}y. \tag{8.28}$$

At this point we encounter the second of the two difficulties mentioned at the outset, that of integrating $\boldsymbol{k}(\xi,\eta)$, a matrix function of ξ and η, over the physical area of the mapped element. Rather than attempting to perform the integration as it stands (in the x–y plane), it is simpler to transform the entire integral to the parent region. This is done by multiplying the integrand by the determinant of the Jacobian matrix — which has already been calculated to obtain the components of $\boldsymbol{B}^{\mathrm{e}}$ — and then integrating with respect to ξ and η. The expression for the stiffness matrix then becomes

$$\boldsymbol{K}^{\mathrm{e}} = \int_{-1}^{+1}\int_{-1}^{+1} \boldsymbol{k}(\xi,\eta)\,|\det \boldsymbol{J}|\,\mathrm{d}\xi\mathrm{d}\eta. \tag{8.29}$$

An advantage of this procedure is that the integration now takes place over the same region in the ξ–η plane (a square of side two centred at the origin) irrespective of the size or shape of the physical element. The integrand itself is a square matrix whose components, although 'computable' for given values of ξ and η, cannot readily be written in closed analytical form. The integration is generally performed numerically (section 8.6). Analogous expressions for the equivalent nodal forces are

$$\boldsymbol{f}_{\mathrm{T}}^{\mathrm{e}} = \int_{-1}^{+1}\int_{-1}^{+1} h\boldsymbol{B}(\xi,\eta)^{\mathrm{eT}}\boldsymbol{D}\boldsymbol{e}_{\mathrm{T}}\,|\det\boldsymbol{J}|\,\mathrm{d}\xi\mathrm{d}\eta, \tag{8.30a}$$

and

$$\boldsymbol{f}_{\mathrm{g}}^{\mathrm{e}} = \int_{-1}^{+1}\int_{-1}^{+1} h\boldsymbol{N}(\xi,\eta)^{\mathrm{eT}}\boldsymbol{g}\,|\det\boldsymbol{J}|\,\mathrm{d}\xi\mathrm{d}\eta. \tag{8.30b}$$

These must also be integrated numerically.

8.5.3 Higher-order elements and curved boundaries

The isoparametric formulation described above can be applied to any of the higher-order elements formulated in sections 8.3 and 8.4; the mapping from the parent element to the physical element being of the same general form as equation 8.23 irrespective of the number of nodes in the element topology.

A useful characteristic of the higher-order isoparametric elements (quadratic and above) is their ability to model curved element boundaries. In higher-order elements, the intermediate node(s) of the mapped element need not lie on a straight line between the corners. Straight sides of the parent element can therefore be mapped onto curved sides in the physical element simply by positioning the intermediate nodes on an arc in the mapped space. This is illustrated in Fig. 8.8 by the mapped shapes of elements (b) and (c), which are based on a quadratic rectangle and triangle. Curved element boundaries of this sort can clearly be used to model physical boundaries arbitrary shape. When isoparametric elements are used in this way, care must be taken to ensure that the intermediate nodes are spaced at roughly equal intervals along the mapped arc. In the case of a quadratic element for example, the midside node should lie at least within the central half of the curved arc according to [4]. If not, the results obtained can deteriorate rapidly. In fact, the accuracy of isoparametric elements is consistently better if they are used in as undistorted a form as possible. The literature abounds with sobering comparisons between orthogonal meshes which give virtually perfect solutions to benchmark problems, and comparable meshes of distorted elements which give appalling results for the same problems (for example section 6.14 of [1]). As we will see shortly, however, the ultimate convergence of an isoparametric mesh *is* guaranteed. Nevertheless, the ability to model arbitrary shapes and curved boundaries should be used sparingly. Wherever possible, meshes should be generated which contain elements which are as undistorted as possible and curved boundaries should not be used except where they are needed to model essential geometrical features.

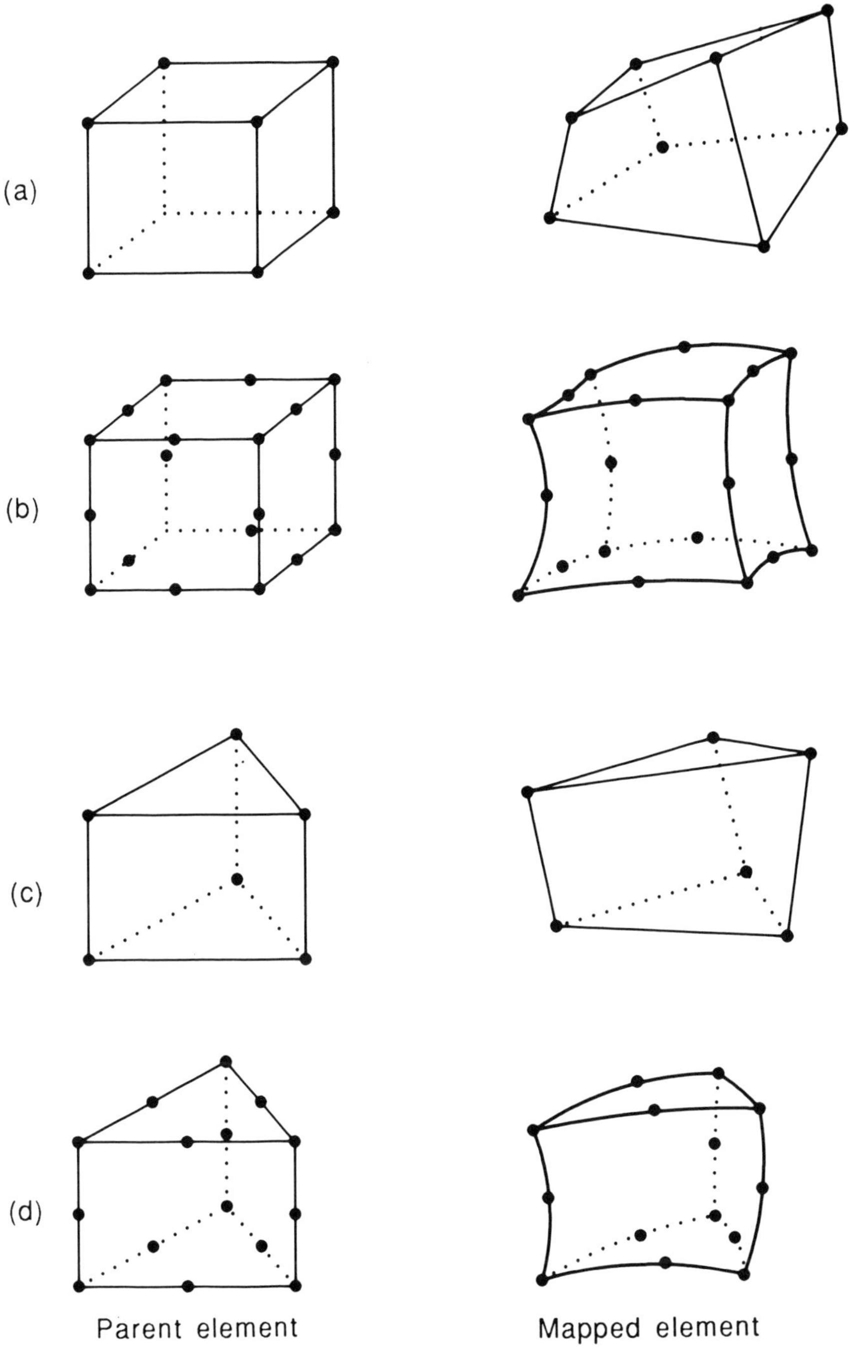

Fig. 8.10 *Some three-dimensional isoparametric elements.*

8.5.4 Three-dimensional elements

The extension of the isoparametric approach to three dimensions is entirely straightforward. The mapping from the parent to the physical element is of the same form as the two-dimensional mapping (equation 8.23) but involves three coordinates instead of two. The integration of element quantities then takes place over a three-dimensional region in the parent space (a cube of side 2 in the case of an isoparametric block). A selection of three-dimensional isoparametric elements is shown in Fig. 8.10. Elements (a) and (b) are distorted blocks based on linear and quadratic serendipity elements. Elements (c) and (d) are prisms based on linear and quadratic right prisms. The parent elements for the prisms have not been discussed here, but are simple to formulate. A derivation of the shape functions for element (c) — these involve natural coordinates in the base plane and a cartesian coordinate in the vertical direction — is posed as an exercise at the end of this chapter (problem 15). Full details both of the parent and distorted elements are to be found in [4].

8.5.5 Convergence and completeness

Isoparametric elements are intrinsically capable of satisfying the requirements of inter-element compatibility. They are, after all, mapped from parent elements which themselves satisfy these requirements in the parent space. Since the mapping on either side of an interface between adjacent mapped elements is identical, inter-element continuity of displacement is therefore assured, provided that the same rules are observed in constructing the isoparametric mesh as would be applied in an equivalent mesh of parent elements; that is, elements are joined so that they match 'side for side' and have the same number of nodes on the common interface. If this is done, the first of the two convergence criteria of section 7.2 is trivially satisfied.

The second requirement, that of completeness, is also satisfied but in a less obvious fashion. The shape functions of an isoparametric element certainly include a complete first order polynomial in ξ and η (and ζ, in the three-dimensional case) but it is no longer clear that they also include a complete first order polynomial in the global coordinates x, y and z, as required for completeness. This is, in fact, the case, as can be demonstrated quite simply using the four-noded isoparametric rectangle of section 8.5.1. We know that the displacements u and v inside this element are given by

$$u = \sum_{i=1}^{4} n_i(\xi, \eta) u_i \text{ and } v = \sum_{i=1}^{4} n_i(\xi, \eta) v_i. \tag{8.31}$$

We know also that the x and y coordinates of any point within the element are mapped from (ξ, η) in the parent space by

$$x = \sum_{i=1}^{4} n_i(\xi, \eta)x_i \text{ and } y = \sum_{i=1}^{4} n_i(\xi, \eta)y_i. \tag{8.32}$$

The question is whether a displacement field of the form,

$$u = \alpha_1 + \alpha_2 x + \alpha_3 y, \quad v = \beta_1 + \beta_2 x + \beta_3 y, \quad (\alpha_i, \beta_i \text{ constants}), \tag{8.33}$$

can be represented *exactly* by expressions in equation 8.31. If this *is* the case, then the completeness criterion of section 7.2 is satisfied.

The nodal values of displacement obtained from equations 8.33 are

$$u_i = \alpha_1 + \alpha_2 x_i + \alpha_3 y_i, \qquad (i = 1, 2, \ldots, 4) \tag{8.34a}$$

and

$$v_i = \beta_2 + \beta_2 x_i + \beta_3 y_i, \qquad (i = 1, 2, \ldots, 4)\,. \tag{8.34b}$$

Substitution of these values into equations 8.31 gives interpolated displacements within the element of

$$u = \alpha_1 \sum_{i=1}^{4} n_i(\xi, \eta) + \alpha_2 \sum_{i=1}^{4} n_i(\xi, \eta)x_i + \alpha_3 \sum_{i=1}^{4} n_i(\xi, \eta)y_i, \tag{8.35a}$$

and

$$v = \beta_1 \sum_{i=1}^{4} n_i(\xi, \eta) + \beta_2 \sum_{i=1}^{4} n_i(\xi, \eta)x_i + \beta_3 \sum_{i=1}^{4} n_i(\xi, \eta)y_i, \tag{8.35b}$$

From equations 8.32, however, the second and third summations on the right hand side of the above expressions are simply x and y. With this simplification, the interpolated values of displacement are

$$u = \alpha_1 \sum_{i=1}^{4} n_i(\xi, \eta) + \alpha_2 x + \alpha_3 y, \tag{8.36a}$$

and

$$v = \beta_1 \sum_{i=1}^{4} n_i(\xi, \eta) + \beta_2 x + \beta_3 y. \tag{8.36b}$$

These are close to the required values (equation 8.33), and would be identical if

$$\sum_{i=1}^{4} n_i(\xi, \eta) = 1. \tag{8.37}$$

This is, in fact, the case provided that the parent element satisfies the completeness condition in the parent space. The argument goes as follows: (a) If the parent element satisfies the completeness criterion, a uniform translation u_0 in the parent space will result in an exact state of zero strain at all points in the parent element, this being a particular case of 'constant' strain; (b) A state of zero strain in the parent element is possible only if the interpolated displacement $u(\xi, \eta)$ is identically equal to u_0 at all points, otherwise nonzero derivatives of $u(\xi, \eta)$ will exist and hence nonzero strains; (c) The interpolated displacement $u(\xi, \eta)$ is identically equal to u_0 at all points in the parent element only if

$$u(\xi,\eta) = \sum_{i=1}^{4} n_i(\xi,\eta)\, u_0 \equiv u_0. \tag{8.38}$$

Division by u_0 in the second part of the above identity then yields the desired result (equation 8.37) and ensures, working backwards to our original supposition, that completeness is satisfied in the mapped element.

The same argument holds for all isoparametric elements and guarantees convergence in the isoparametric formulation as a whole provided that the parent elements satisfy the completeness criterion of Chapter 7, and provided also that the mesh of mapped elements is formed in such a way that inter-element compatibility would be preserved in an equivalent mesh of parent elements.

8.6 NUMERICAL INTEGRATION

In the isoparametric formulation, area or volume integrals must be evaluated in the parent space to give the element stiffnesses and equivalent nodal forces (expressions 8.29 and 8.30). The integrands however are no longer polynomials and analytic integration is therefore difficult if not impossible. Integration is performed in most instances using 'numerical' techniques. The most appropriate numerical schemes are found to be those based on Gauss–Legendre integration, also termed 'Gauss quadrature'.

8.6.1 Numerical integration of a line integral, Gauss–Legendre integration

Let us look first at numerical schemes for the one-dimensional integral,

$$I = \int_{-1}^{+1} f(\xi)\, d\xi. \tag{8.39}$$

By definition, this is the limit, as the length of each subinterval tends to zero, of the finite summation

$$\sum_{i=1}^{n} f(\xi_i)\, \Delta\xi_i, \tag{8.40}$$

where $\Delta\xi_i$ is the length of a subinterval and ξ_i is an integration point which lies within it (Fig. 8.11). In physical terms, the limit of the summation, and hence

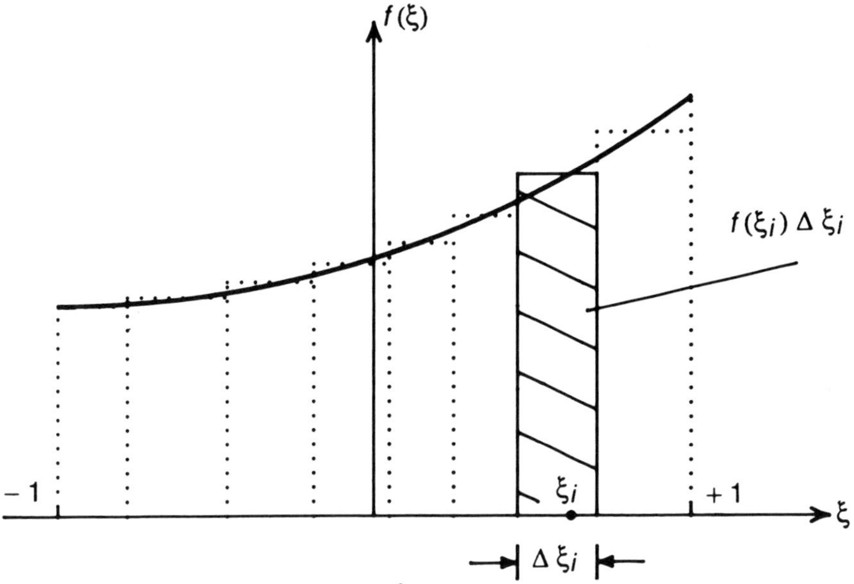

Fig. 8.11 *Graphical representation of* $\int f(\xi)\,d\xi$ *and* $\sum f(\xi)\,\Delta\xi$.

the integral, corresponds to the area under the f–ξ curve. The selection of a specific set of integration points and subintervals then defines an integration scheme for evaluating the integral. The extent to which the finite summation approximates the integral limit is a measure of the accuracy of the scheme.

Many strategies exist for determining sets of integration points and interval lengths [5]. The most straightforward are the Newton–Cotes formulae, which approximate the integrand by polynomial segments of a particular order. The trapezoidal rule, for example, uses a linear interpolation for $f(\xi)$ over each subinterval, Simpson's rule uses quadratic interpolation spanning two adjacent subintervals, and so on. Algorithms of this type are not particularly efficient, however, in integrating high-order (but smooth) polynomial-like functions. A more effective method in such instances is to use Gauss quadrature. The notation used in Gauss quadrature differs slightly from that of equation 8.40 in that the lengths of the subintervals, $\Delta\xi_i$, become 'weights', W_i over a standard interval. These represent factors by which the discrete values of the function, evaluated at the integration points ξ_i, are multiplied to give an approximate value of the integral. In the present instance, if the standard interval is taken as $[-1, +1]$, expression 8.34 becomes

$$I \simeq \sum_{i=1}^{n} f(\xi_i)\, W_i. \qquad (8.41)$$

In the Gauss–Legendre approach, the weights and integration points are then selected so that the above expression *exactly* duplicates the analytic integral for a polynomial of a given order. The derivation of Gaussian integration points and weights for specific polynomial orders will not be repeated here. Details are to be found in [5]. Suffice to say, a scheme with n integration points integrates exactly a polynomial of order $(2n - 1)$. The Gaussian scheme with two integration points $(n = 2)$, for example, integrates exactly an arbitrary cubic polynomial, that with three points an arbitrary quintic polynomial and so on. Gauss points and weights for one, two, three and four-point schemes are given in Table 8.1 and a more comprehensive list can be found in [4]. Although designed specifically for polynomial integrands, Gauss quadrature is effective for most integrands which have a 'polynomial-like' appearance. Isoparametric stiffness and force integrands fall into this category and are invariably integrated using this method.

Table 8.1 *Gaussian integration points and weights for the interval* $[-1, +1]$

Order n	Location ξ_i	Weight w_i
1	0.0	2.0
2	± 0.57735 02691 89626	1.00000 00000 00000
3	± 0.77459 66629 41483	0.55555 55555 55556
	0.00000 00000 00000	0.88888 88888 88889
4	± 0.86113 63115 94053	0.34785 48451 37454
	± 0.33998 10435 84856	0.65214 51548 62546

8.6.2 Gauss quadrature for area and volume integrals; matrix integrands

The calculation of element stiffnesses and equivalent nodal forces in the isoparametric formulation requires the evaluation of integrals over the area or volume of the parent element. In the case of a serendipity rectangle or block, the region in question is a rectangle or cube of side 2. The area or volume integral can therefore be replaced by repeated line integration between the fixed limits -1 and $+1$, and this can be evaluated by repeated application of Gauss quadrature in the ξ, η and ζ directions. In the case of a two-dimensional isoparametric rectangle, for example, the integral for each stiffness component takes the form

$$\int_{-1}^{+1} \int_{-1}^{+1} f(\xi, \eta)\, d\xi d\eta. \tag{8.42}$$

If n-point Gauss quadrature is used in the ξ and η directions, this reduces to

$$\sum_{i,j=1}^{n} (W_i W_j) f(\xi_i, \eta_j), \tag{8.43}$$

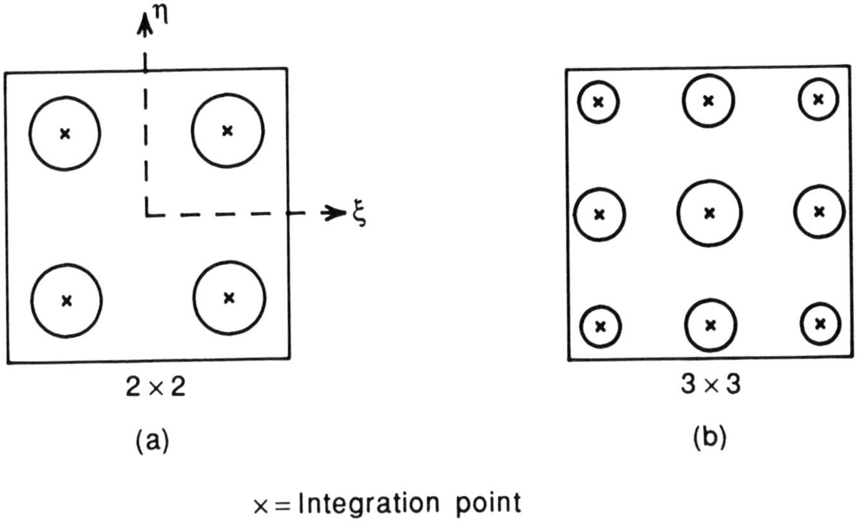

Fig. 8.12 *Gauss schemes for area integration. (a)* 2 × 2, *(b)* 3 × 3.

where ξ_i, η_j, W_i and W_j are taken from Table 8.1. The double summation can, of course, be interpreted as a single summation over n^2 terms with the product $(W_i W_j)$ regarded as a single weighting factor attached to the two-dimensional integration point (ξ_i, η_j). This interpretation is illustrated in Fig. 8.12 which shows the integration points and weights for a square of side 2 using two-point (2×2) and three-point (3×3) quadrature in each direction. The area of the circular region surrounding each integration point is proportional to the combined weight $W_i W_j$.

The discussion of numerical integration so far, has been confined to scalar integrands. The schemes described apply equally well however to matrix quantities. In the case of the four-noded isoparametric rectangle, for example, if the stiffness matrix (given by integral 8.29) is evaluated using n-point Gauss quadrature in the ξ and η directions, we obtain

$$\boldsymbol{K}^{\mathrm{e}} \simeq \sum_{i,j=1}^{n} (W_i W_j)\, \boldsymbol{k}(\xi_i, \eta_j) \left| \det \boldsymbol{J}(\xi_i, \eta_j) \right|, \tag{8.44}$$

where W_i, W_j, ξ_i and η_j are Gauss weights and integration points taken from Table 8.1, and where $\boldsymbol{k}(\xi_i, \eta_j)$ is the integrand, $h(\boldsymbol{B}^{\mathrm{eT}}\boldsymbol{D}\,\boldsymbol{B}_{\mathrm{e}})$, evaluated at the integration point (ξ_i, η_j).

The treatment of three-dimensional integrals follows the same pattern with expressions of the form

$$\int_{-1}^{+1}\int_{-1}^{+1}\int_{-1}^{+1} \boldsymbol{k}(\xi, \eta, \zeta)\, |\det \boldsymbol{J}|\, \mathrm{d}\xi \mathrm{d}\eta \mathrm{d}\zeta \tag{8.45}$$

being represented by triple summations

$$\sum_{i,j,k=1}^{n} (W_i\, W_j\, W_k)\, \boldsymbol{k}\,(\xi_i, \eta_j, \zeta_k)\, |\det \boldsymbol{J}|\,. \tag{8.46}$$

Analogous integration formulae are available for triangles and tetrahedra. In the case of triangular elements, the integration points are specified in terms of natural coordinates. In the case of tetrahedral elements, 'volume' coordinates are used. Gauss points and weights for both cases are to be found in tables 8.2 and 8.3 of [4]. The question which must then be asked in practice is, how many integration points are needed to produce acceptable results for integrands of the type encountered in the isoparametric formulation?

8.6.3 Required order of integration, reduced quadrature

The order of Gauss quadrature which is required to ensure exact integration of the stiffness terms in an isoparametric element can be estimated without much difficulty when the element is undistorted. The Jacobian matrix is then constant and the required number of integration points can be deduced from the polynomial order of terms which occur in the stiffness integrand ($\boldsymbol{B}^{e\mathrm{T}}\boldsymbol{D}\,\boldsymbol{B}^{e}$). In the case of a four-noded serendipity rectangle, for example, the terms in $\boldsymbol{B}^{e}$ are linear in ξ and η. Quadratic terms therefore exist in the product ($\boldsymbol{B}^{e\mathrm{T}}\boldsymbol{D}\,\boldsymbol{B}^{e}$) and a 2×2 scheme of the type illustrated in Fig. 8.12(a) ensures an *exact* result for the stiffness integral. In the case of an eight-noded rectangle, quadratic terms exist in $\boldsymbol{B}^{e}$, giving quartic terms in the stiffness integrand and a 3×3 scheme is therefore required. Similarly, a 4×4 scheme is required in the cubic element and so on. The same arguments cannot strictly be applied to distorted elements but the scheme required for an equivalent undistorted case gives a good indication of what is needed in the mapped case also, provided that the distortion is not excessive.

It is important however to distinguish between the order of integration required for an exact (or nearly exact) evaluation of element stiffness and the minimum order required to produce a convergent solution. In the latter case, all that is needed in a compatible element — following comments made in section 7.2 — is that a state of constant strain be exactly integrable. In other words, the required order of integration is that which can exactly integrate a constant over the area or volume of the physical element. In the case of an undistorted element, single-point integration is therefore sufficient. In the general case, however, the integration of a constant over the physical area or volume of the element is synonymous with the integration of $|\det \boldsymbol{J}|$ over the parent region. The exact integration of this quantity therefore becomes a sufficient condition for convergence. In an isoparametric, four-noded quadrilateral, for example, $|\det \boldsymbol{J}|$ is linear in ξ, and η, and a single Gauss point is therefore adequate to ensure convergence. In the case of an eight-noded rectangle, $|\det \boldsymbol{J}|$ is quadratic in ξ, and η, and a 2×2 scheme is therefore required. A 3×3 scheme is needed for the twelve-noded, cubic element and so on. These orders of integration are one order less than those indicated by an exact integration of the undistorted stiffness. It can be argued that they can be reduced even further in the light of the patch test in its 'weak' form. The argument goes as follows: (a) The exact representation of constant strain is required only in the limit as the element is repeatedly subdivided. (b) Repeatedly subdivided elements have asymptotically straight sides irrespective of the order of their mapping functions. (c) The determinant of the Jacobian of a straight-sided element contains only linear terms in ξ, η and ζ and therefore requires only single-point quadrature. The logic is irrefutable but avoids the important practical issue of how rapidly the solutions will converge as the mesh is refined. After all, there is little point in reducing the order of integration to a single point per element if ten times as many elements are required to produce the same degree of accuracy. In practice, some degree of reduced integration — 'reduced' in the sense that it is less than that required for an exact result — is generally beneficial, yielding results which are more, rather than less, accurate than those obtained using the 'correct' integration scheme. The recommended levels of integration for the isoparametric, quadrilaterals of Fig. 8.8, for example, are: 1-point integration for element (a) and 2×2 integration for element (b). Similar recommendations apply to three-dimensional elements. A justification of these statements lies beyond the scope of the present treatment but can be found in [4], [6] and [7].

At a heuristic level, it is not difficult to explain why 'reduced integration' often gives an overall solution which is more accurate than that obtained from an exact evaluation of the element integrals. In effect, 'higher-order' displacement modes are 'filtered' from the strain energy integral when it is sampled at a reduced number of points. These modes do not then contribute to the total energy of the system, and compensate, in some sense, for the over-estimate of total energy which is inevitable in any discrete model which is based on the location of an approximate minimum.

The use of reduced integration is not entirely without pitfalls. In particular, care must be taken to ensure that the total number of unconstrained degrees of freedom does not exceed the number of independent relations supplied at the integration points of the system. Singularity of the stiffness matrix will then result. The number of independent relations contributing to the stiffness equations is equal to the order of $\boldsymbol{D}$ multiplied by the total number of integration points. A singular stiffness matrix can therefore arise when reduced integration is used for coarse, lightly constrained meshes. This can be avoided in practice by adding extra integration points where necessary. This phenomenon is discussed in some detail in section 8.11 of [4].

REFERENCES

[1] Cook, R. D., Malkus D. S. and Plesha M. E. (1989) *Concepts and Applications of Finite Element Analysis*, 3rd edn. John Wiley & Sons, New York, Chapter 6.

[2] Taylor, R. L., Beresford, P. J. and Wilson, E. L. (1976) A non-conforming element for stress analysis. *International Journal for Numerical Methods in Engineering*, **10**, 1211–20.

[3] Norrie, D. H. and De Vries G. (1978) *An Introduction to Finite Element Analysis*, Academic Press, New York, Chapter 9.

[4] Zienkiewicz, O. C. and Taylor, R. L. (1990) *The Finite Element Method*, 4th edn, McGraw-Hill, London, Chapters 7 and 8.

[5] Al-Khafaji, A. W. and Tooley J. R. (1986) *Numerical Methods in Engineering Practice*, CBS publishing, Tokyo, Chapter 11.

[6] Fried, I. (1973) Accuracy and condition of curved isoparametic elements. *Journal of Sound and Vibration*, **31**, 345–55.

[7] Fried, I. (1974) Numerical integration in the finite element method. *Journal of Computers and Structures*, **4**, 921–32.

PROBLEMS

1. Confirm that the shape functions for the two-noded bar element of Chapter 4 (see expressions 4.5) take the value of unity at their own node and zero at others as required by the condition in equation 8.5. Show that this is true also of the three-noded, quadratic element whose shape functions are given in problem two of Chapter 4.

2. Nodes 1, 2 and 3 of a linear triangle are located at points (0,0), (1,0) and (0,1) in the x–y plane. Calculate the shape function n_1 in terms of x and y (use equations 6.5 and 6.7) and confirm that it takes the value of unity at node 1 and zero at nodes 2 and 3, as required by condition 8.5.

3. Use expressions 6.5 and 6.7 of Chapter 6 to show that the shape function $n_1(x, y)$ of a linear triangle can be written as a determinant

$$n_1(x, y) = (1/2A) \begin{vmatrix} 1 & x & y \\ 1 & x_2 & y_2 \\ 1 & x_3 & y_3 \end{vmatrix} .$$

Hence confirm that in the general case, $n_1\,(x, y)$ takes the value of unity at (x_1, y_1) and zero at (x_2, y_2) and (x_3, y_3).

4. A linear tetrahedron has the geometry and topology shown in problem 11 of Chapter 6. Confirm that the shape function $n_1(x, y, z)$ satisfies condition 8.5.

5. The four-noded serendipity rectangle of Fig. 8.4(a) is to be used for the analysis of plane stress. Write down the components of the strain–displacement matrix $\boldsymbol{B}^e$ for this element and confirm that the 1–1 term in the element stiffness matrix is

$$\left[\frac{Eh}{3(1-\nu^2)}\right]\left[\frac{b}{a}+\frac{1}{2}(1-\nu)\frac{a}{b}\right],$$

where E, ν and h are the Young's modulus, Poisson's ratio and thickness, respectively. Show also that the equivalent thermal loads which must be applied at node 1 to simulate a uniform temperature rise T° are; $E\alpha hTb/(1-\nu)$ and $E\alpha hTa/(1-\nu)$ in the negative x and y directions respectively (α is the coefficient of thermal expansion).

6. A uniform body force $\mathbf{g}_0 = (g_{0x}, g_{0y})$ acts on the four-noded rectangle of problem 5. Show that the equivalent nodal forces which must be applied at node 1 to simulate the distributed load are; $g_{0x}abh$ and $g_{0y}abh$ in the x and y directions, respectively. What proportion does this represent of the net resultant load acting on the element?

7. Calculate suitable shape functions for a typical corner node and a typical intermediate node (say nodes 1 and 5) of the twelve-noded, cubic, serendipity element of figure 8.4(c). The terms in the polynomial interpolation for the element are indicated in Fig. 8.5. (Suggestion: investigate the suitability of expressions of the form: *constant* $\times (1-x'/a)(1-y'/b)\ ((x'/a)^2 + (y'/b)^2 - 10/9)$ for shape function n_1; and *constant* $\times (1-x'/a)(1+x'/a)(1/3-x'/a)(1-y'/b)$ for shape function n_5).

8. A nine-noded Lagrangian element has the topology shown on (a) (overleaf). Interpolation within the element is of the form

$$\alpha_1 + \alpha_2 x' + \alpha_3 y' + \alpha_4 x'^2 + \alpha_5 x'y' + \alpha_6 y'^2 + \alpha_7 x'^2 y' + \alpha_8 x'y'^2 + \alpha_9 x'^2 y'^2,$$

where x' and y' are local cartesian coordinates whose origin is at the centre of the element. Obtain expressions for the shape functions n_1, n_5 and n_9. Explain carefully why the element can be compatibly matched to an eight-noded serendipity rectangle as shown in (b), but *not* to two four-noded rectangles as indicated in (c). (Suggestion: investigate the suitability of expressions of the form: *constant* $\times (x'/a)(1-x'/a)(y'/b)(1-y'/b)$ for shape function n_1; *constant* $\times (1-x'/a)(1+x'/a)(1-y'/b)(y'/b)$ for shape function n_5; and *constant* $\times (1-x'/a)(1+x'/a)(1-y'/b)(1+y'/b)$ for shape function n_9.)

9. The Lagrangian element of problem eight is used for plane stress and is of constant thickness. Show that an in-plane body force, uniformly distributed over the volume of the element, is consistently modelled by the following nodal forces: 1/36 of the net resultant load at each corner node, 1/9 at each midside node and 4/9 at the centroid. In what proportion should the same load be distributed among the eight nodes of a quadratic serendipity element?

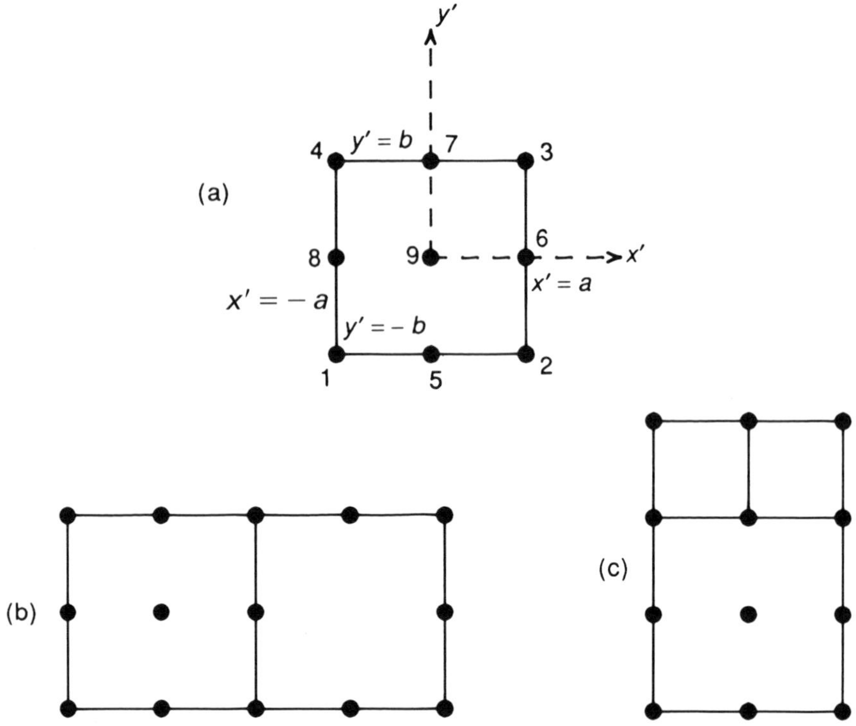

Problem 8.

10. A six-noded rectangle (as shown below) is formulated using an interpolation of the form

$$\alpha_1 + \alpha_2 x' + \alpha_3 y' + \alpha_4 x'^2 + \alpha_5 x'y' + \alpha_6 x'^2 y'.$$

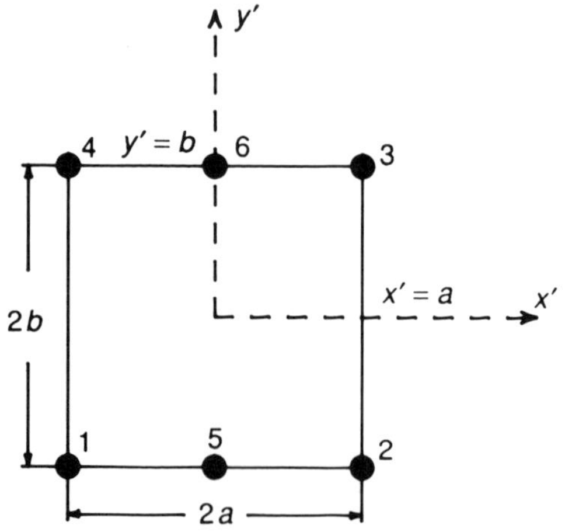

Problem 10.

Confirm that the element can be compatibly matched to a quadratic element along sides 1–2 and 4–3, and to a linear element along sides 1–4 and 2–3. Obtain expressions for the shape functions n_1 and n_5 in terms of x' and y'. (Suggestion: investigate the suitability of expressions of the form: *constant* $\times (1 - x'/a)(1 - y'/b)(x'/a)$ for shape function n_1, and *constant* $\times (1 - x'/a)(1 + x'/a)(1 - y'/b)$ for shape function n_5.)

11. Verify that the triangular subareas A_i $(i = 1, 2, 3)$ shown in Fig. 8.6 are given in terms of cartesian coordinates by expressions 8.12. (Suggestion: show that A_1 is equal to $\frac{1}{2}ab\sin\theta$ (see figure) and hence that A_1 is also equal to the magnitude of the vector product $\frac{1}{2}\boldsymbol{a}\times\boldsymbol{b}$, where $\boldsymbol{a}$ and $\boldsymbol{b}$ are vectors joining the point $(x_2\ y_2)$ to (x_3, y_3) and (x, y) respectively. Evaluate $\boldsymbol{a}\times\boldsymbol{b}$ using cartesian components to obtain the required result.)

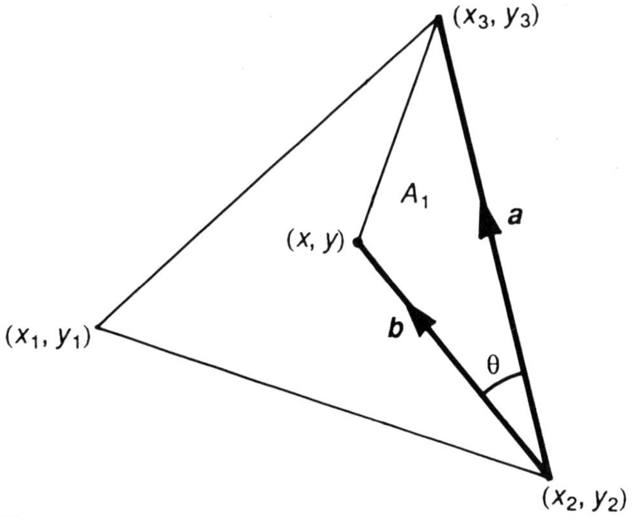

Problem 11.

12. Derive the shape functions for a four-noded triangle with linear variation of displacement along two sides and quadratic variation along the third as shown overleaf. (Suggestion: use natural coordinates and assume an interpolation of the form, $\alpha_1 L_1 + \alpha_2 L_2 + \alpha_3 L_3 + \alpha_4 L_1 L_2$.)

13. Derive the shape functions for nodes 1, 4 and 10 of the ten-noded triangle of Fig. 8.7(c). The central node has natural coordinates (1/3,1/3,1/3). (Suggestion: assume a complete cubic interpolation within the element (see equation 8.17) and investigate the suitability of the expressions: *constant* $\times L_1\,[L_1 - (1/3)(L_1 + L_2 + L_3)]\,[L_1 - (2/3)(L_1 + L_2 + L_3)]$ for shape function n_1, *constant* $\times L_1L_2\,[L_1 - (1/3)(L_1 + L_2 + L_3)]$ for shape function n_4 and *constant* $\times L_1L_2L_3$ for shape function n_{10}.)

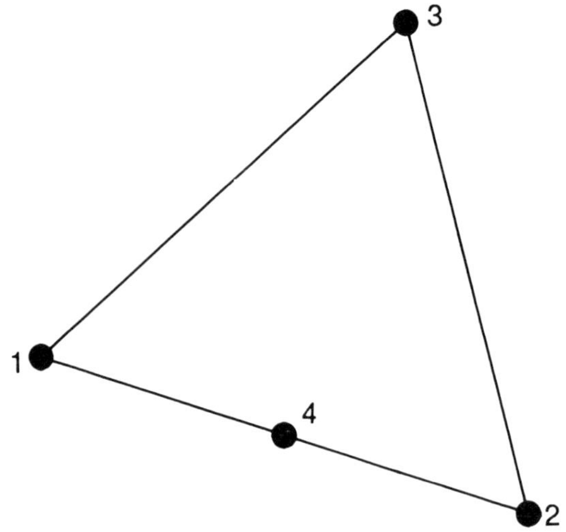

Problem 12.

14. Show that the proportions of a uniformly distributed body force which must be placed at nodes 1 and 4 of the six-noded triangle of Fig. 8.7(b) are

$$\frac{1}{A}\int_A n_i \, \mathrm{d}x\mathrm{d}y, \text{ and } \frac{1}{A}\int_A n_4 \, \mathrm{d}x\mathrm{d}y, \text{ respectively,}$$

where n_1 and n_4 are given by equations 8.19. Evaluate these integrals and confirm that the equivalent nodal forces act entirely at the midside nodes of the element (Suggestion: use the identity of expression 8.22).

15. Derive shape functions $n_1, \ldots, n_6$ for a six-noded, right triangular prism shown. In what proportion should a uniformly distributed body force acting on such an element be applied at each node? (Suggestion: use natural coordinates L_1, L_2, L_3 in the x–y plane, and investigate the suitability of an expression of the form, *constant* $\times L_1 (1 - z/h)$ for the shape function n_1.)

16. A four-noded isoparametric quadrilateral has nodes 1, 2, 3 and 4 at points $(-1, 0)$, $(0, -1)$, $(1, 1)$ and $(0, 2)$ in the x–y plane. Sketch the element and obtain explicit equations for the isoparametric transformation. Show that the Jacobian matrix is a constant and calculate the components of the strain–displacement matrix $\boldsymbol{B}^{\mathrm{e}}$ in terms of the parent coordinates (assume a state of plane stress). If the element is of thickness 0.1 m. and is subject to a uniform temperature rise of 100°C, what forces must be applied at the nodes of the element to correctly simulate the thermal load? ($E = 210$ GPa, $\nu = 0.33$ and $\alpha = 0.000011$ per deg C. Nodal coordinates are given in metres.)

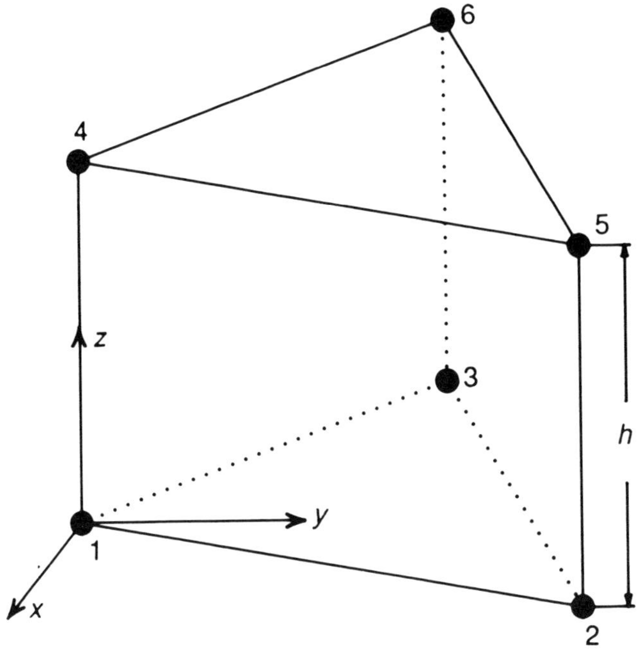

Problem 15.

17. A four-noded isoparametric rectangle has nodes at (0,0), (2,0), (3,3) and (0, 2) as shown. Obtain an explicit expression for the isoparametric mapping and confirm that the determinant of the Jacobian varies linearly with

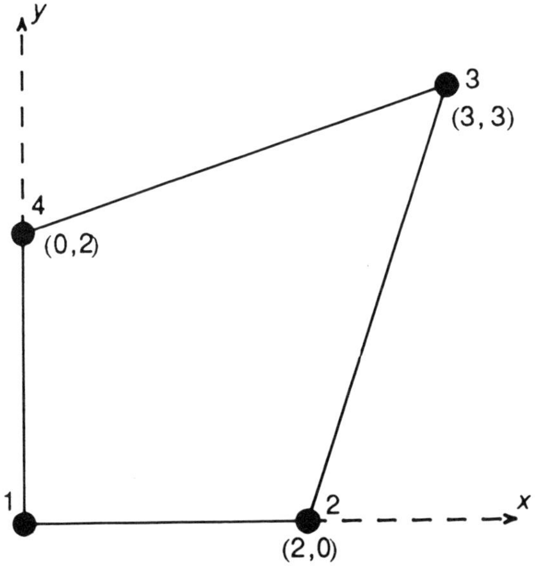

Problem 17.

ξ and η. If a body force is distributed uniformly over the element, show that 2/9 of the resultant load should be applied to node 1 as an equivalent nodal force.

18. Sketch the following functions of ξ on the interval [– 1, + 1] and compute their integrals using single-point, two-point and three-point Gauss quadrature. Compare with the exact value in each case and comment.

(i) $f(\xi) = 1 + \xi + \xi^2 + \xi^3$,

(ii) $f(\xi) = \xi^2(1 - \xi^2)$,

(iii) $f(\xi) = \cos\left(\dfrac{\pi\xi}{2}\right)$,

(iv) $f(\xi) = \dfrac{1}{\sqrt{(1 + \xi)}}$

(v) $f(\xi) = -\xi, \; -1 \leqslant \xi < 0,$
$\quad\;\; = \xi, \quad\; 0 \leqslant \xi \leqslant 1.$

19. A four-noded serendipity rectangle with sides parallel to global x and y-axes is used for the analysis of plane stress. The element is of variable thickness $t(x, y)$. What is the minimum number of Gauss points which will exactly integrate the element stiffness for the three cases:

(i) $t(x, y)$ constant over the element
(ii) $t(x, y)$ varies linearly with x *and* y
(iii) $t(x, y)$ is constant with x and varies quadratically with y.

20. What is the minimum number of integration points which will ensure that the element of problem 19 is convergent? (Investigate cases (i), (ii) and (iii) separately.)

21. What orders of Gauss quadrature are required in the x' and y' directions to exactly integrate all stiffness contributions for the six-noded element of problem ten? What is the minimum order required for convergence?

9

Beam and frame elements

9.1 INTRODUCTION

In formulating the elements discussed in previous chapters, the full continuum equations of linear elasticity were taken as a starting point. Elements were then developed using trial functions for the primitive displacement components. These were assumed to behave independently but were chosen so that they satisfied the compatibility requirements within each element.

The same general approach can be applied to the formulation of beam and plate elements, but here the trial displacements are chosen so that they already satisfy the fundamental axiom of beam and plate theory; that plane sections which are initially perpendicular to the neutral axis (or plane) remain plane after deformation. If these sections also remain orthogonal to the neutral axis, the resulting elements model the behaviour of Bernoulli–Euler beams or Kirchhoff plates as discussed in Chapter 2 (section 2.5). Beam elements of various types are presented in the remainder of this chapter. Plate elements follow in Chapter 10.

In the classical analysis of beams and plates, the variation of stress and displacement through the depth of the beam (or the thickness of the plate) is assumed to be linear. Stress and displacement are therefore defined at all points in terms of the displacement of the neutral axis (or plane). The same assumptions are made in finite element idealizations of such entities. In other words, it is only the displacement of the neutral axis — rather than the displacement field throughout the element — which need be represented in such models. The economies of computational effort which can be achieved by adopting this approach are substantial.

By way of illustration, Fig. 9.1 shows two finite element models for a cantilever beam of thin rectangular section. The first is a two-dimensional continuum model in which the beam is represented by ten, plane-stress, serendipity rectangles. This model is shown as Fig. 9.1(a). The distribution of stress in each element is approximately linear and the model uses a total of 80 degrees of freedom. This gives an effectively 'converged' solution, in the sense that further mesh refinement does not sensibly alter the predicted values of the centreline displacement.

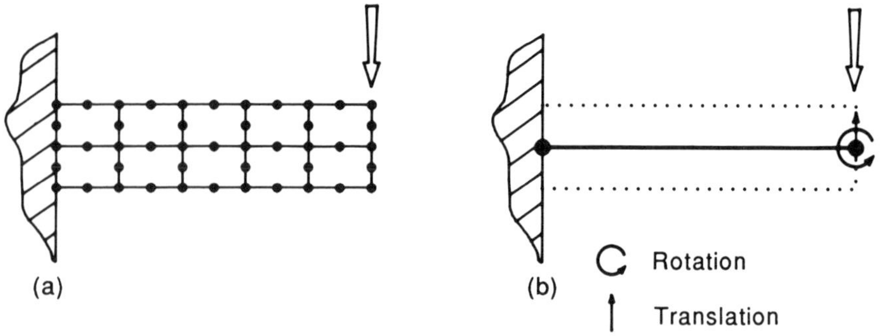

Fig. 9.1 *Finite element models of a cantilever plate loaded in its own plane: (a) continuum model, (b) beam model.*

The second model, Fig. 9.1(b), is a structural idealization in which the beam is represented by a single 'beam' element of a type to be described shortly. The lateral displacement of the neutral axis is the only quantity which is directly interpolated in such an element and the nodal degrees of freedom are a lateral translation and an angular rotation at each end. Both are constrained at the left hand end to simulate a built-in support. The entire structure is therefore modelled using two degrees of freedom, the displacement and rotation at the free end.

The discrete solution obtained using model (b) is exact in terms of simple beam theory. In other words, it reproduces exactly the analytic solution obtained by integrating the beam deflection equation and applying suitable end conditions. It is not, of course, 'exact' in the continuum sense since beam theory itself is only an approximation to the two or three-dimensional problem. The difference is small, however, in the current instance, the discrepancy between the centreline displacements predicted by models (a) and (b) being of the order of a few percent. This is clearly quite acceptable in many engineering calculations, and the economies of computational effort which can be achieved by using the two-degree of freedom model of (b), rather than the 80-degree of freedom model of (a) are then self-evident.

9.2 SIMPLE BEAM ELEMENT

A prismatic element is now developed for planar bending. The element incorporates the assumptions of Bernoulli–Euler beam theory as outlined in Chapter 2 (section 2.5.1). The element is of length L, cross-sectional area A, flexural rigidity EI (in the x–z plane) and lies parallel to the x-axis of a global cartesian system (Fig. 9.2). Its undeformed geometry is indistinguishable from that of the two-noded bar element of Chapter 4. In formulating the bar element, however, it was assumed that no lateral load was carried by the member; its strain energy being derived entirely from axial tension or compression. This assumption is now reversed. In other words, we assume that no loads are carried by the element in tension or compression and hence that its strain energy derives entirely from bending. In practice both effects, bending and tension/compression, are present in structural members, and both are subsequently included in the finite element idealization when a 'frame' element is developed later in this chapter. For the time being, however, let us assume that the deformation of the beam consists entirely of a lateral displacement $w(x')$ where x' is a local axial coordinate. Note that the subscript '0' which was used in Chapter 3 to distinguish between displacements on the neutral axis and those at a general point, is now dropped. In all that follows, the unsubscripted variable, w, therefore refers to the lateral displacement of the neutral axis.

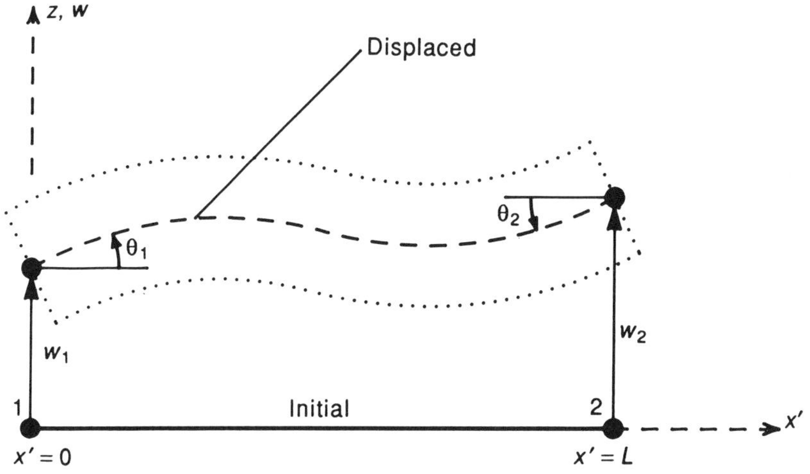

Fig. 9.2 *The simple beam element, geometry and displaced shape.*

The next step in the formulation is the choice of an interpolation for w. Before attempting this, let us look briefly at the implications of the compatibility criterion of Chapter 7 for the current element. The compatibility condition requires that the displacement field is continuous throughout the model. This must hold on the neutral axis but also at points above and below it. It is not sufficient, therefore, to insist that the displacement is continuous on the neutral axis itself, since this permits incompatible displacements of the type shown in Fig. 9.3(a). These can occur at inter-element nodes even when the displacement of the neutral axis is specified as a nodal variable common to adjacent elements. Such behaviour is prevented only when the rotation of the point is specified as an additional nodal parameter, as indicated in Fig. 9.3(b). The simplest, compatible two-noded beam element therefore has four degrees of freedom. These are the displacements at each node, w_1 and w_2, and the accompanying rotations θ_1 and θ_2 (Fig. 9.2). We will place these in a nodal displacement vector $\boldsymbol{d}^e$ for future reference, where

$$\boldsymbol{d}^e = [w_1, \theta_1, w_2, \theta_2]^T. \tag{9.1}$$

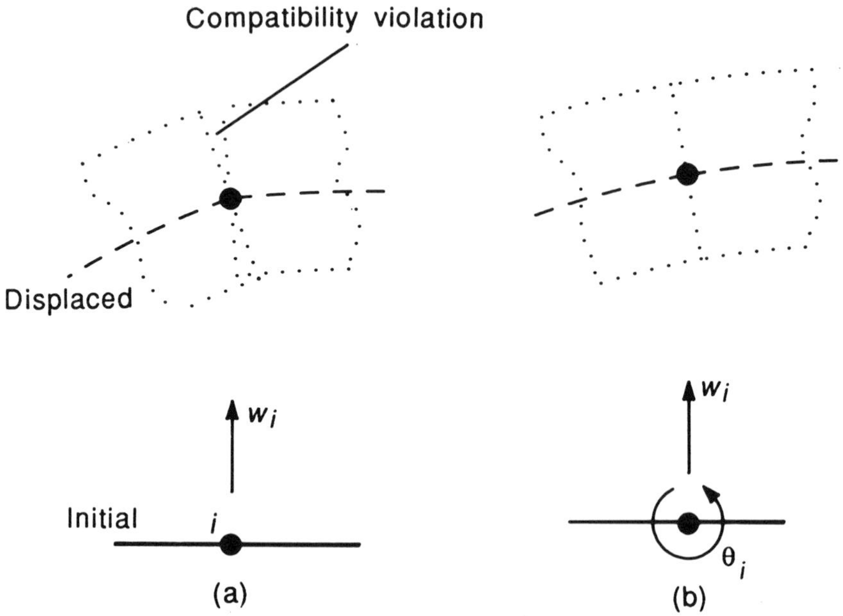

Fig. 9.3 *Matching of beam elements at a common node: (a) incompatible, (b) compatible*

9.2.1 Element shape relationship

Given that the beam element of Fig. 9.2 has four degrees of freedom, the natural choice for an interpolating function is a cubic polynomial in x' with four unknown constants. This gives a trial displacement of the form,

$$w(x') = \alpha_1 + \alpha_2 x' + \alpha_3 x'^2 + \alpha_4 x'^3. \tag{9.2}$$

Consistent with this interpolation for w, the rotation of the neutral axis, $\theta(x')$, is given by

$$\theta(x') = \frac{dw}{dx} = \alpha_2 + 2\alpha_3 x' + 3\alpha_4 x'^2. \tag{9.3}$$

The constants $\alpha_1, \ldots, \alpha_4$ are then determined by equating w and θ to nodal values at $x' = 0$ and $x' = L$. This gives, after some manipulation,

$$\alpha_1 = w_1, \quad \alpha_2 = \theta_1,$$

$$\alpha_3 = -\frac{3}{L^2} w_1 - \frac{2}{L} \theta_1 + \frac{3}{L^3} w_2 - \frac{1}{L} \theta_2,$$

and

$$\alpha_4 = \frac{2}{L^3} w_1 + \frac{1}{L^2} \theta_1 - \frac{2}{L^3} w_2 + \frac{1}{L^2} \theta_2. \tag{9.4}$$

When substituted into the original interpolation 9.3, these give the shape relationship

$$w(x') = n_1(x')w_1 + n_2(x')\theta_1 + n_3(x')w_2 + n_4(x')\theta_2, \tag{9.5}$$

where

$$n_1(x') = 1 - 3\left(\frac{x'}{L}\right)^2 + 2\left(\frac{x'}{L}\right)^3,$$

$$n_2(x') = x'\left[1 - 2\left(\frac{x'}{L}\right) + \left(\frac{x'}{L}\right)^2\right]$$

$$n_3(x') = 3\left(\frac{x'}{L}\right)^2 - 2\left(\frac{x'}{L}\right)^3,$$

$$n_4(x') = x'\left[-\left(\frac{x'}{L}\right) + \left(\frac{x'}{L}\right)^2\right] \tag{9.6}$$

The functions n_i $(i = 1, \ldots, 4)$ are the shape functions of the element. They define the lateral displacement within the element in terms of its nodal degrees of freedom which now include rotations as well as translations. Equation 9.6 can also be written in the standard form of Chapter 4 as,

$$\boldsymbol{u} = \boldsymbol{N}^e \boldsymbol{d}^e, \tag{9.7}$$

where $\boldsymbol{d}^e$ is the nodal displacement vector, defined by equation 9.1, $\boldsymbol{u}$ is the vector of displacements at a point, which in this case contains the single component w, and $\boldsymbol{N}^e$ is a shape matrix with components

$$\boldsymbol{N}^e = [n_1(x'), n_2(x'), n_3(x'), n_4(x')]. \tag{9.8}$$

9.2.2 Strain energy, 'generalized' stress and strain

In the absence of axial tension/compression, the strain energy per unit length of the beam (equation 3.23) is given by

$$\text{SE/length} = \frac{1}{2} M \left(-\frac{d^2 w}{dx^2}\right), \tag{9.9}$$

where

$$M = EI\left(-\frac{\mathrm{d}^2 w}{\mathrm{d}x^2}\right). \tag{9.10}$$

Both of these expressions can be recast in a form which closely resembles that of the continuum formulation, if we regard the bending moment M as a 'generalized' stress, and the curvature $(-\mathrm{d}^2w/\mathrm{d}x^2)$ as a 'generalized' strain. This is purely a notational change but allows us to define a 'generalized' stress vector $\boldsymbol{s}$, and a 'generalized' strain vector $\boldsymbol{e}$ as

$$\boldsymbol{s} = [M] \text{ and } \boldsymbol{e} = \left[-\frac{\mathrm{d}^2 w}{\mathrm{d}x^2}\right]. \tag{9.11}$$

Expression 9.9 then becomes

$$\text{SE/length} = \tfrac{1}{2}(\boldsymbol{s}^{\mathrm{T}}\,\boldsymbol{e}), \tag{9.12}$$

which is identical, in form at least, to the expression for the strain energy density of an elastic continuum. Moreover, the moment–curvature relationship 9.10 can be rewritten as a 'generalized' stress–strain relationship by defining a constant matrix, $\boldsymbol{D}$, which contains the flexural rigidity of the beam as its only component. In other words, equation 9.10 can be rewritten

$$\boldsymbol{s} = \boldsymbol{D}\,\boldsymbol{e}, \tag{9.13}$$

where $\boldsymbol{D} = [EI]$.

Equations 9.12 and 9.13 then mimic the expression for the strain energy density and the statement of Hooke's law which lie at the heart of the continuum formulation. This is an extremely useful device since many of the matrix and vector expressions which were derived for continuum elements can now be extended to beam and plate elements provided that stresses and strains are 'reinterpreted' as moments and curvatures.

Pursuing this analogy, we can regard the curvature–displacement relationship of a beam as a 'generalized' strain–displacement relationship and can write it in the standard discrete form introduced in Chapter 4. This is done by differentiating $w(x')$ twice using expression 9.5 and rearranging the resulting terms to give

$$\boldsymbol{e} = \boldsymbol{B}^{\mathrm{e}}\,\boldsymbol{d}^{\mathrm{e}} \tag{9.14}$$

where

$$\boldsymbol{B}^{\mathrm{e}} = \left[-\frac{12x'}{L^3}+\frac{6}{L^2},\; -\frac{6x'}{L^2}+\frac{4}{L},\; -\frac{6}{L^2}+\frac{12x'}{L^3},\; -\frac{6x'}{L^2}+\frac{2}{L}\right]. \tag{9.15}$$

Since the components of $\boldsymbol{B}^{\mathrm{e}}$ are linear in x', we conclude that the element is capable of modelling linear variations of curvature and bending moment without approximation.

9.2.3 Stiffness matrix

The strain energy of the element is obtained by integrating equation 9.1 with respect to x'. This gives

$$U^{\mathrm{e}} = \int_0^L \tfrac{1}{2}(\boldsymbol{s}^{\mathrm{T}}\boldsymbol{e})\,\mathrm{d}x'.$$

Substitution of equations 9.12, 9.13 and 9.14 then yields the 'standard' result, namely

$$U^{\mathrm{e}} = \tfrac{1}{2}\boldsymbol{d}^{\mathrm{eT}}\boldsymbol{K}^{\mathrm{e}}\boldsymbol{d}^{\mathrm{e}}, \tag{9.16}$$

where $\boldsymbol{K}^{\mathrm{e}}$ is given by

$$\boldsymbol{K}^{\mathrm{e}} = \int_0^L \boldsymbol{B}^{\mathrm{eT}}\boldsymbol{D}\boldsymbol{B}^{\mathrm{e}}\,\mathrm{d}x'. \tag{9.17}$$

This is virtually identical to the expression for the stiffness matrix of a continuum element (equation 4.39) except that the integration is performed along the neutral axis of the element, rather than over its physical volume, and the nodal displacement vector contains rotations as well as displacements. Expression 9.17 can be integrated explicitly, using the components of $\boldsymbol{B}^{\mathrm{e}}$ from equation 9.15, to give

$$\boldsymbol{K}^{\mathrm{e}} = a\begin{bmatrix} 12 & 6L & -12 & 6L \\ 6L & 4L^2 & -6L & 2L^2 \\ -12 & -6L & 12 & -6L \\ 6L & 2L^2 & -6L & 4L^2 \end{bmatrix}, \tag{9.18}$$

where $a = EI/L^3$. The algorithm for assembling stiffness matrices of this type to form a model is then identical to that already described for continuum elements. The only practical difference is that the degrees of freedom of the assembled system now include rotations as well as translations. Both must be treated without distinction when the degrees of freedom of the system are numbered prior to assembly.

9.2.4 Assembly and solution, a simple example

The assembly and solution of a simple beam model is illustrated using a two-element idealization for a uniform beam of length $2L$. The problem to be solved and the model used are shown in Fig. 9.4. The beam is of flexural rigidity EI, is built-in at each end and carries a concentrated load P and at its midpoint. The finite element model is formed from two identical elements of the type described above and has six degrees of freedom before constraints are applied. Since the ends of the beam are 'built-in', the displacements and rotations at these points are constrained to zero when the final stiffness equations are formed, leaving two unconstrained degrees of freedom at the central node. These are numbered δ_1 and δ_2 as shown. Although it is not strictly necessary to do so, the constrained degrees of freedom are also numbered, $\delta_3, \ldots, \delta_6$, and are included in the assembly of the stiffness matrix for the purposes of illustration, even though they play no part in the final solution.

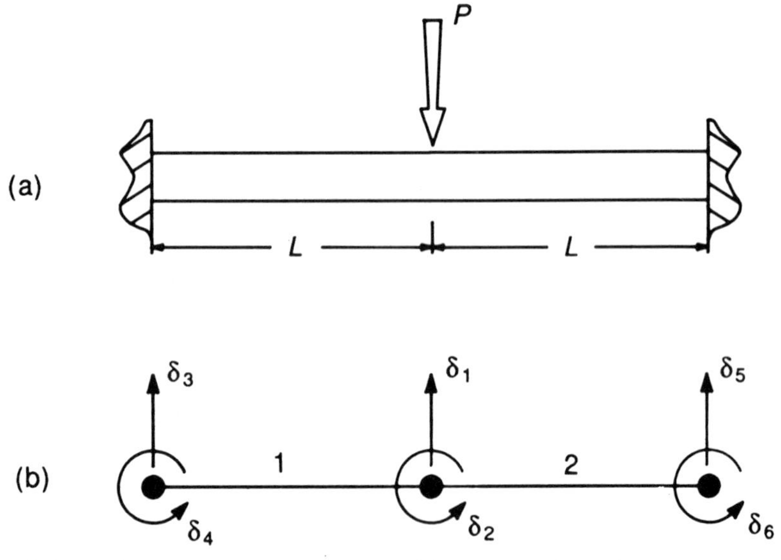

Fig. 9.4 *Worked example: (a) geometry and loading, (b) finite element model.*

The stiffness matrices for elements 1 and 2 are identical and are both given by expression 9.18. The nodal displacement vectors are

$$\boldsymbol{d}^{(1)} = \begin{bmatrix} \delta_3 \\ \delta_4 \\ \delta_1 \\ \delta_2 \end{bmatrix} \text{ and } \boldsymbol{d}^{(2)} = \begin{bmatrix} \delta_1 \\ \delta_2 \\ \delta_5 \\ \delta_6 \end{bmatrix}.$$

The stiffness matrix for the assembled model is obtained in the usual way by inserting the stiffness contributions from each element into an initially zero 6×6 array using the degree of freedom numbers in $\boldsymbol{d}^{(1)}$ and $\boldsymbol{d}^{(2)}$ to determine the row and column locations. After assembly of the first element, the assembled matrix therefore has components

$$\begin{bmatrix} 12a & -6aL & -12a & -6aL & 0 & 0 \\ -6aL & 4aL^2 & 6aL & 2aL^2 & 0 & 0 \\ -12a & 6aL & 12a & 6aL & 0 & 0 \\ -6aL & 2aL^2 & 6aL & 4aL^2 & 0 & 0 \\ 0 & 0 & 0 & 0 & 0 & 0 \\ 0 & 0 & 0 & 0 & 0 & 0 \end{bmatrix}.$$

where $a = EI/L^3$. After assembly of the second element, this becomes

$$\begin{bmatrix} 24a & 0 & -12a & -6aL & -12a & 6aL \\ 0 & 8aL^2 & 6aL & 2aL^2 & -6aL & 2aL^2 \\ -12a & 6aL & 12a & 6aL & 0 & 0 \\ -6aL & 2aL^2 & 6aL & 4aL^2 & 0 & 0 \\ -12a & -6aL & 0 & 0 & 12a & -6aL \\ 6aL & 2aL^2 & 0 & 0 & -6aL & 4aL^2 \end{bmatrix}. \tag{9.19}$$

Since there are only two elements in the current model, this completes the assembly process. The external loads consist of a concentrated force P acting in the negative δ_1 direction. This gives a single component, $-P$, in the first row of a 6×1 force vector for the system.

The stiffness equations are now formed by partitioning $\boldsymbol{K}$ and $\boldsymbol{f}$ to eliminate rows 3–6 which correspond to the constrained degrees of freedom $\delta_3, \ldots, \delta_6$. This leaves us with the reduced equation,

$$\begin{bmatrix} 24a & 0 \\ 0 & 8aL^2 \end{bmatrix} \begin{bmatrix} \delta_1 \\ \delta_2 \end{bmatrix} = \begin{bmatrix} -P \\ 0 \end{bmatrix},$$

which is trivially solved to give $\delta_1 = -PL^3/24EI$ and $\delta_2 = 0$. These values of the displacement and rotation at midspan are, in fact, identical to those obtained by integrating the beam deflection equation and applying 'built-in' boundary conditions at each end. The finite element model therefore yields the 'correct' solution and does so, not only at the nodes of the model but also at all points within each element. This complete agreement between the analytic and discrete solutions is not unexpected. The analytic solution itself takes the form of a cubic expressions in each half of the beam, and can therefore be modelled, without approximation, by cubic interpolation within each element. The same argument applies to any assembly of such elements which involves concentrated loads applied at nodal points. In such instances, the finite element model will always reproduce the exact result. The same is not necessarily true when distributed loads are applied.

9.2.5 Distributed loads, equivalent nodal forces and moments

A distributed load $q(x')$ per unit length acting in the direction of the displacement w, contributes a quantity, V^{e}, to the potential energy of the element, where

$$V^{\mathrm{e}} = -\int_0^L w(x')q(x')\,\mathrm{d}x.$$

Substitution of the interpolation relationship 9.8 into the above integral then gives

$$V^{\mathrm{e}} = -\boldsymbol{d}^{\mathrm{eT}}\boldsymbol{f}_q^{\mathrm{e}}$$

where

$$\boldsymbol{f}_q^{\mathrm{e}} = \int_0^L \boldsymbol{N}^{\mathrm{eT}}\,q(x')\,\mathrm{d}x'. \tag{9.20}$$

This result is analogous to that obtained for a distributed load acting on a continuum element. It tells us that the load can be replaced by an equivalent set of nodal forces given by the components of $\boldsymbol{f}_q^{\mathrm{e}}$. In the case of the beam element, however, the nodal 'forces' comprise not only concentrated forces in the usual sense but also concentrated moments, which may be regarded as 'generalized' forces acting in the 'direction' of the rotations θ_1 and θ_2. The terms in rows one and three of $\boldsymbol{f}_q^{\mathrm{e}}$, for example, are concentrated forces, while those in rows two and four are concentrated moments. The magnitudes of these quantities are obtained by evaluating the integral in equation 9.20 for a specific load distribution. In the case of a uniformly distributed load q_0 acting vertically downwards (Fig. 9.5) the integration gives

$$\boldsymbol{f}_q^{\mathrm{e}} = \int_0^L \begin{bmatrix} n_1 \\ n_2 \\ n_3 \\ n_4 \end{bmatrix} (-q_0)\,\mathrm{d}x'$$

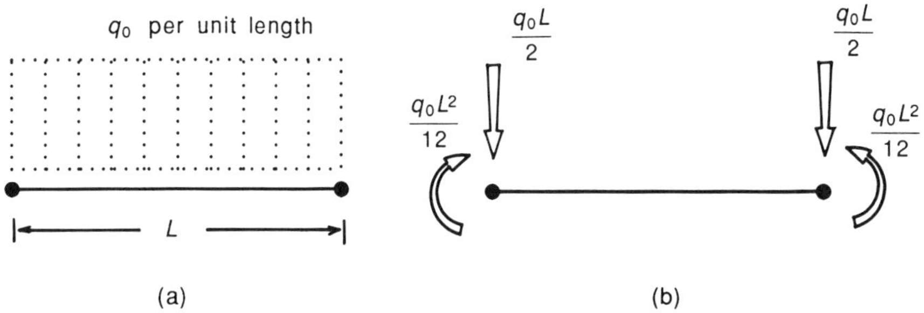

Fig. 9.5 *Nodal forces and moments due to a uniformly distributed load: (a) loading, (b) equivalent nodal forces.*

$$= -\int_0^L q_0 \begin{bmatrix} 1 - 3(x'/L)^2 + 2(x'/L)^3 \\ x'[1 - 2(x'/L) + (x'/L)^2] \\ 3(x'/L)^2 - 2(x'/L)^3 \\ x'[-(x'/L) + (x'/L)^2] \end{bmatrix} \mathrm{d}x' = \begin{bmatrix} -q_0L/2 \\ -q_0L^2/12 \\ -q_0L/2 \\ +q_0L^2/12 \end{bmatrix}. \tag{9.21}$$

In other words, the distributed load is equivalent to nodal forces, $-q_0L/2$, acting in the directions of w_1 and w_2, and nodal moments $-q_0L^2/12$ and $+q_0L^2/12$ acting in the 'directions' of θ_1 and θ_2. These are shown in Fig. 9.5(b). Some exercises which involve the assembly of such contributions for uniform and non-uniform loads.are to be found at the end of this chapter (see problems 3, 6 and 9).

9.3 EFFECTS OF SHEAR DEFORMATION, MODIFIED ELEMENTS FOR DEEP BEAMS

In formulating the element of the previous section, the only stresses present in the beam were assumed to be direct stresses due to bending. The effects of bending shear stress were therefore ignored, as indeed they are in the analytic development of simple beam theory (section 2.5.1). In the case of slender beams, the neglect of shear deformation is quite acceptable, the errors incurred being of the order of $(h/L)^2$, where h is the depth of the beam and L its length. In situations where the beam is relatively short or deep, however, the effect can be significant and should be included both in analytical and numerical treatments. A simple approach is presented here for the incorporation of shear deformation within a modified beam element. This produces a robust and widely used 'Timoshenko' element which is to be found in many finite element codes.

9.3.1 Shear deformation, analytic considerations

In situations where bending shear strain contributes significantly to the deflection of the beam, the Bernoulli–Euler equations of Chapter 2 and the resulting expression for strain energy per unit length must be modified as follows (a more complete derivation of these relationships can be found in [1] and [2]).

First, the assumption that cross-sections remain perpendicular to the neutral axis after deformation is replaced by the less restrictive assumption that they remain plane but experience a shear strain γ, as indicated in Fig. 9.6. The rotation of each cross-section, θ say, is then given by

$$\theta = \mu - \gamma. \tag{9.22}$$

θ then replaces $\mu(= \mathrm{d}w/\mathrm{d}x)$ in equation 2.27 and in all subsequent expressions of Chapter 2. In particular, the beam deflection equation becomes

$$\frac{\mathrm{d}\theta}{\mathrm{d}x} = -\frac{M}{EI} \quad \text{replacing} \quad \frac{\mathrm{d}^2 w}{\mathrm{d}x^2} = -\frac{M}{EI} \tag{9.23}$$

and the expression for the strain energy due to the direct bending stress is

$$\text{SE/length (bending)} = \frac{1}{2} EI \left(\frac{\mathrm{d}\theta}{\mathrm{d}x}\right)^2. \tag{9.24}$$

This latter quantity is supplemented, however, by an energy contribution which derives from shearing deformation. The general expression for strain

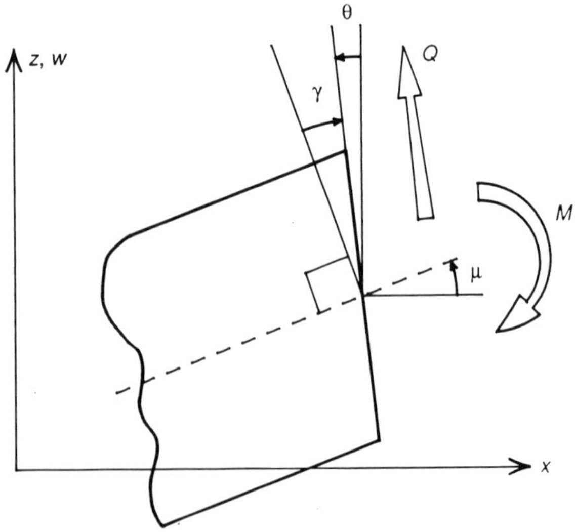

Fig. 9.6 *Shear deformation of a Timoshenko beam.*

energy due to shear is simply $\frac{1}{2}\tau\gamma$ per unit volume (equation 3.19b). Applying this to the beam and assuming constant average values of τ and γ over each cross-section, we obtain a shear strain energy contribution of magnitude $\frac{1}{2}A(\tau\gamma)$ per unit length. In practice, this must be multiplied by a non-dimensional 'shear coefficient' κ which relates what is effectively an integrated value of $\frac{1}{2}(\tau\gamma)$ to appropriate averaged or centreline values of τ and γ. The values of κ derived by Cowper [3] are generally regarded as the most satisfactory (a table of these is to be found in [3]). With this modification the combined bending and shearing strain energy of the beam is given by

$$\text{SE/length} = \frac{1}{2}EI\left(\frac{\mathrm{d}\theta}{\mathrm{d}x}\right)^2 + \frac{1}{2}\kappa AG\gamma^2. \qquad (9.25)$$

Consistent with the above expression, the shear strain γ is related to the shear force Q (Fig. 9.6), by $\gamma = Q/\kappa AG$. The deflection equation of the beam, equation 9.23, then becomes

$$\frac{\mathrm{d}\theta}{\mathrm{d}x} = \frac{\mathrm{d}}{\mathrm{d}x}\left(\frac{\mathrm{d}w}{\mathrm{d}x} - \frac{Q(x)}{\kappa AG}\right)$$
$$= -\frac{M(x)}{EI}. \qquad (9.26)$$

This can be solved analytically in simple cases to yield estimates for the deflections of 'deep' beams including the effect of shear deformation. A simple problem of this type which will be used later to test the accuracy of the Timoshenko element, is that of a cantilever beam of length L which is fixed at one end and carries a concentrated load P at the other. The problem is shown in Fig. 9.7. A simple finite element model formed from a single 'Timoshenko' beam element of a type to be described shortly is also shown. An exact solution

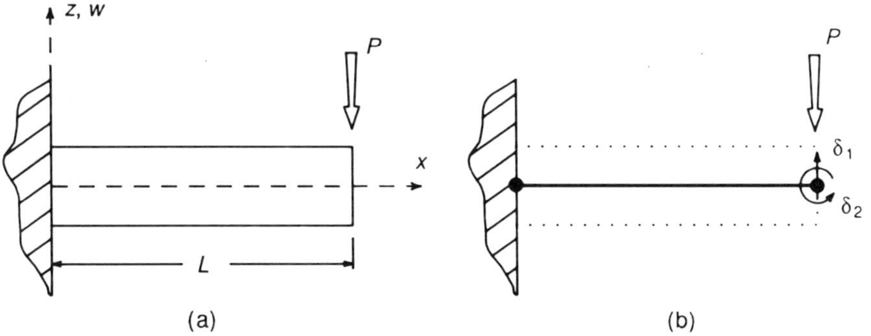

Fig. 9.7 *Test case for Timoshenko model: (a) geometry and loading, (b) discrete model.*

— 'exact' in terms of Timoshenko's theory — is obtained by integrating equation 9.26 setting $Q = -P$ and $M = P(L - x)$. The constants of integration are then selected so that $w = \theta = 0$ at the left hand end. Note that the 'built-in' condition now requires that the rotation of the cross-section, rather than the slope of the neutral axis, is zero, these quantities no longer being synonymous. This gives a solution

$$w(x) = -\frac{P}{EI}\left(\frac{x^3}{6} - \frac{Lx^2}{2}\right) - \frac{Px}{\kappa AG}, \tag{9.27a}$$

and

$$\theta(x) = -\frac{P}{EI}\left(\frac{x^2}{2} - Lx\right). \tag{9.27b}$$

The deflection and rotation at the free end are then

$$w(L) = -\frac{PL^3}{3EI}\left(1 + \frac{\varepsilon}{4}\right), \quad \text{where } \varepsilon = \frac{12EI}{\kappa AGL^2}, \tag{9.28}$$

and

$$\theta(L) = -\frac{PL^2}{2EI}. \tag{9.29}$$

Note that as L, the length of the beam, increases, the non-dimensional parameter ε approaches zero and the solution converges to the conventional Euler–Bernoulli result for a slender beam.

9.3.2 The Timoshenko beam element

A beam element is now proposed which incorporates independent variation for w and γ. A number of elements of this type have been developed and many are reviewed in [4] and [5]. The one presented here is the simplest. It is a two-noded element similar in appearance to the conventional element of Fig. 9.2, but with the subtle difference that the rotational degrees of freedom, θ_1 and θ_2, define rotations of the end planes rather than rotations of the neutral axis. The latter are denoted by μ_1 and μ_2. The topology, deformed shape and degrees of freedom of the element are shown in Fig. 9.8. The strain energy is obtained by integrating expression 9.25 along the axis of the element. Eliminating θ, using equation 9.22, we obtain

$$U^e = \int_0^L \frac{1}{2}\left[EI\left(\frac{d^2w}{dx^2} - \frac{d\gamma}{dx}\right)^2 + \kappa AG\gamma^2\right]dx. \tag{9.30}$$

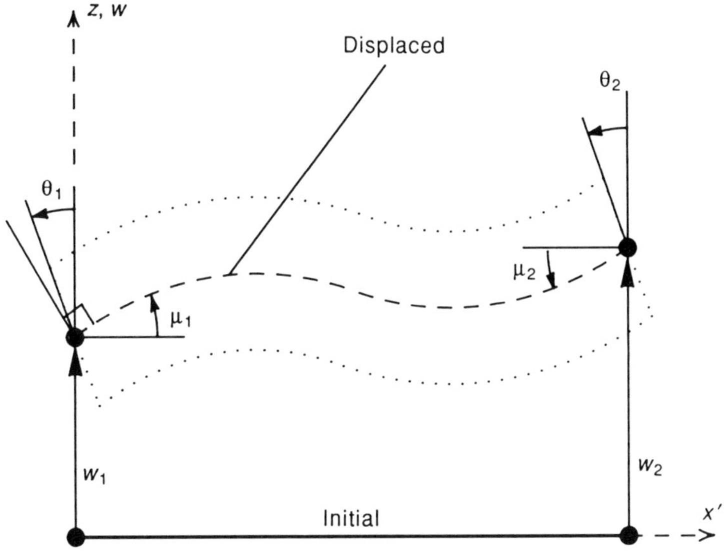

Fig. 9.8 *The Timoshenko beam element.*

Shape functions are now specified for w and γ. The same cubic interpolation is used for w as was used in the standard element of Fig. 9.2. The shape relationship is therefore identical to that of the previous element but with θ replaced by μ. This gives

$$w(x') = n_1(x')w_1 + n_2(x')\mu_1 + n_3(x')w_2 + n_4(x')\mu_2, \tag{9.31}$$

where the shape functions n_i are defined by expressions comprising equation 9.6. The shape relationship for γ can be formulated in a number of ways. The most straightforward approach, and that adopted here, is to use zeroth order interpolation. We assume, in other words, that γ is a constant, equal to γ_0 say, within each element. This implies also a constant shear stress and shear force within the element.

The introduction of γ_0 as an independent variable gives the element a fifth degree of freedom. This is apparent when equation 9.31 is rewritten in terms of w_i, θ_i and γ_0. Using equation 9.22 to rewrite μ_i in terms of θ_i and γ_0, we obtain

$$w(x') = n_1(x')w_1 + n_2(x')(\theta_1 + \gamma_0) + n_3(x')w_2 + n_4(x')(\theta_2 + \gamma_0), \tag{9.32}$$

which, after rearrangement of terms, becomes

$$w(x') = n_1(x')w_1 + n_2(x')\theta_1 + n_3(x')w_2 + n_4(x')\theta_2 + n_5(x')\gamma_0, \tag{9.33a}$$

where $n_5(x') = n_2(x') + n_4(x')$. This can also be written in the standard form as a matrix shape relationship,

$$w(x') = [n_1(x'), n_2(x'), n_3(x'), n_4(x'), n_5(x')] \begin{bmatrix} w_1 \\ \theta_1 \\ w_2 \\ \theta_2 \\ \gamma_0 \end{bmatrix} = \boldsymbol{N}_\gamma^e \, \boldsymbol{d}_\gamma^e, \qquad (9.33b)$$

where, $\boldsymbol{d}_\gamma^e$, contains the displacements and rotations at the ends of the element, *plus* the shear strain γ_0. Substitution into equation 9.30 then gives, after further manipulation,

$$U^e = \tfrac{1}{2} \boldsymbol{d}_\gamma^{eT} \boldsymbol{K}_\gamma^e \boldsymbol{d}_\gamma^e, \qquad (9.34)$$

where the stiffness matrix $\boldsymbol{K}_\gamma^e$ has components

$$\boldsymbol{K}_\gamma^e = \begin{bmatrix} 12a & 6aL & -12a & 6aL & 12aL \\ 6aL & 4aL^2 & -6aL & 2aL^2 & 6aL^2 \\ -12a & -6aL & 12a & -6aL & -12aL \\ 6aL & 2aL^2 & -6aL & 4aL^2 & 6aL^2 \\ 12aL & 6aL^2 & -12aL & 6aL^2 & (12aL^2 + b) \end{bmatrix}, \qquad (9.35)$$

where $a = EI/L^3$ and $b = \kappa AGL$. The first four rows and columns of this matrix are then identical to those in the stiffness matrix for the conventional beam element (expression 9.18). The additional terms in row and column five result from shear deformation and are associated with the 'shear'' degree of freedom γ_0.

There are now two ways to proceed. We can, for example, assemble the stiffness matrix, $\boldsymbol{K}_\gamma^e$, as it stands, having assigned a degree of freedom number to the additional degree of freedom γ_0. The Timoshenko element must then be treated as a five-degree of freedom element and cannot, for example, be used in place of a conventional two-noded element without modification to the mesh and degree of freedom numbering systems of the overall model.

A more common course of action is to eliminate the new degree of freedom *before* assembling the element into a finite element model. This can be done relatively easily since γ_0 is unique to a particular element, unlike w_i and θ_i which are common to adjacent elements. The elimination (or 'condensation') of γ_0 as an internal degree of freedom leaves us with an element which has the same topology and as its Bernoulli–Euler equivalent and can therefore be inserted into a standard mesh without modification. The condensation of γ_0 is performed in the following way.

First, we write the equation which would have resulted from the 'γ_0' row of the assembled stiffness matrix, had the element been assembled in its five-degree of freedom state. This is easily done since γ_0 is connected only to the four other components of its 'own' element displacement vector $\boldsymbol{d}^{\mathrm{e}}_{\gamma}$. Moreover, there are no forces on the right hand side of the equation provided that loads are applied only at the nodal points. The coefficients of the 'γ_0' equation in the assembled system are therefore extracted from the bottom row of the element stiffness matrix to give

$$(12aL)w_1 + (6aL^2)\theta_1 - (12aL)w_2 + (6aL^2)\theta_2 + (12aL^2 + b)\gamma_0 = 0, \quad (9.36)$$

or

$$\gamma_0 = \frac{1}{12aL^2 + b}[-(12aL)w_1 - (6aL^2)\theta_1 + (12aL)w_2 - (6aL^2)\theta_2]. \quad (9.37)$$

This relationship can be incorporated into a transformation between the degrees of freedom with and without γ_0. In other words, we can write,

$$\begin{bmatrix} w_1 \\ \theta_1 \\ w_2 \\ \theta_2 \\ \gamma_0 \end{bmatrix} = \begin{bmatrix} 1 & 0 & 0 & 0 \\ 0 & 1 & 0 & 0 \\ 0 & 0 & 1 & 0 \\ 0 & 0 & 0 & 1 \\ -2c & -cL & 2c & -cL \end{bmatrix} \begin{bmatrix} w_1 \\ \theta_1 \\ w_2 \\ \theta_2 \end{bmatrix},$$

where $c = 6aL/(12aL^2 + b)$. In more concise notation this becomes

$$\boldsymbol{d}^{\mathrm{e}}_{\gamma} = \boldsymbol{T}\,\boldsymbol{d}^{\mathrm{e}}, \quad (9.38)$$

where $\boldsymbol{d}^{\mathrm{e}}$ is the nodal displacement vector for a four-degree of freedom element and $\boldsymbol{T}^{\mathrm{e}}$ is a 5×4 transformation matrix. The stiffness matrix, $\boldsymbol{K}^{\mathrm{e}}$, for the four-degree of freedom element is obtained by substituting equation 9.38 into equation 9.34 to give

$$U^{\mathrm{e}} = \tfrac{1}{2}\,\boldsymbol{d}^{\mathrm{eT}}\,\boldsymbol{K}^{\mathrm{e}}\,\boldsymbol{d}^{\mathrm{e}},$$

where

$$\boldsymbol{K}^{\mathrm{e}} = \boldsymbol{T}^{\mathrm{T}}\,\boldsymbol{K}^{\mathrm{e}}_{\gamma}\,\boldsymbol{T}. \quad (9.39)$$

Evaluation of this matrix triple product gives the components

$$\boldsymbol{K}^{\mathrm{e}} = \left(\frac{a}{1+\varepsilon}\right) \begin{bmatrix} 12 & 6L & -12 & 6L \\ 6L & (4+\varepsilon)L^2 & -6L & (2-\varepsilon)L^2 \\ -12 & -6L & 12 & -6L \\ 6L & (2-\varepsilon)L^2 & -6L & (4+\varepsilon)L^2 \end{bmatrix}, \quad (9.40)$$

where $\varepsilon = 12EI/(\kappa AGL^2)$. This is the same nondimensional parameter which occurred in expression 9.28. Clearly, as the beam becomes slender, that is as ε approaches zero, the 'Timoshenko' matrix, expression 9.40, converges to the conventional stiffness matrix of expression 9.18.

As a check on the accuracy of the formulation, let us calculate the end deflection for the test problem of Fig. 9.7 using a Timoshenko element to model the beam. The model is shown in Fig. 9.7(b). The resulting stiffness equation, after eliminating the displacement and rotation at the left hand end, is

$$\left(\frac{a}{1+\varepsilon}\right)\begin{bmatrix} 12 & -6L \\ -6L & (4+\varepsilon)L^2 \end{bmatrix}\begin{bmatrix} \delta_1 \\ \delta_2 \end{bmatrix} = \begin{bmatrix} -P \\ 0 \end{bmatrix}, \qquad (9.41)$$

giving a solution; $\delta_1 = -(P/3a)(1+\varepsilon/4)$ and $\delta_2 = -P/(2aL)$. These agree exactly with the analytic result obtained at the beginning of this discussion (equations 9.28 and 9.29).

In deriving the Timoshenko stiffness matrix, it was assumed that loads and moments were applied only at the nodes. No discrete forces were therefore associated with 'shear' degree of freedom γ_0. This is not generally the case when distributed loads are applied to the element. The equivalent nodal force vector (given by integral 9.20 with $\boldsymbol{N}^{\mathrm{e}}$ replaced by $\boldsymbol{N}_{\gamma}^{\mathrm{e}}$) then includes a component associated with γ_0. This 'generalized' force is somewhat difficult to visualize but must be included on the right hand side of the ensuing stiffness relationship if a consistent solution is to be obtained. It is not difficult, however, to show this force is zero for the particular case of a *uniformly* distributed load, which can therefore be modelled without approximation by the condensed, four-degree of freedom version of the Timoshenko element provided that the same equivalent nodal forces and moments are applied at the nodes of the model as would be applied to a similar model formed from simple beam elements (problem 10 at the end of this chapter).

9.4 FRAME ELEMENTS IN TWO AND THREE DIMENSIONS

Elements are now derived for the analysis of two and three-dimensional frameworks. The modifications required to convert a beam element into a frame element are analogous to those used to transform a bar element into a two-dimensional truss element (section 4.5). In the case of the frame element however, the notion of simply rotating the element in space is frustrated by the coupling which exists in the structure as a whole between bending, tension/compression and torsion.

Consider, for example, the structural connection shown in Fig. 9.9. Structural members (a), (b) and (c) are joined at the point A, which experiences cartesian translations u, v, w and rotations ϕ, θ, and ψ. Consider the nature of one of the translations, u say, as it appears to each member. To members (b) and (c), it is a lateral (bending) displacement in the x–z and x–y planes, respectively. To member (a), however, it is an axial (tensile) displacement. The translation v is similarly percieved as a bending displacement in members (a) and (b) but a compression in member (c). The situation with respect to rotations is equally convoluted; θ, for example, acts as a bending rotation in the x–z plane for members (a) and (b), but as a torsional rotation of member (c) about its own axis. Clearly, it is no longer possible to regard bending, torsion and tension/compression as distinct modes of deformation which can be uncoupled throughout the structure. A satisfactory finite element model therefore requires an element which incorporates the effects of all three types of deformation.

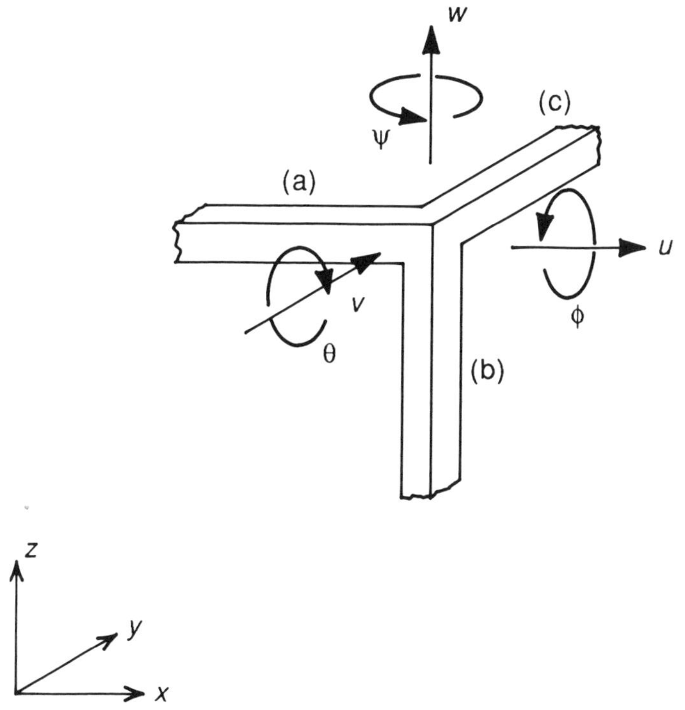

Fig. 9.9 *Displacement of a joint in a three-dimensional framework.*

The formulation of such elements is simplified in the case of planar systems which deform only in their own plane. No torsional deformations then occur

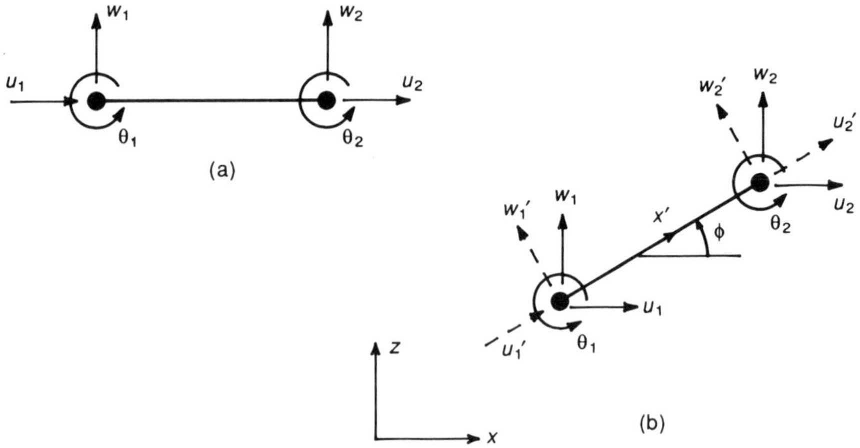

Fig. 9.10 *The planar frame element: (a) local orientation, (b) global orientation.*

and coupling exists only between bending and tension/compression. This case is discussed first.

9.4.1 Six-degree of freedom beam element with axial tension

Consider now an element similar to that of Fig. 9.2, but which permits axial deformation in addition to bending. The element is shown in Fig. 9.10(a). Nodal displacements u_1 and u_2 are specified in addition to the existing degrees of freedom w_1, θ_1, w_2 and θ_2, giving a six-degree of freedom element. The axial component of displacement within the element is then approximated using the same (linear) interpolation as was employed in the two-noded bar element of Chapter 4. The strain energy of the element is given by (equation 3.24)

$$U^e = \int_0^L \left[\frac{1}{2} EI \left(\frac{d^2 w}{dx^2} \right)^2 dx + \frac{1}{2} EA \left(\frac{du}{dx} \right)^2 \right] dx. \tag{9.42}$$

The first term in this integral represents a bending contribution and the second a contribution due to axial tension/compression. Each has been encountered separately in our formulation of the beam element (section 9.2) and the bar element (section 4.2). The two contributions are uncoupled within the element itself so that U^e may be written

$$U^e = U_b^e + U_m^e, \tag{9.43}$$

where U_b^e and U_m^e are the strain energies of an equivalent beam and bar, respectively. U^e can therefore be written in matrix form as

$$U^{\mathrm{e}} = \frac{1}{2}[w_1, \theta_1, w_2, \theta_2]\begin{bmatrix} 12a & 6aL & -12a & 6aL \\ 6aL & 4aL^2 & -6aL & 2aL^2 \\ -12a & -6aL & 12a & -6aL \\ 6aL & 2aL^2 & -6aL & 4aL^2 \end{bmatrix}\begin{bmatrix} w_1 \\ \theta_1 \\ w_2 \\ \theta_2 \end{bmatrix} + \frac{1}{2}[u_1, u_2]\begin{bmatrix} k & -k \\ -k & k \end{bmatrix}\begin{bmatrix} u_1 \\ u_2 \end{bmatrix}, \tag{9.44}$$

where $a = EI/L^3$ and $k = EA/L$. Alternatively, both contributions can be included in a single triple product:

$$U^{\mathrm{e}} = \frac{1}{2}[u_1, w_1, \theta_1, u_2, w_2, \theta_2]\begin{bmatrix} k & 0 & 0 & -k & 0 & 0 \\ 0 & 12a & 6aL & 0 & -12a & 6aL \\ 0 & 6aL & 4aL^2 & 0 & -6aL & 2aL^2 \\ -k & 0 & 0 & k & 0 & 0 \\ 0 & -12a & -6aL & 0 & 12a & -6aL \\ 0 & 6aL & 2aL^2 & 0 & -6aL & 4aL^2 \end{bmatrix}\begin{bmatrix} u_1 \\ w_1 \\ \theta_1 \\ u_2 \\ w_2 \\ \theta_2 \end{bmatrix}. \tag{9.45}$$

The 6×6 array at the centre of the above expression then defines the stiffness matrix for the six-degree of freedom, frame element. Note that the same expression can be obtained by regarding the frame element itself as a simple two-element model formed from separate beam and bar elements occupying the same physical location in space. In either event, we obtain an element which incorporates the effects both of bending and of tension/compression. Note also that although the conventional beam element of section 9.2 has been used as the 'bending' component of the current frame element, the Timoshenko version can equally well be used if the final element is intended for problems in which shear deformation is significant.

9.4.2 Planar frame element

The element of Fig. 9.10(a) is now rotated in the x–z plane so that it is orientated at an angle ϕ to the global x-axis. Nodal displacements, u_1, w_1, θ_1, u_2, w_2 and θ_2 are then defined in the global coordinate system (Fig. 9.10(b)) and are placed in a 'global', element displacement vector, $\boldsymbol{d}^{\mathrm{e}}$. Analogous components in the local system, whose x'-axes lies along the neutral axis of the element, are denoted by u_1', w_1' etc. These are placed in a 'local' displacement vector, $\boldsymbol{d}^{\mathrm{e}'}$. The global and local components of displacement are related by the equations

$$u_i' = u_i \cos\phi + w_i \sin\phi,\ w_i' = w_i \cos\phi - u_i \sin\phi,\ \theta_i' = \theta_i \quad (i = 1, 2).$$

These can be formed into a single matrix transformation:

$$d^{e'} = T^e d^e, \tag{9.46}$$

where

$$T^e = \begin{bmatrix} \cos\phi & \sin\phi & 0 & 0 & 0 & 0 \\ -\sin\phi & \cos\phi & 0 & 0 & 0 & 0 \\ 0 & 0 & 1 & 0 & 0 & 1 \\ 0 & 0 & 0 & \cos\phi & \sin\phi & 0 \\ 0 & 0 & 0 & -\sin\phi & \cos\phi & 0 \\ 0 & 0 & 0 & 0 & 0 & 1 \end{bmatrix}. \tag{9.47}$$

The strain energy of this element has already been calculated, however, in terms of the local displacements. It is given by

$$U^e = \tfrac{1}{2} d^{e'T} K^{e'} d^{e'},$$

where $K^{e'}$ is the 'local' stiffness matrix at the centre of expression 9.45. Substitution of transformation 9.46 then gives

$$U^e = \tfrac{1}{2} d^{eT} K^e d^e, \tag{9.48}$$

where

$$K^e = T^{eT} K^{e'} T^e. \tag{9.49}$$

In this form, expression 9.48 defines a global stiffness matrix, K^e, for the arbitrarily orientated element. This can be assembled in the usual way to model planar structures with rigid joints.

9.4.3 Assembly of frame elements: a simple example

The assembly of frame elements to form a planar structure is demonstrated by the following example. Two identical members, each of length L, flexural

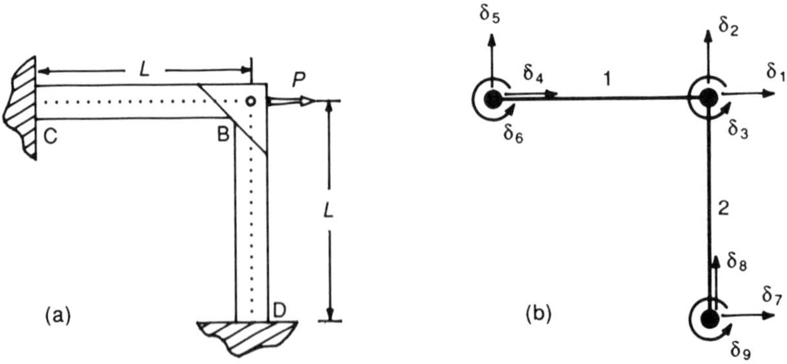

Fig. 9.11 *Plane framework, worked example: (a) geometry and loading, (b) discrete model.*

rigidity EI and cross-section A, are rigidly joined at B (Fig. 9.11(a)). The structure is built-in at C and D, and is loaded at B with a horizontal force P. Each member is modelled by a planar frame element of the type described in the preceding section.

The degrees of freedom of the assembled structure are numbered $\delta_1, \ldots, \delta_9$ as shown (Fig. 9.11(b)). The unconstrained displacements and rotations at node B are numbered first. The nodal displacement vectors, $\boldsymbol{d}^{(1)}$ and $\boldsymbol{d}^{(2)}$, for elements 1 and 2 are then

$$\boldsymbol{d}^{(1)} = \begin{bmatrix} \delta_4 \\ \delta_5 \\ \delta_6 \\ \delta_1 \\ \delta_2 \\ \delta_3 \end{bmatrix} \text{ and } \boldsymbol{d}^{(2)} = \begin{bmatrix} \delta_7 \\ \delta_8 \\ \delta_9 \\ \delta_1 \\ \delta_2 \\ \delta_3 \end{bmatrix}.$$

The element stiffness matrices, $\boldsymbol{K}^{(1)}$ and $\boldsymbol{K}^{(2)}$, are given by equation 9.48 with the transformation matrix $\boldsymbol{T}^{e}$ calculated by putting ϕ equal to 0° and 90°, respectively. In the case of element 1, $\boldsymbol{T}^{(1)}$ is simply the identity matrix and $\boldsymbol{K}^{(1)}$ is then identical to the local stiffness matrix of expression 9.45. In the case of element 2, the transformation matrix $\boldsymbol{T}^{(2)}$ has components

$$\boldsymbol{T}^{(2)} = \begin{bmatrix} 0 & 1 & 0 & 0 & 0 & 0 \\ -1 & 0 & 0 & 0 & 0 & 0 \\ 0 & 0 & 1 & 0 & 0 & 0 \\ 0 & 0 & 0 & 0 & 1 & 0 \\ 0 & 0 & 0 & -1 & 0 & 0 \\ 0 & 0 & 0 & 0 & 0 & 1 \end{bmatrix},$$

and $\boldsymbol{K}^{(2)}$ is given by

$$\boldsymbol{K}^{(2)} = \begin{bmatrix} 12a & 0 & -6aL & -12a & 0 & -6aL \\ 0 & k & 0 & 0 & -k & 0 \\ -6aL & 0 & 4aL^2 & 6aL & 0 & 2aL^2 \\ -12a & 0 & 6aL & 12a & 0 & 6aL \\ 0 & -k & 0 & 0 & k & 0 \\ -6aL & 0 & 2aL^2 & 6aL & 0 & 4aL^2 \end{bmatrix}.$$

Assembly of $\boldsymbol{K}^{(1)}$ and $\boldsymbol{K}^{(2)}$ into an initially zero 9×9 matrix then yields a stiffness matrix with components:

$$K=\begin{bmatrix} (k+12a) & 0 & 6aL & -k & 0 & 0 & -12a & 0 & 6aL \\ 0 & (k+12a) & -6aL & 0 & -12a & -6aL & 0 & -k & 0 \\ 6aL & -6aL & 8aL^2 & 0 & 6aL & 2aL^2 & -6aL & 0 & 2aL^2 \\ -k & 0 & 0 & k & 0 & 0 & 0 & 0 & 0 \\ 0 & -12a & 6aL & 0 & 12a & 6aL & 0 & 0 & 0 \\ 0 & -6aL & 2aL^2 & 0 & 6aL & 4aL^2 & 0 & 0 & 0 \\ -12a & 0 & -6aL & 0 & 0 & 0 & 12a & 0 & -6aL \\ 0 & -k & 0 & 0 & 0 & 0 & 0 & k & 0 \\ 6aL & 0 & 2aL^2 & 0 & 0 & 0 & -6aL & 0 & 4aL^2 \end{bmatrix}.$$

All but the first three degrees of freedom are constrained, however, since points C and D are rigidly fixed. Deleting all but the first three rows of the above matrix and inserting a load P acting in the direction of degree of freedom δ_1, we obtain a reduced stiffness relationship of the form

$$\begin{bmatrix} (k+12a) & 0 & 6aL \\ 0 & (k+12a) & -6aL \\ 6aL & -6aL & 8aL^2 \end{bmatrix}\begin{bmatrix} \delta_1 \\ \delta_2 \\ \delta_3 \end{bmatrix}=\begin{bmatrix} P \\ 0 \\ 0 \end{bmatrix}.$$

Solving, we obtain

$$\delta_1=\frac{\mathrm{P}\,[10a+(4/3)k]}{[4a+(4/3)k]\,[k+12a]},$$

$$\delta_2=\frac{-6aP}{[4a+(4/3)k]\,[k+12a]},$$

and

$$\delta_3=-\frac{P}{L\,[4a+(4/3)k]}.$$

Although the details are not included here, it can be shown with some difficulty that these values are, in fact, identical to the those obtained from an exact 'equilibrium' solution. This is not unexpected, since the trial displacement field within each element — a cubic polynomial for bending displacement and a linear polynomial for axial deformation — is capable of modelling the 'true', equilibrium displacement field without approximation.

9.4.4 Three-dimensional frame element

The development of a three-dimensional frame element follows closely that of its planar equivalent. First, a frame element is formed which lies along the

x-axis, but is capable of displacing in three dimensions, in other words, is capable of sustaining bending deformation in two planes, torsion about its own axis and axial tension/compression. A simple transformation then rotates this element arbitrarily in space.

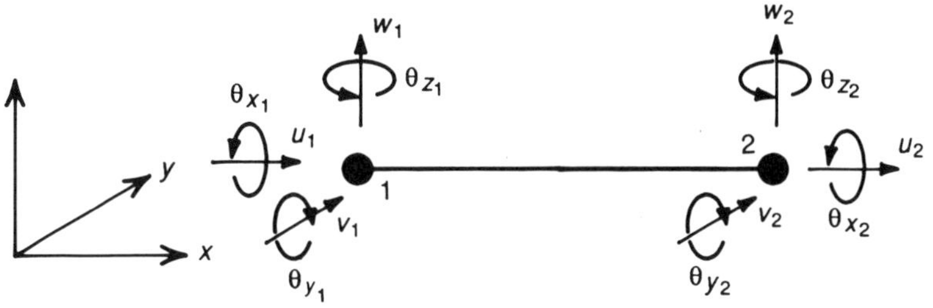

Fig. 9.12 *The three-dimensional frame element, local degrees of freedom.*

The first stage in this procedure is accomplished by summing the strain energy contributions due to bending, torsion and tension/compression. This is done by formulating an element which lies along the x-axis and whose displaced state is defined by nodal displacements u_i, v_i, w_i and rotations $\theta_{xi}, \theta_{yi}, \theta_{zi}$ (Fig. 9.12). Note that the nodal rotations are defined by the right hand rule applied to the coordinate directions. This is consistent in the x–y plane with the previous definition of rotation as an axial derivative of transverse displacement. The same is not true, however, in the x–z plane, where the previous rotation $\theta\ (=\partial w/\partial x)$ is now equal to $-\theta_y$. This change of sign must be taken into account when forming an expression for the strain energy due to bending.

The strain energy of the element comprises four distinct contributions and may be written

$$U^e = U^e_{xz} + U^e_{xy} + U^e_m + U^e_t. \tag{9.50}$$

The first two terms in the above summation derive from bending in the x–z and x–y planes. They are analogous to the bending energy U^e_b in the formulation of the planar element, and are given by

$$U^e_{xz} = \frac{1}{2}[w_1, \theta_{y1}, w_2, \theta_{y2}]\begin{bmatrix} 12a & -6aL & -12a & -6aL \\ -6aL & 4aL^2 & 6aL & 2aL^2 \\ -12a & 6aL & 12a & 6aL \\ -6aL & 2aL^2 & 6aL & 4aL^2 \end{bmatrix}\begin{bmatrix} w_1 \\ \theta_{y1} \\ w_2 \\ \theta_{y2} \end{bmatrix}, \tag{9.51a}$$

and

$$U_{xy}^{e} = \frac{1}{2}[v_1, \theta_{z1}, v_2, \theta_{z2}] \begin{bmatrix} 12b & 6bL & -12b & 6bL \\ 6bL & 4bL^2 & -6bL & 2bL^2 \\ -12b & -6bL & 12b & -6bL \\ 6bL & 2bL^2 & -6bL & 4bL^2 \end{bmatrix} \begin{bmatrix} v_1 \\ \theta_{z1} \\ v_2 \\ \theta_{z2} \end{bmatrix}, \quad (9.51b)$$

where $a = EI_1/L^3$ and $b = EI_2/L^3$ (I_1 and I_2 are second moments of area about the y and z-axes, respectively, assumed to be principal axes of the cross-section).

The third contribution, U_m^e, is the membrane strain energy due to tension/compression. This has already been used in the planar frame element and is given by

$$U_m^e = \frac{1}{2}[u_1, u_2] \begin{bmatrix} k & -k \\ -k & k \end{bmatrix} \begin{bmatrix} u_1 \\ u_2 \end{bmatrix}. \quad (9.51c)$$

The last contribution, U_t^e, is the strain energy due to torsion. It arises from a contribution (section 3.5.2) of $\frac{1}{2}C\,(d\theta_x/dx)^2$ to the strain energy per unit length, where C is the torsional rigidity of the member. If a linear distribution is assumed for θ_x, it is simple to show that U_t^e is given by an expression similar to that obtained for the strain energy due to tension/compression but with u replaced by θ_x and with the axial stiffness k replaced by the torsional stiffness $c\ (=C/L)$. In other words

$$U_t^e = \tfrac{1}{2}[\theta_{x1}, \theta_{x2}] \begin{bmatrix} c & -c \\ -c & c \end{bmatrix} \begin{bmatrix} \theta_{x1} \\ \theta_{x2} \end{bmatrix}. \quad (9.51d)$$

Assembly of the contributions, expressed in equations 9.51a-d, into an initially zero 12×12 array then gives a 'local' stiffness matrix, $\boldsymbol{K}^{e'}$ say, for the composite element. If the nodal degrees of freedom are ordered $u_1, v_1, w_1, \theta_{x1}, \theta_{y1}, \theta_{z1}, u_2, \ldots$, the components of $\boldsymbol{K}^{e'}$ are

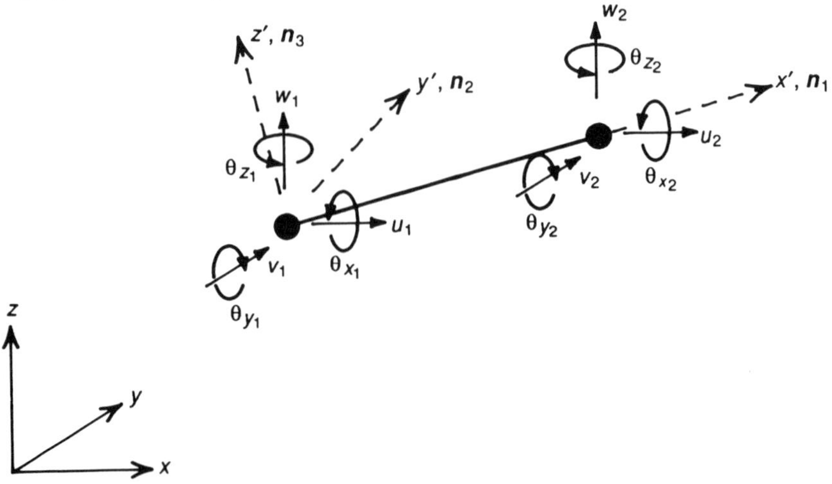

Fig. 9.13 *The three-dimensional frame element, global degrees of freedom.*

$$
\boldsymbol{K}^{e'} = \begin{bmatrix}
k & 0 & 0 & 0 & 0 & 0 & -k & 0 & 0 & 0 & 0 & 0 \\
0 & 12b & 0 & 0 & 0 & 6bL & 0 & -12b & 0 & 0 & 0 & 6bL \\
0 & 0 & 12a & 0 & -6aL & 0 & 0 & 0 & -12a & 0 & -6aL & 0 \\
0 & 0 & 0 & c & 0 & 0 & 0 & 0 & 0 & -c & 0 & 0 \\
0 & 0 & -6aL & 0 & 4aL^2 & 0 & 0 & 0 & 6aL & 0 & 2aL^2 & 0 \\
0 & 6bL & 0 & 0 & 0 & 4bL^2 & 0 & -6bL & 0 & 0 & 0 & 2bL^2 \\
-k & 0 & 0 & 0 & 0 & 0 & k & 0 & 0 & 0 & 0 & 0 \\
0 & -12b & 0 & 0 & 0 & -6bL & 0 & 12b & 0 & 0 & 0 & -6bL \\
0 & 0 & -12a & 0 & 6aL & 0 & 0 & 0 & 12a & 0 & 6aL & 0 \\
0 & 0 & 0 & -c & 0 & 0 & 0 & 0 & 0 & c & 0 & 0 \\
0 & 0 & -6aL & 0 & 2aL^2 & 0 & 0 & 0 & 6aL & 0 & 4aL^2 & 0 \\
0 & 6bL & 0 & 0 & 0 & 2bL^2 & 0 & -6bL & 0 & 0 & 0 & 4bL^2
\end{bmatrix}. \tag{9.52}
$$

The above matrix relates to a 'local' coordinate system aligned with the centroidal axis of the element. A global version of the same matrix is obtained by rotating the element in three-dimensional space. Consider, therefore, an identical element orientated so that the x, y and z-axes of Fig. 9.12 become the x', y' and z' axes of Fig. 9.13. Their orientation, relative to a global axis set, Oxyz, is defined by unit vectors $\boldsymbol{n}_1, \boldsymbol{n}_2$ and $\boldsymbol{n}_3$, where $\boldsymbol{n}_i = (l_i, m_i, n_i)$.

The global degrees of freedom for the rotated element comprise displacements u_i, v_i and w_i, and corresponding rotations θ_{xi}, θ_{yi} and θ_{zi} in the global x, y and z directions. The equivalent 'local' degrees of freedom — not shown in Fig. 9.13 — bear the same orientation to the element as the translations and rotations of Fig. 9.12. They are denoted now by the 'dashed' variables, u_i', v_i', w_i', θ'_{xi}, θ'_{yi} and $\theta'_{zi} (i = 1, 2)$.

The transformation between the local and global degrees of freedom at each node can then be written

$$
\begin{bmatrix} u_i' \\ v_i' \\ w_i' \\ \theta'_{xi} \\ \theta'_{yi} \\ \theta'_{zi} \end{bmatrix} = \begin{bmatrix}
l_1 & m_1 & n_1 & 0 & 0 & 0 \\
l_2 & m_2 & n_2 & 0 & 0 & 0 \\
l_3 & m_3 & n_3 & 0 & 0 & 0 \\
0 & 0 & 0 & l_1 & m_1 & n_1 \\
0 & 0 & 0 & l_2 & m_2 & n_2 \\
0 & 0 & 0 & l_3 & m_3 & n_3
\end{bmatrix} \begin{bmatrix} u_i \\ v_i \\ w_i \\ \theta_{xi} \\ \theta_{yi} \\ \theta_{zi} \end{bmatrix}, \; (i = 1, 2), \tag{9.53}
$$

or

$$d_i^{e\prime} = T\, d_i^{e},\ (i = 1, 2) \tag{9.54}$$

The procedure for forming a stiffness matrix for the global element is now analogous to that which led to expression 9.48. That is to say, the strain energy U^{e}, currently defined in terms of the local displacements by

$$U^{e} = \tfrac{1}{2}\, d^{e\prime T}\, K^{e\prime}\, d^{e\prime},$$

where $K^{e\prime}$ is the stiffness matrix of expression 9.52. This is rewritten, using equation 9.54, as

$$U^{e} = \tfrac{1}{2}\, d^{eT}\, K^{e}\, d^{e}, \tag{9.55}$$

where

$$K^{e} = T^{eT}\, K^{e\prime}\, T^{e} \text{ and } T^{e} = \begin{bmatrix} T & 0 \\ 0 & T \end{bmatrix}. \tag{9.56}$$

The global matrix K^{e} defined by equation 9.56 can then be assembled in the usual way. Structures formed from any number of rigidly connected and arbitrarily orientated members can be modelled in this way; 'built-in' or 'simple' supports being simulated by restraining combinations of rotational and translational degrees of freedom.

9.4.5 Rigid offsets, modelling the shear centre

The two and three-dimensional frame elements described here can be used to model most framed structures. Care must be exercised, however, in modelling correctly the joints between elements. Intrinsic to the formulation as it stands is the assumption that elements are joined so that their neutral axes intersect at a common node. In many instances this is not the case. That is to say, members are joined so that their centroidal axes are rigidly offset. A typical arrangement is shown in Fig. 9.14(a). Here members AB and CD are rigidly but eccentrically connected between B and C.

A similar problem arises when the centroid C and shear centre B of a structural member do not coincide (Fig. 9.14(b)). If such a member is modelled using a frame element of the type described above, the correct locations for the nodes of the element are ambiguous. Should they be placed at the centroid of area or at the centre of shear? Neither location is in fact correct for *all* of the degrees of freedom. The evaluation of membrane strain energy, for example, involves displacement of the centroid, whereas calculation of the strain and potential energies in bending involves displacement of the shear centre. Before the element can be compatibly joined to others of the same type,

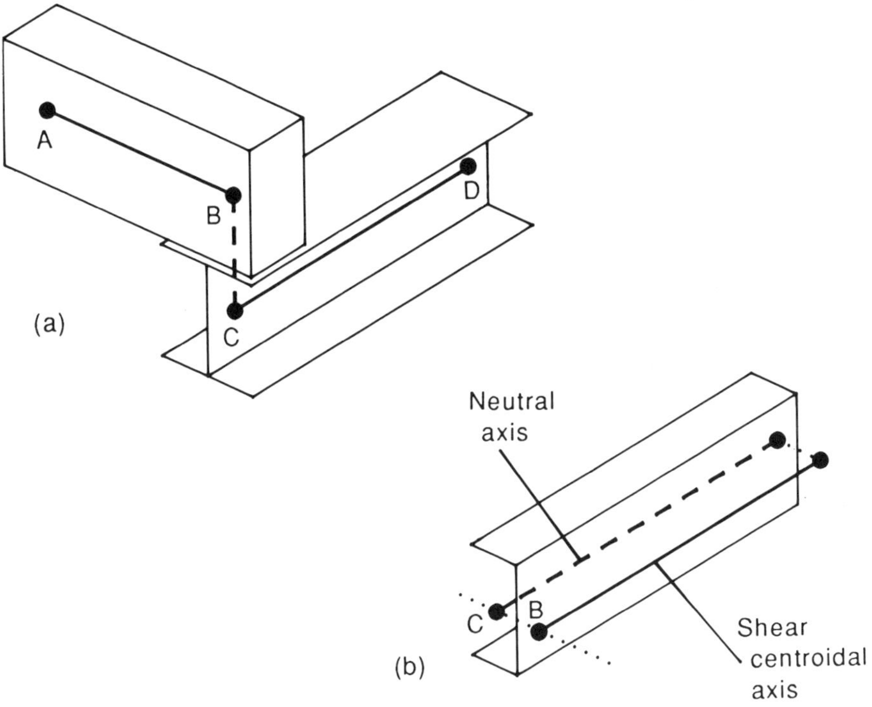

Fig. 9.14 Rigid offsets: (a) eccentrically connected members, (b) non-coincident centroid and shear centre.

all degrees of freedom must therefore be referenced to a common set of local axes at each node. This can be done by inserting an imaginary rigid connection which defines the necessary kinematic relationships between centroidal and shear-centroidal degrees of freedom. The mathematics involved is then similar to that required to define the rigid offset of Fig. 9.14(a). The procedure is relatively straightforward in both instances. Further details are not included here but a brief description of the procedure is to be found in [6], and the analysis involved in adding rigid links to each end of a planar frame element is included as an exercise at the end of this chapter (problem 12).

REFERENCES

[1] Timoshenko, S. P. (1921) On the correction for shear of the differential equation for transverse vibrations of prismatic bars. *Philosophical Magazine*, Series 6, **41**, 744–6.
[2] Dym, C. L. and Shames, I. H. (1973) *Solid Mechanics, a Variational Approach*, McGraw-Hill, New York, Chapter 4.
[3] Cowper, G. R. (1966) The shear coefficient in Timoshenko beam theory. *Journal of Applied Machanics*, **33**, 335–40.

[4] Dawe, D. J. (1978) A finite element for the vibration analysis of Timoshenko beams. *Journal of Sound and Vibration*, **60** (1), 11–20.
[5] Bakr, E. M. and Shabana, A. A. (1987) Timoshenko beams and flexible multibody system dynamics. *Journal of Sound and Vibration*, **116** (1), 89–107.
[6] Cook, R. D. (1981) *Concepts and Applications of Finite Element Analysis*, 2nd edn, John Wiley & Sons, New York, Chapter 6, pp. 158–160.

PROBLEMS

1. A beam of length $2L$ and flexural rigidity EI is fixed at one end and simply supported at the other. It carries a concentrated load P at its centre and is modelled by two simple beam elements (as shown). Show that the stiffness relationship for the discrete model is

$$\begin{bmatrix} 24a & 0 & 6aL \\ 0 & 8aL^2 & 2aL^2 \\ 6aL & 2aL^2 & 4aL^2 \end{bmatrix} \begin{bmatrix} \delta_1 \\ \delta_2 \\ \delta_3 \end{bmatrix} = \begin{bmatrix} -P \\ 0 \\ 0 \end{bmatrix},$$

where $a = EI/L^3$. Confirm that this yields the correct value of the central deflection ($7PL^3/96EI$ downwards).

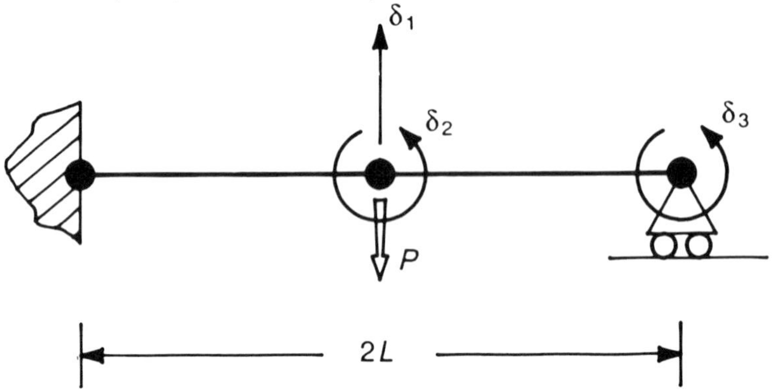

Problem 1.

2. Show that the displacement at the centre of the beam element of Fig. 9.2 is

$$w(L/2) = \frac{1}{2}w_1 + \frac{1}{8}L\theta_1 + \frac{1}{2}w_2 - \frac{1}{8}L\theta_2. \quad \text{(i)}$$

Hence show that a concentrated load P acting vertically downward at the centre of the element can be consistently modelled by applying concentrated forces and moments of magnitude $P/2$ and $PL/8$, respectively, as shown. Repeat the analysis of problem one using a single element to model the beam. Solve for the rotation at the simply supported end and obtain an estimate for the central deflection using (i) above. Compare to the exact solution. Why would you not expect this model to give an exact result?

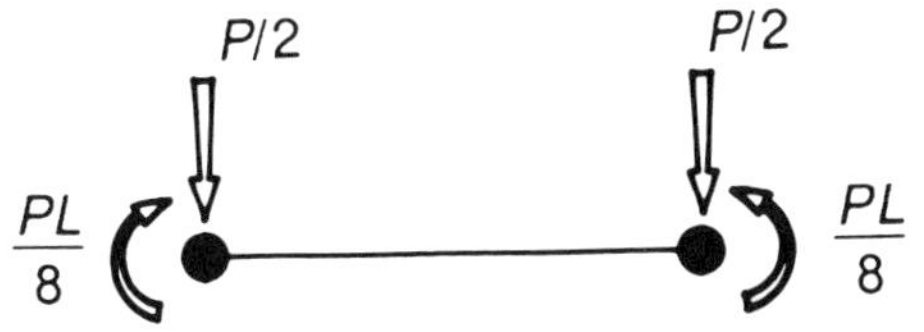

Problem 2.

3. The beam element shown in Fig. 9.2 is subject to a distributed load $q(x')$ per unit length which acts vertically downward and varies linearly from zero at the left hand end to q_0 at the right (as shown in (a) below). Show that this is consistently modelled by the nodal forces and moments shown in (b). A single element of this type is used to model a cantilever beam of length L and flexural rigidity EI subject to a linearly varying load depicted in (c). Calculate the deflection at the free end and compare with the exact solution.

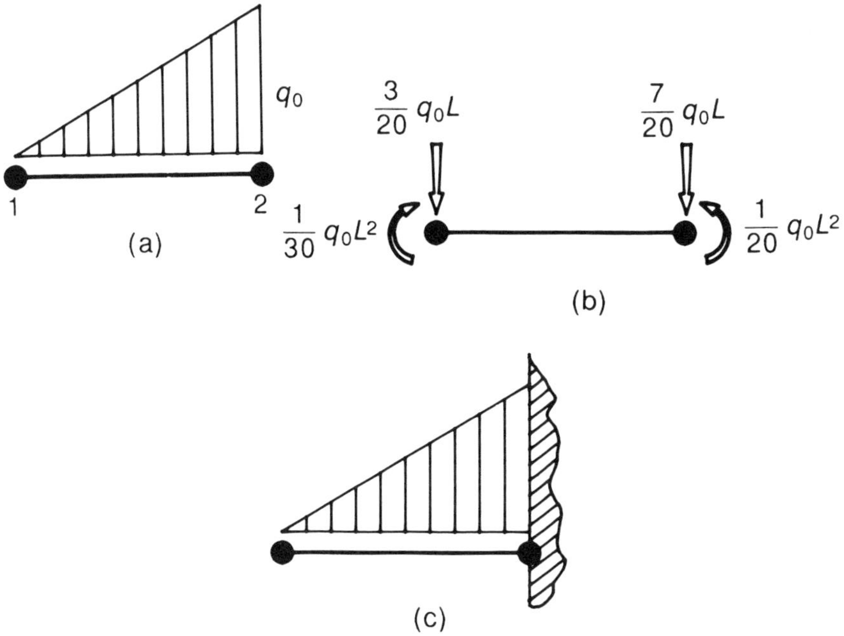

Problem 3.

4. A tapered beam element of length L is to be used to model a beam whose second moment of area varies along its length. The topology and shape functions of the element are identical to those of the element shown in

Fig. 9.2, but the second moment of area $I(x')$ varies linearly from I_1 at node 1 to I_2 at node 2. Show that the stiffness matrix for the element is

$$\boldsymbol{K}^{\mathrm{e}} = \left(\frac{E}{L^3}\right)\begin{bmatrix} 6(I_1+I_2) & (5I_1+I_2)L & -6(I_1+I_2) & (I_1+5I_2) \\ (5I_1+I_2)L & (3I_1+I_2)L^2 & -(5I_1+I_2)L & (I_1+I_2)L^2 \\ -6(I_1+I_2) & -(5I_1+I_2)L & 6(I_1+I_2) & -(I_1+5I_2) \\ (I_1+5I_2)L & (I_1+I_2)L^2 & -(I_1+5I_2)L & (I_1+3I_2)L^2 \end{bmatrix}.$$

(Suggestion: put $I(x') = I_1 + (x'/L)(I_2 - I_1)$, and evaluate the integral in equation 9.17 with $\boldsymbol{D} = [EI(x')]$.)

5. A beam of flexural rigidity EI and length $2L$ is simply supported at each end and carries a concentrated load P at its centre. The beam is modelled using two simple beam elements. The unconstrained degrees of freedom are numbered $\delta_1, \ldots, \delta_4$ as shown. Show that the stiffness relationship for the model is

$$\begin{bmatrix} 24a & 0 & -6aL & 6aL \\ 0 & 8aL^2 & 2aL^2 & 2aL^2 \\ -6aL & 2aL^2 & 4aL^2 & 0 \\ 6aL & 2aL^2 & 0 & 4aL^2 \end{bmatrix}\begin{bmatrix}\delta_1\\ \delta_2\\ \delta_3\\ \delta_4\end{bmatrix} = \begin{bmatrix}-P\\ 0\\ 0\\ 0\end{bmatrix}, \quad (a = EI/L^3).$$

Solve for δ_1 and confirm that the model yields an exact result ($PL^3/6EI$) for the central deflection (assume from symmetry that δ_2 is zero and $\delta_3 = -\delta_4$).

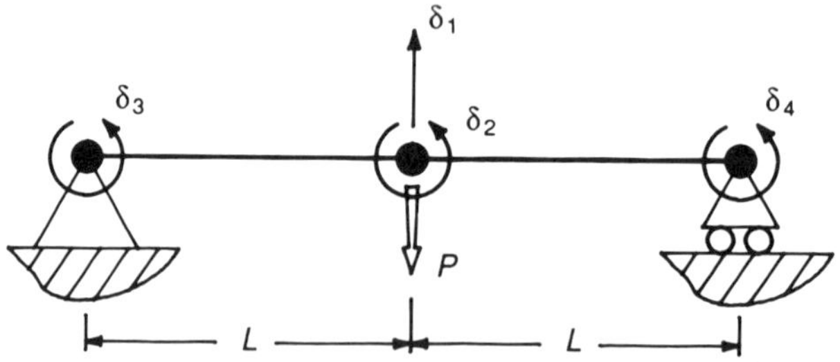

Problem 5.

6. Repeat problem five replacing the concentrated load by a uniformly distributed load q_0 acting downwards along the length of the beam. Show that the finite element model yields an exact result ($5q_0L^4/24EI$) for the central deflection. Is the solution exact at all points along the beam?

7. A beam at length $2L$ is fixed at each end and carries a vertical load P at its centre. The flexural rigidity of the beam is EI_a to the left of the load, and

EI_b to the right. The structure is modelled using two beam elements. Show that the stiffness equation for the system is

$$\begin{bmatrix} 12E(I_a + I_b)/L^3 & 6E(I_b - I_a)/L^2 \\ 6E(I_b - I_a)/L^2 & 12E(I_a + I_b)/L \end{bmatrix} \begin{bmatrix} \delta_1 \\ \delta_2 \end{bmatrix} = \begin{bmatrix} -P \\ 0 \end{bmatrix}, \quad (a = EI/L^3).$$

The degrees of freedom are numbered as in Fig. 9.4(b). Does this yield an exact or an approximate result for the central deflection?

8. A quartic beam element of length $2L$ is defined by three nodes, one at each end and one at the centre (as shown). It has five degrees of freedom: w_1, θ_1, w_2, θ_2 and w_3. Confirm that a suitable shape relationship is

$$w(\xi) = n_1(\xi)w_1 + n_2(\xi)\theta_2 + n_3(\xi)w_2 + n_4(\xi)\theta_2 + n_5(\xi)w_3,$$

where $\xi = x'/L$, and

$$n_1(\xi) = -\left(\frac{1}{2}\xi^2 + \frac{3}{4}\xi\right)(1-\xi)^2,$$

$$n_2(\xi) = -\frac{L}{4}\xi(\xi+1)(\xi-1)^2,$$

$$n_3(\xi) = -\left(\frac{1}{2}\xi^2 - \frac{3}{4}\xi\right)(1+\xi)^2,$$

$$n_4(\xi) = \frac{L}{4}\xi(\xi-1)(\xi+1)^2,$$

$$n_5(\xi) = (1-\xi^2)^2.$$

The degrees of freedom are placed in the nodal displacement vector in the order w_1, θ_1, w_2, θ_2, w_3. Show that the 5–5 term in the element stiffness matrix is $128EI/5L^3$. Use a single element of this type to model the beam of Fig. 9.4(a) and show that this model does *not* give an exact result for the central deflection. Comment on the discrepancy.

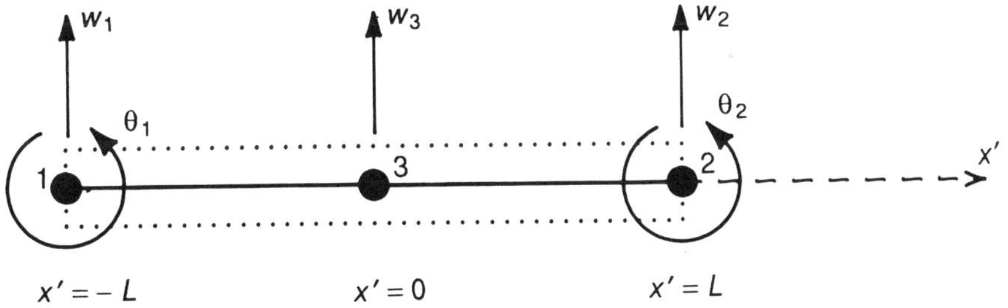

Problem 8.

9. What nodal forces and moments are required to model a uniformly distributed load q_0 acting on the upper surface of the quartic beam of problem eight? Use a single element of this type to model a beam which is built-in at both ends and carries a uniformly distributed load along its entire length. Confirm that the discrete model gives an exact solution for the deflection of the beam.

10. The five-degree of freedom Timoshenko element whose stiffness matrix is given by expression 9.35 is subject to a distributed load $q(x')$ per unit length. Show that the vector of equivalent nodal forces f_q^e, has components

$$f_q^e = \begin{bmatrix} F_1 \\ F_2 \\ F_3 \\ F_4 \\ F_4 + F_2 \end{bmatrix},$$

where $F_1 \dots, F_4$ are the forces and moments which would be applied to the four-degree of freedom, Bernoulli–Euler element to model the same load. In the case of a uniformly distributed load q_0, show that no nodal force is associated with the fifth degree of freedom γ_0 and that the equivalent nodal forces which must be applied to the Bernoulli–Euler and Timoshenko elements are therefore identical.

11. A plane framework ABCD, is rigidly fixed at A and D, and carries a vertical load P at the midpoint of the horizontal span BC. A distributed load q_0 acts on the upper surface of the same member (as shown in (a)). All members are of flexural rigidity EI and axial stiffness k. A finite element solution is sought by modelling one half of the structure and constraining the horizontal displacement and in-plane rotation to be zero on the plane of symmetry. Two frame elements are used and the unconstrained degrees of freedom are numbered $\delta_1, \dots, \delta_4$ as shown in (b). Show that the stiffness equations which determine the behaviour of the model are

$$\begin{bmatrix} k+12a & 0 & 6aL & 0 \\ 0 & k+12a & -6aL & -12a \\ 6aL & -6aL & 8aL^2 & 6aL \\ 0 & -12a & 6aL & 12a \end{bmatrix} \begin{bmatrix} \delta_1 \\ \delta_2 \\ \delta_3 \\ \delta_4 \end{bmatrix} = \begin{bmatrix} 0 \\ -\frac{1}{2} q_0 L \\ q_0 L^2/12 \\ -\frac{1}{2} P - \frac{1}{2} q_0 L \end{bmatrix},$$

where $a = EI/L^3$.

12. A planar six-degree of freedom frame element BC of length L is connected to rigid links AB and CD of length e (as shown). The displacements

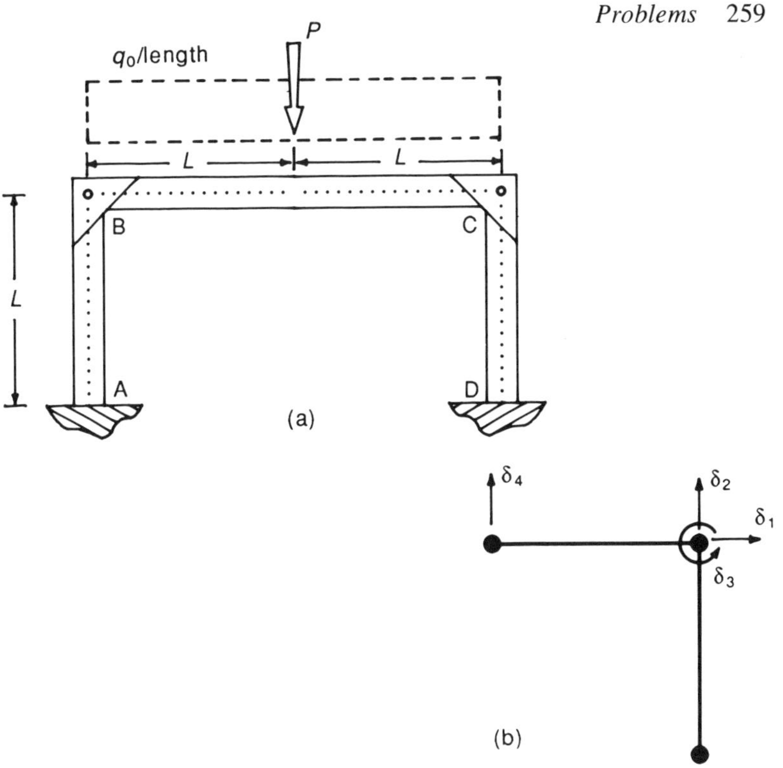

Problem 11.

and rotations at B and C are $\delta_1, \ldots, \delta_6$ and those at A and D are $d_1, \ldots, d_6$. Confirm that they are related by the transformation

$$\begin{bmatrix} \delta_1 \\ \delta_2 \\ \delta_3 \\ \delta_4 \\ \delta_5 \\ \delta_6 \end{bmatrix} = \begin{bmatrix} 1 & 0 & e & 0 & 0 & 0 \\ 0 & 1 & 0 & 0 & 0 & 0 \\ 0 & 0 & 1 & 0 & 0 & 0 \\ 0 & 0 & 0 & 1 & 0 & e \\ 0 & 0 & 0 & 0 & 1 & 0 \\ 0 & 0 & 0 & 0 & 0 & 1 \end{bmatrix} \begin{bmatrix} d_1 \\ d_2 \\ d_3 \\ d_4 \\ d_5 \\ d_6 \end{bmatrix}. \tag{i}$$

Hence show that the stiffness matrix $\boldsymbol{K}_{AD}$ of the element AD which has degrees of freedom $d_1, \ldots, d_6$ and comprises the frame BC *and* the rigid end attachments AB and CD, is given by

$$\boldsymbol{K}_{AD} = \begin{bmatrix} k & 0 & ek & -k & 0 & -ek \\ 0 & 12a & 6aL & 0 & -12a & 6aL \\ ek & 6aL & 4aL^2 + e^2k & -ek & -6aL & 2aL^2 - e^2k \\ -k & 0 & -ek & k & 0 & ek \\ 0 & -12a & -6aL & 0 & 12a & -6aL \\ -ek & 6aL & 2aL^2 - e^2k & ek & -6aL & 4aL^2 + e^2k \end{bmatrix}$$

where $a = EI/L^3$ and $k = EA/L$. (Suggestion: write transformation (i) as, $\boldsymbol{d}_{BC} = \boldsymbol{T}\,\boldsymbol{d}_{AD}$, and show that the strain energy of the frame element BC, given by $\frac{1}{2}\boldsymbol{d}_{BC}^{T}\boldsymbol{K}_{BC}\,\boldsymbol{d}_{BC}$ where $\boldsymbol{K}_{BC}$ is stiffness matrix of expression 9.45, can be rewritten as, $\frac{1}{2}\boldsymbol{d}_{AD}^{T}\boldsymbol{K}_{AD}\,\boldsymbol{d}_{AD}$, where $\boldsymbol{K}_{AD} = \boldsymbol{T}^{T}\boldsymbol{K}_{BC}\,\boldsymbol{T}$. Evaluate to obtain the required result.)

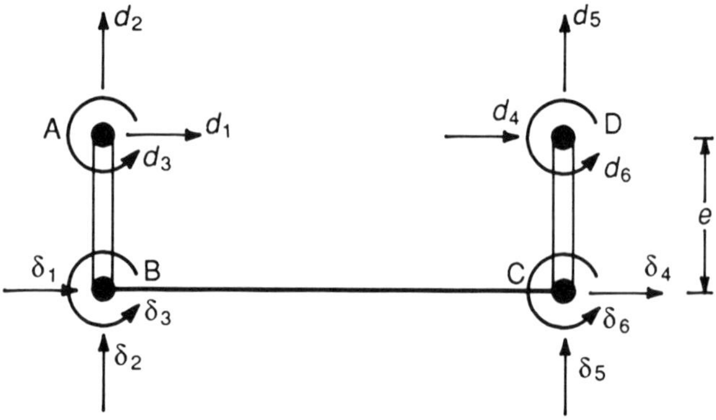

Problem 12.

10

Plate and shell elements

10.1 INTRODUCTION

The formulation of finite elements for plates and shells closely parallels that for beams and frames. Plate elements, which carry lateral loads in bending, are two-dimensional analogues of the beam elements described in sections 9.2 and 9.3. Shell elements, which are capable of sustaining loads both in bending and in in-plane (membrane) action, are analogous to the frame elements of section 9.4.

Finite elements for plates and shells were among the earliest to be developed [1–3] and yet today are still the subject of intense research. The continued interest in these elements after three decades of development, is an interesting commentary on the unsatisfactory nature of many of the more straightforward or 'obvious' formulations. There is, in fact, no such thing as an entirely straightforward plate element. The compatibility requirements, in particular, are much more difficult to enforce in plate and shell formulations than in the continuum problem. This has led to the development of non-conforming elements which do not satisfy the obvious compatibility criteria, but still produce convergent solutions. The same compatibility problem has led also to elements which are based on variational or residual principles quite different to those which have been adopted as a basis for the continuum formulations of previous chapters. The ingenuity which has been expended in resolving the difficulties encountered in plate formulations has inevitably spilled over into other areas of finite element application, and in this sense the continuing development of plate elements still lies at the forefront of finite element research, even though a large number of reliable and accurate elements are now available for routine calculations.

The diversity of plate formulations in common use is enormous. This renders any discussion of them at an introductory level necessarily selective. Discussion is limited here to simple examples of a few major categories:

(a) Kirchhoff displacement elements;
(b) Mindlin elements;
(c) discrete Kirchhoff elements.

Specific elements of each type will be discussed in the remainder of this chapter and will be used to illustrate the essential features of each formulation.

Topology	Ref.	Description	Interpolation
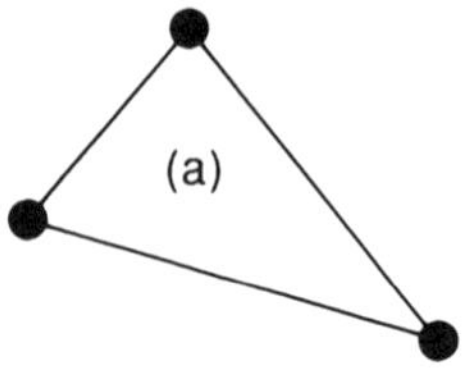	[3]	Kirchhoff displacement element, nonconforming	w — incomplete cubic
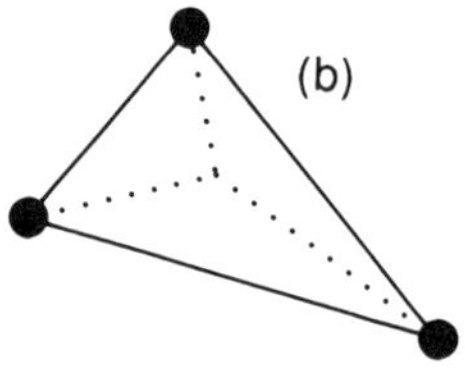	[2]	Kirchhoff displacement element, conforming	w — cubic, three subdomains
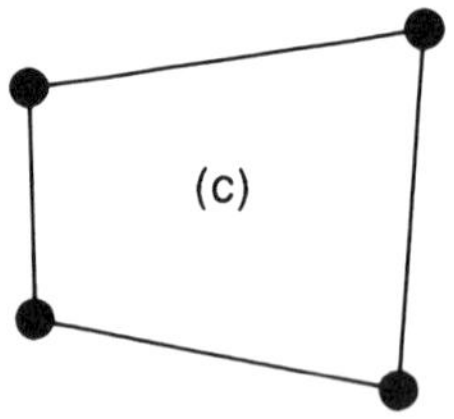	[4]	Mindlin rectangle	w, θ_x, θ_y — bilinear
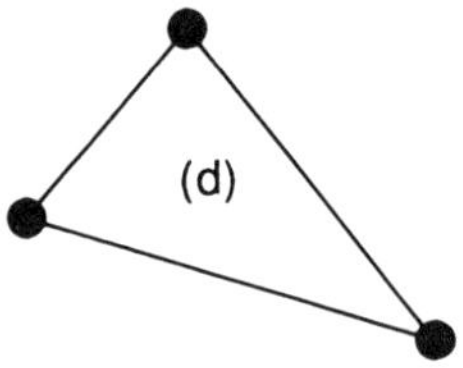	[5]	Discrete Kirchhoff triangle	w — cubic $\theta_x \theta_y$ — quadratic

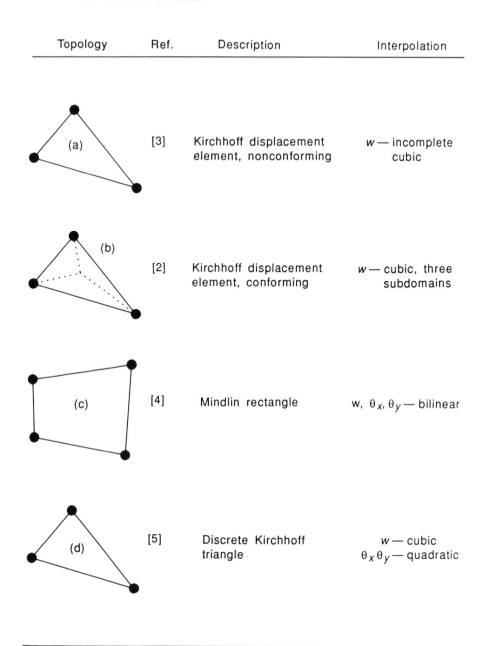

Note: all nodes have three degrees of freedom, w, θ_x and θ_y.

Fig. 10.1 *A selection of plate bending elements.*

The elements to be discussed are listed with brief, descriptive comments in Fig. 10.1. The sequence progresses roughly from early (simple) formulations to later (more complex) ones. It is in no sense complete and excludes all elements which are based on principles other than the principle of minimum total energy. 'Hybrid' and 'mixed' formulations, in particular, are significant omissions. Readers who seek to remedy this deficiency will find a simple introduction to these formulations in [6] and a more comprehensive treatment in [7]

10.2 KIRCHHOFF DISPLACEMENT ELEMENTS, THE GENERAL FORMULATION

The Kirchhoff displacement approach is the most straightforward of the thin plate formulations. Elements formulated in this way are based on a single interpolation for the lateral displacement of the neutral plane. Shear strain energy is ignored and the resulting elements are direct analogues of the simple beam element of section 9.2.

10.2.1 Geometry and notation

In all that follows, the neutral plane of the plate is taken to lie in the x–y plane of a right-handed cartesian coordinate system (Fig. 10.2). The lateral displacement of the neutral plane in the direction of z is denoted by w. As in the beam formulations of Chapter 9, the subscript '0' used in Chapter 2 to distinguish between the displacement of the neutral plane and displacements at other points within the plate has been removed. Rotations of the neutral plane about the x and y-axes are defined as θ_x and θ_y, their directions being determined by the right hand rule. Some care must be taken here to distinguish between 'right handed' rotations of the neutral plane and the slopes, $\partial w/\partial x$ and $\partial w/\partial y$. The relationship between these quantities is (Fig. 10.2)

$$\theta_x = \frac{\partial w}{\partial y} \quad \text{and} \quad \theta_y = -\frac{\partial w}{\partial x} . \tag{10.1}$$

Either θ_x and θ_y or $\partial w/\partial x$ and $\partial w/\partial y$ can be used as nodal variables. The choice of which pair to use varies from author to author and can cause confusion if a convention is not clearly established. We will use θ_x and θ_y.

10.2.2 Compatibility requirements for a Kirchhoff element

A major problem in the formulation of a Kirchhoff plate element is the difficulty encountered in satisfying compatibility requirements at inter-element boundaries. These are more demanding than in the continuum case, for reasons

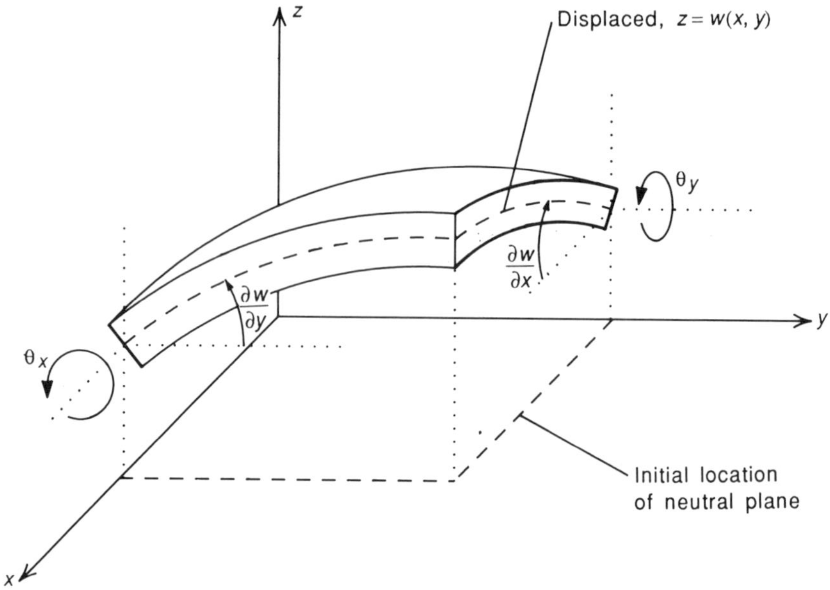

Fig. 10.2 *The deformation of a thin plate.*

touched upon already in relation to beam elements. That is to say, continuity of displacement at *all* points above and below the neutral plane, requires not only that w is continuous but that the derivatives of w with respect to x and y are continuous as well. Otherwise, physical discontinuities can occur which are two-dimensional versions of those illustrated in Fig. 9.3(a). In the beam formulation of the previous chapter, compatibility was assured by selecting θ, the slope of the neutral axis, as an additional nodal variable. This ensured that the rotation of the neutral axis took the same value in adjacent elements, guaranteeing a smooth transition from one element to another (Fig. 9.3(b)). The same philosophy can be applied to the formulation of plate elements by selecting θ_x and θ_y as nodal variables. This is less immediately successful however in ensuring full, inter-element compatibility. It does, admittedly, guarantee that the deformed surface is 'smooth' *at* the nodes themselves — since the values of θ_x and θ_y are then common to adjacent elements — but offers no assurance of a smooth transition at all points along a common boundary. This is illustrated in Fig. 10.3, which shows the displaced shape of the neutral plane on either side of a boundary between two rectangular plate elements. The terms w_i, θ_{xi} and θ_{yi} are specified as nodal values at each end of the boundary and the displaced neutral plane must clearly be smooth *at* the nodes themselves where the derivatives $\partial w/\partial x\ (=\theta_y)$ and $\partial w/\partial y\ (=-\theta_x)$ take common values in both elements. Let us assume also that the displacement itself is continuous across the interface, as indicated by the continuity of the displaced surface in Fig. 10.3. Neither of these conditions is any guarantee however that the slope of the surface normal to the interface is also continuous. A 'kink' is therefore permitted in the displaced surface as indicated in Fig. 10.3. A deformation of this type violates continuity of displacement at points above and below the neutral plane and is clearly unacceptable in terms of the full compatibility requirements.

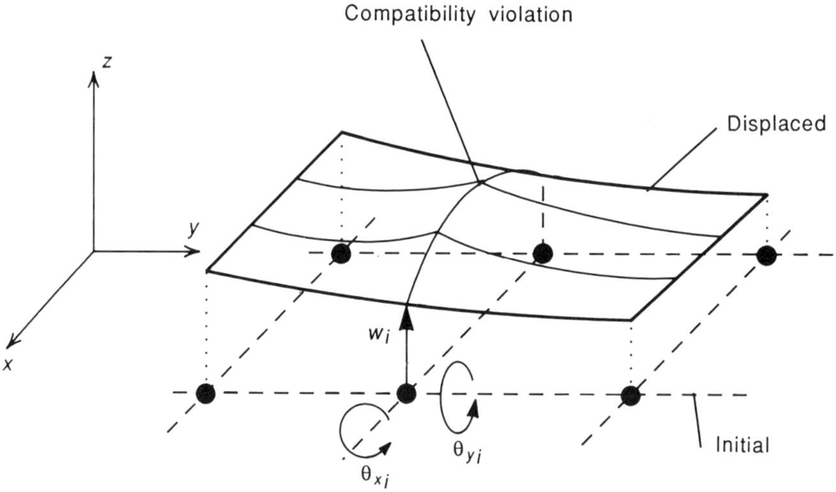

Fig. 10.3 *Incompatible matching of adjacent plate elements.*

Full compatibility is, in fact, extremely difficult to achieve using polynomial interpolation with simple nodal variables (displacements and rotations). Fortunately, elements which nearly satisfy the compatibility requirements are relatively easy to formulate and some of these perform very well. They are 'nonconforming', in the terminology of Chapter 7. Specific examples of successful and not so successful conforming and nonconforming Kirchhoff triangles are presented in sections 10.3 and 10.4.

10.2.3 Some basic relationships

Before looking at specific Kirchhoff elements, it is useful to review some of the relationships of Kirchhoff plate theory which were established in Chapter 2, and to recast them in suitable matrix form for finite element analysis.

Kirchhoff theory is based on the assumption that the bending stresses within the plate are determined entirely by the curvatures of the neutral plane. In the most general case, these are supplemented by 'membrane' stresses due to in-plane loads. These in-plane effects are disregarded in our initial formulation of the plate element, but are re-introduced later when shell elements are discussed in section 10.8.

The bending stresses in the plate vary linearly with depth and are statically equivalent to distributed bending moments M_x, M_y and M_{xy} (Fig. 2.15(b)). These in turn are related to the curvatures of the neutral plane, $-\partial^2 w/\partial x^2$, $-\partial^2 w/\partial y^2$ and $-\partial^2 w/\partial x\partial y$, through the moment–curvature equations 2.45. It is convenient to rewrite these now in matrix form as

$$\begin{bmatrix} M_x \\ M_y \\ M_{xy} \end{bmatrix} = \begin{bmatrix} D & \nu D & 0 \\ \nu D & D & 0 \\ 0 & 0 & D(1-\nu)/2 \end{bmatrix} \begin{bmatrix} -\partial^2 w/\partial x^2 \\ -\partial^2 w/\partial y^2 \\ -2\partial^2 w/\partial y\partial x \end{bmatrix}, \tag{10.2a}$$

where $D = Et^3/[12(1-\nu^2)]$. In the notation of Chapter 4 this relationship can also be written

$$\boldsymbol{s} = \boldsymbol{D}\boldsymbol{e}, \tag{10.2b}$$

where $\boldsymbol{s}$ and $\boldsymbol{e}$ are generalized stress and strain vectors which contain moments and curvatures, respectively, and where $\boldsymbol{D}$ is a constant matrix. The strain energy of the plate, obtained from expression 3.32 of Chapter 3 with the membrane contribution omitted, is given by

$$\text{SE/area} = \frac{1}{2}\left[M_x\left(-\frac{\partial^2 w}{\partial x^2}\right) + M_y\left(-\frac{\partial^2 w}{\partial y^2}\right) + M_{xy}\left(-2\frac{\partial^2 w}{\partial x\partial y}\right)\right] = \frac{1}{2}\boldsymbol{s}^{\mathrm{T}}\boldsymbol{e}. \tag{10.3}$$

Equations 10.2b and 10.3 now mimic the stress–strain relationship and strain energy density of an elastic continuum in much the same way that equations 9.12 and 9.13 mimic the same relationships in the case of a beam. Once again, this is a useful notational device since it allows us to apply the continuum expressions derived in Chapter 4 to plate formulations, provided that stresses and strains are reinterpreted as moments and curvatures.

10.2.4 Element stiffness matrix

A general expression for the stiffness matrix for a Kirchhoff plate element is now derived. The first step is to select an interpolation for the lateral displacement of the neutral plane. Without specifying a precise topology for the element, let us assume that it has p degrees of freedom, $\delta_1, \delta_2, \ldots, \delta_p$. These will generally include rotations as well as translations; indeed, in particular instances, they may also include higher order quantities, such as nodal curvatures. Whatever the nature of the nodal variables, the interpolation within the element will be defined by shape functions $n_1(x, y), n_2(x, y), \ldots, n_p(x, y)$, where

$$w(x, y) = [n_1(x, y), n_2(x, y), \ldots, n_p(x, y)] \begin{bmatrix} \delta_1 \\ \delta_2 \\ \vdots \\ \delta_p \end{bmatrix} = \boldsymbol{N}^{\mathrm{e}}\boldsymbol{d}^{\mathrm{e}}. \tag{10.4}$$

The components of a generalized 'strain' vector, $\boldsymbol{e}$, are then obtained by differentiating equation 10.4 twice with respect to x and y to give consistent expressions for the curvatures, $-\partial^2 w/\partial x^2$, $-\partial^2 w/\partial y^2$ and $-2\partial^2 w/\partial x\partial y$. These define a discrete form of the 'strain–displacement' relationship (in reality, it is now a curvature–displacement relationship) which can be written in the notation of Chapter 4 as

$$\boldsymbol{e} = \boldsymbol{B}^{\mathrm{e}}\,\boldsymbol{d}^{\mathrm{e}}, \qquad (10.5\mathrm{a})$$

where $\boldsymbol{B}^{\mathrm{e}}$, has components

$$\boldsymbol{B}^{\mathrm{e}} = \begin{bmatrix} -\partial^2 n_1/\partial x^2 & -\partial^2 n_2/\partial x^2 & \cdots & -\partial^2 n_p/\partial x^2 \\ -\partial^2 n_1/\partial y^2 & -\partial^2 n_2/\partial y^2 & \cdots & -\partial^2 n_p/\partial y^2 \\ -2\partial^2 n_1/\partial x\partial y & -2\partial^2 n_2/\partial x\partial y & \cdots & -2\partial^2 n_p/\partial x\partial y \end{bmatrix}. \qquad (10.5\mathrm{b})$$

The strain energy of the element is obtained by integrating expression 10.3 over the area of the element. Substitution of equations 10.5(b) and 10.2 then gives the familiar result

$$U^{\mathrm{e}} = \tfrac{1}{2}\,\boldsymbol{d}^{\mathrm{eT}}\,\boldsymbol{K}^{\mathrm{e}}\,\boldsymbol{d}^{\mathrm{e}}, \qquad (10.6)$$

where

$$\boldsymbol{K}^{\mathrm{e}} = \int_A \boldsymbol{B}^{\mathrm{eT}}\boldsymbol{D}\,\boldsymbol{B}^{\mathrm{e}}\,\mathrm{d}A, \qquad (10.7)$$

where A is the area of the element. This expression is virtually identical to that for the stiffness matrix of a continuum element (equations 4.39) except that the integration is now performed over the area of the neutral plane rather than over the physical volume of the element, and the matrices $\boldsymbol{B}^{\mathrm{e}}$ and $\boldsymbol{D}$ have been defined in a somewhat different manner.

10.2.5 Distributed loads, equivalent forces and moments

The potential energy, V^{e}, of a distributed load $q(x, y)$ per unit area acting in the direction of z is

$$V^{\mathrm{e}} = -\int_A w(x, y)\, q(x, y)\, \mathrm{d}A.$$

Substitution of the shape relationship 10.4 and some rearrangement of terms then yields the familiar result

$$V^{\mathrm{e}} = -\,\boldsymbol{d}^{\mathrm{eT}}\boldsymbol{f}_q^{\mathrm{e}},$$

where the equivalent nodal force vector, $\boldsymbol{f}_q^{\mathrm{e}}$, is given by

$$\boldsymbol{f}_q^{\mathrm{e}} = \int_A \boldsymbol{N}^{\mathrm{eT}}\, q(x, y)\, \mathrm{d}A. \tag{10.8}$$

Again, this is similar to the analogous expression for a continuum element, except that the integration is performed over the area of the element rather than its volume. Also, since the degrees of freedom now include rotations as well as translations, the components of the nodal 'force' vector, $\boldsymbol{f}_q^{\mathrm{e}}$, will include nodal moments as well as nodal forces.

10.3 NON-CONFORMING TRIANGLES

Two triangular elements with ten and nine degrees of freedom, respectively, are shown in Fig. 10.4. They are based on complete (element (a)) and incomplete (element (b)) cubic interpolation. Both elements are nonconforming to the extent that neither satisfies continuity of normal slope at inter-element boundaries.

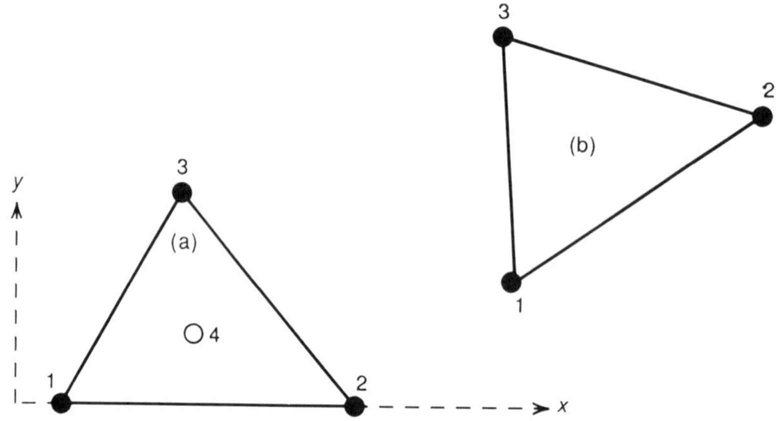

Fig. 10.4 *Geometry and topology of nonconforming plate elements: (a) complete cubic, (b) incomplete cubic (BCIZ). Key* (•)ω_i, θ_{x_i}, θ_{y_i}; (○)ω_i.

Element (a), although quite simple to formulate, is profoundly unsatisfactory in terms of its overall performance. However, the completeness of its interpolating function is a useful feature in discussing the compatibility requirements for elements of this type. The element has three degrees of freedom at each vertex, w_i, θ_{xi} and θ_{yi} ($i = 1, 2, 3$) and one at the central node (w_4). A complete cubic polynomial with ten unknown constants is used as an interpolating function. This can be written either in terms of cartesian coordinates, x and y, or natural coordinates L_1, L_2 and L_3. We will use cartesian coordinates here. We will assume also that the 1–2 side defines the x-axis of the cartesian system. Since the interpolating polynomial is complete, and since a complete polynomial of any order has the same algebraic form in *any* orthogonal coordinate system, no loss of generality is incurred by taking the 1–2 side to lie parallel to a particular coordinate direction. The interpolation for w then takes the form,

$$w(x, y) = \alpha_1 + (\alpha_2 x + \alpha_3 y) + (\alpha_4 x^2 + \alpha_5 xy + \alpha_6 y^2) + (\alpha_7 x^3 + \alpha_8 x^2 y + \alpha_9 xy^2 + \alpha_{10} y^3), \quad (10.9)$$

and the rotations θ_x and θ_y are given by

$$\theta_x = \frac{\partial w}{\partial y} = \alpha_3 + \alpha_5 x + 2\alpha_6 y + \alpha_8 x^2 + 2\alpha_9 xy + 3\alpha_{10} y^2, \quad (10.10a)$$

$$\theta_y = -\frac{\partial w}{\partial x} = -(\alpha_2 + 2\alpha_4 x + \alpha_5 y + 3\alpha_7 x^2 + 2\alpha_8 xy + \alpha_9 y^2). \quad (10.10b)$$

The unknown constants α_i $(i = 1, \ldots, 10)$ are obtained by equating w, θ_x and θ_y to the ten nodal values of w_i, θ_{xi} and θ_{yi}. The specific shape functions which are generated in this way need not concern us here (they are given as expressions 12.35 in [6] if required). We will confine our attention solely to the question of whether the interpolation so defined satisfies inter-element compatibility along the 1–2 side. Because of the peculiar geometry of the element, the interpolation for w along this side is obtained by setting y equal to zero in equation 10.9 to give

$$w(x, 0) = \alpha_1 + \alpha_2 x + \alpha_4 x^2 + \alpha_7 x^3.$$

Therefore w varies cubically with x. Moreover, w itself and its derivative with respect to $x(= -\theta_y)$ are specified at nodes 1 and 2 so that a unique cubic variation is defined at all points along the interface. Continuity of displacement is therefore assured if the element is matched to another of the same type. The same is not true, however, of the normal slope $\partial w/\partial y\ (= \theta_x)$. An expression for the variation of this quantity along the 1–2 side is obtained by setting $y = 0$ in equation 10.10(a) to give

$$\theta_x = \alpha_3 + \alpha_5 x + \alpha_8 x^2. \quad (10.11)$$

θ_x therefore varies quadratically with x but is specified only by *two* nodal values, θ_{x1} and θ_{x2}. Since a unique specification of a quadratic function requires *three* values, we conclude that θ_x is not uniquely determined by the nodal parameters at nodes 1 and 2. When two elements of this type are joined together, therefore, the normal derivative is not necessarily continuous across the interface and the displacement field is incompatible in the manner shown in Fig. 10.3. The same argument applies to the other two sides of the element.

At first sight, it appears that this deficiency could be remedied by specifying an extra 'nodal' value of $\partial w/\partial n$ along each side. This could be done, for example, by treating the normal derivative of w at the midpoint of each side as an additional degree of freedom. In doing so, however, we introduce three new variables, and therefore require three additional coefficients in the interpolating polynomial. This, in turn, introduces quartic terms into equation 10.9 and destroys the unique cubic variation of w along each side. Any attempt to produce a compatible element by introducing additional translations and/or rotations in this way results in a circular argument of this type and is similarly self-defeating. A compatible element can be formed but only by increasing the number of coefficients to 21 and introducing curvatures as nodal degrees of freedom. Such an element is described in [8]. This 'quintic' element, although accurate, cannot readily be integrated with conventional structural elements and is not widely used.

Incompatibility is not in itself an obstacle to satisfactory performance and nonconforming elements are quite capable of producing accurate and convergent results. Unhappily, the complete cubic element described above does not fall into this category. It generates extremely poor solutions for even the simplest of test cases and is quite unsatisfactory for practical calculations.

A more effective, non-conforming element — element (b) of Fig. 10.4 — is obtained by discarding the fourth node at the element centroid and developing an element with nine degrees of freedom. Various incomplete polynomials with nine rather than ten unknown coefficients can then be used. The most successful of these gives the 'BCIZ' triangle, an element whose acronym derives from the names of its inventors Bazeley, Cheung, Irons and Zienkiewicz [3]. It is non-conforming in exactly the same way as the complete cubic element described above (continuity of normal slope is not achieved along element boundaries) does not (quite) pass the patch test, but nonetheless performs well and can be regarded as a reliable element for straightforward plate calculations.

The interpolating function in the BCIZ element is a particular, incomplete polynomial in the area coordinates, L_1, L_2 and L_3. It takes the form,

$$w(L_1, L_2, L_3) = \beta_1 L_1 + \beta_2 L_2 + \beta_3 L_3 + \beta_4 (L_1^2 L_2 + \tfrac{1}{2} L_1 L_2 L_3) + \beta_5 (L_1^2 L_3 + \tfrac{1}{2} L_1 L_2 L_3) + \ldots + \beta_9 (L_3^2 L_2 + \tfrac{1}{2} L_1 L_2 L_3). \quad (10.12)$$

The logic behind this particular grouping of terms is explained in more detail in [9]. Briefly, the first three terms represent a rigid body displacement, and the remaining six combine to represent arbitrary states of constant curvature. This latter combination is significant, ensuring that completeness at least is satisfied if not compatibility.

The coefficients β_i are evaluated by equating, w, $\partial w/\partial y$ and $-\partial w/\partial x$ to nodal values of w, θ_x and θ_y. Upon substitution into equation 10.12 and rearrangement of terms, this yields the shape relationship:

$$w(L_1, L_2, L_3) = n_1(L_1, L_2, L_3)w_1 + n_2(L_1, L_2, L_3)\theta_{x1} + n_3(L_1, L_2, L_3)\theta_{y1} + \ldots + n_9(L_1, L_2, L_3)\theta_{y3},$$

where the first three shape functions are

$$\begin{aligned} n_1(L_1, L_2, L_3) &= L_1 + L_1^2L_2 + L_1^2L_3 - L_2^2L_1 - L_3^2L_1, \\ n_2(L_1, L_2, L_3) &= b_2(L_1^2L_3 + \tfrac{1}{2}L_1L_2L_3) + b_3(L_1^2L_2 + \tfrac{1}{2}L_1L_2L_3), \\ n_3(L_1, L_2, L_3) &= c_2(L_1^2L_3 + \tfrac{1}{2}L_1L_2L_3) - c_3(L_1^2L_2 + \tfrac{1}{2}L_1L_2L_3), \end{aligned} \tag{10.13}$$

and the remainder follow by cyclical rotation of the subscripts 1, 2 and 3 (b_i and c_i are the geometrical constants defined in Chapter 6 (expressions 6.5)). The components of the strain–displacement matrix $\boldsymbol{B}^e$ are readily calculated as second derivatives of these expressions, and the stiffness matrix is then obtained by evaluating the integral in equation 10.7. This can be done numerically or analytically. Analytic calculation of the 1–1 term for an element of this type is included as an exercise at the end of this chapter (problem 10).

It would be reassuring to be able to report that the BCIZ element passed the patch test in spite of its failure to satisfy the full, compatibility requirements. Unfortunately, this is not the case, although the patch test is passed for all meshes formed by parallel and equally spaced lines. Although not absolutely convergent, the element generally performs well, however, giving solutions which are often more accurate than those obtained from simple conforming elements of the same topology.

10.4 A CONFORMING KIRCHHOFF TRIANGLE

Two approaches may be used to generate conforming, nine-degree of freedom triangles by the direct Kirchhoff approach. Both result in shape functions which have nonunique second derivatives at the vertices. The first involves the use of rational functions, rather than polynomials, and is discussed in Chapter 1 of [9]. The second, which is discussed here, is associated with the names of Clough and Tocker [2] and involves matched polynomial interpolation within separate subdomains of the element.

The topology of the Clough–Tocker triangle is shown in Fig. 10.5. It is similar in external appearance to the nonconforming triangle of Fig. 10.4(b), but is divided internally into three fixed subdomains. Within each subdomain, a complete cubic interpolation is assumed for the displacement $w(x, y)$. Let us denote these by $w_a(x, y)$, $w_b(x, y)$ and $w_c(x, y)$ in subdomains (a), (b) and (c),

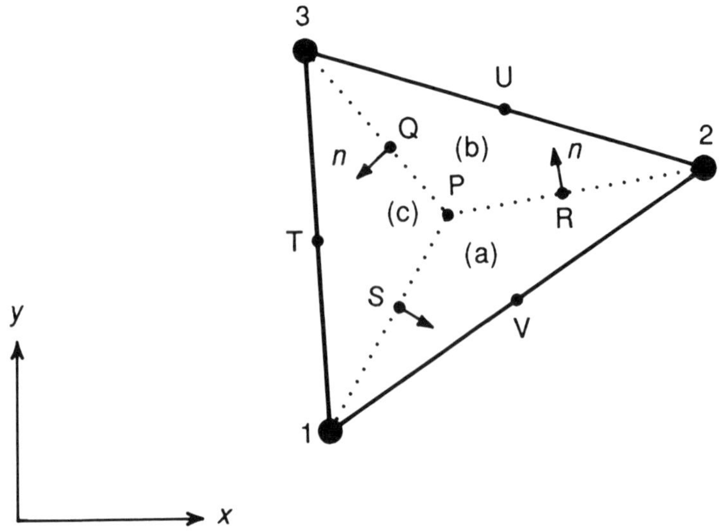

Fig. 10.5 *The Clough-Tocker, conforming triangle.*

respectively. Each involves ten unknown coefficients, giving a total of 30 unknown constants which must be eliminated in favour of the nine nodal degrees of freedom, w_i, θ_{xi} and θ_{yi} $(i = 1, 2, 3)$. This is done by forming and solving 30 constraint equations. These are obtained in the following way:·

(i) At node 1: $(w_a, \partial w_a/\partial y, -\partial w_a/\partial x)$ and $(w_c, \partial w_c/\partial y, -\partial w_c/\partial x)$ are equated to $(w_1, \theta_{x1}, \theta_{y1})$, (6 equations).

(ii) At node 2: $(w_a, \partial w_a/\partial y, -\partial w_a/\partial x)$ and $(w_b, \partial w_b/\partial y, -\partial w_b/\partial x)$ are equated to $(w_2, \theta_{x2}, \theta_{y2})$, (6 equations).

(iii) At node 3: $(w_b, \partial w_b/\partial y, -\partial w_b/\partial x)$ and $(w_c, \partial w_c/\partial y, -\partial w_c/\partial x)$ are equated to $(w_3, \theta_{x3}\ \theta_{y3})$, (6 equations).

(iv) At the central point P: $(w_a, \partial w_a/\partial x, \partial w_a/\partial y)$ is equated to $(w_b, \partial w_b/\partial x, \partial w_b/\partial y)$ and $(w_c, \partial w_c/\partial x, \partial w_c/\partial y)$, (6 equations).

(v) At the subdomain interfaces: $\partial w_a/\partial n$ is equated to $\partial w_b/\partial n$ at R, $\partial w_b/\partial n$ is equated to $\partial w_c/\partial n$ at Q, $\partial w_c/\partial n$ is equated to $\partial w_a/\partial n$ at S (3 equations).

The above 27 constraints — three short of the required 30 — ensure that the three representations are compatibly matched along each internal interface (the conditions imposed at the ends of each internal division *plus* the normal derivative condition at the midpoint determine unique cubic and quadratic variations of w and $\partial w/\partial n$, respectively). Also, each subdomain representation is consistent with the nodal values of w, θ_x and θ_y at the vertices of the triangle. We still require three additional equations or constraints. These are formed in either of the following ways:

(vi) by defining the normal derivatives of w at midside nodes T, U and V to be additional degrees of freedom for the element and then equating them to the normal derivatives of w_a, w_b and w_c to provide three additional equations,

or

(vii) by constraining the normal derivatives at T, U and V to be the average of their values at the ends of each side, also providing three additional relationships.

The first of these approaches gives an element with twelve degree of freedom; three at each vertex, and one (the normal derivative of w) at the midpoint of each side. The normal slope of w is then uniquely defined as a quadratic function along each side. The second approach gives a nine-degree of freedom element which has a unique linear variation of the normal slope along each side. In both cases, the cubic variation of displacement is uniquely specified along each edge and both elements are therefore fully conforming. The second approach is generally more popular, being somewhat simpler to implement since it assigns the same number of degrees of freedom (three) to all nodes. Neither element is particularly accurate, although the first is a good deal better than the second (further comments, section 10.7).

10.5 MINDLIN PLATE ELEMENTS

10.5.1 Theoretical considerations

Kirchhoff plate theory can be modified to accommodate shear deformation in much the same way that Bernoulli–Euler beam theory is adjusted to give the Timoshenko beam (section 9.3). In both instances, the requirement that fibres normal to the midsurface remain normal after deformation is no longer enforced. In the case of a Mindlin plate, these fibres are assumed to remain straight, but shear deformation is permitted in the x–z and y–z planes. This is illustrated in Fig. 10.6. The rotations of such a fibre are denoted by the variables ϕ and ψ, no longer synonymous with the slopes $\partial w/\partial x$ and $\partial w/\partial y$. Nonzero shear strains $\gamma_{xz}\,(=\partial w/\partial x - \phi)$ and $\gamma_{yz}(=\partial w/\partial y - \psi)$ are therefore generated and must be included in our estimate of the strain energy of the plate. Note that the rotation ϕ is no longer right-handed with respect to the y-axis. This discrepancy can be accommodated, if required, within the 'right-handed' sign convention established earlier for rotational degrees of freedom, by using $(-\phi)$ and ψ as nodal variables rather than ϕ and ψ.

The effect of the above modification to the assumed displacement field is to substitute ϕ and ψ for $\partial w/\partial x$ and $\partial w/\partial y$ in many of the relationships derived in Chapter 2. The curvatures $\partial^2 w/\partial x^2$, $\partial^2 w/\partial y^2$ and $\partial^2 w/\partial x\partial y$, for example, are replaced by $\partial\phi/\partial x$, $\partial\psi/\partial y$ and $(\partial\phi/\partial y) + (\partial\psi/\partial x)$; the moment curvature relationship 10.2(a) becomes

$$\begin{bmatrix} M_x \\ M_y \\ M_{xy} \end{bmatrix} = \begin{bmatrix} D & \nu D & 0 \\ \nu D & D & 0 \\ 0 & 0 & D(1-\nu)/2 \end{bmatrix} \begin{bmatrix} -\partial\phi/\partial x \\ -\partial\psi/\partial y \\ -\partial\psi/\partial y - \partial\psi/\partial x \end{bmatrix}. \tag{10.14a}$$

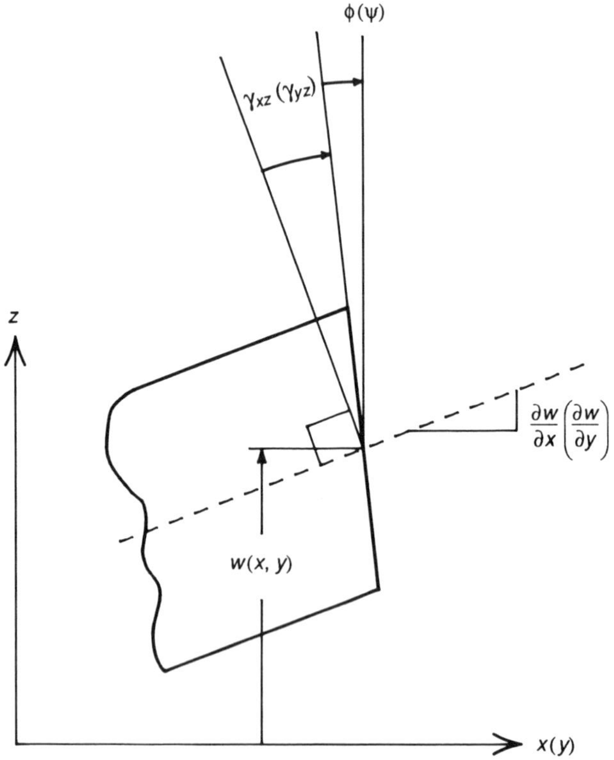

Fig. 10.6 *Deformation of Mindlin plate in the x–z plane (corresponding quantities in the y–z plane show in parentheses).*

This may still be written, using the previous notation, as

$$\boldsymbol{s} = \boldsymbol{D}\,\boldsymbol{e}, \tag{10.14b}$$

provided that the 'generalized' strain vector, $\boldsymbol{e}$, is now defined as

$$\boldsymbol{e} = \begin{bmatrix} -\partial\phi/\partial x \\ -\partial\psi/\partial y \\ -\partial\phi/\partial y - \partial\psi/\partial x \end{bmatrix}. \tag{10.14c}$$

With this modification, the bending strain energy of the plate is given once again by the inner product of $\boldsymbol{s}$ and $\boldsymbol{e}$, that is

$$\text{SE(bending)/area} = \frac{1}{2}\left[M_x\left(-\frac{\partial\phi}{\partial x}\right) + M_y\left(-\frac{\partial\psi}{\partial y}\right) + M_{xy}\left(-\frac{\partial\phi}{\partial y} - \frac{\partial\psi}{\partial x}\right)\right] = \frac{1}{2}\boldsymbol{s}^{\mathrm{T}}\boldsymbol{e}. \tag{10.15}$$

This quantity is supplemented in the Mindlin formulation by the strain energy due to shear. This contributes an additional term,

$$\text{SE(shear)/area} = \tfrac{1}{2}\,Gt\kappa(\gamma_{xz}^2 + \gamma_{yz}^2), \tag{10.16}$$

where κ is a shear coefficient, customarily taken to be 6/5, and t is the thickness of the plate. It is convenient at this stage to place the shear strains in a separate vector, $\boldsymbol{g}$ say, with components

$$\boldsymbol{g} = \begin{bmatrix}\gamma_{xz}\\ \gamma_{yz}\end{bmatrix} = \begin{bmatrix}\partial w/\partial x - \phi\\ \partial w/\partial y - \psi\end{bmatrix}. \tag{10.17}$$

By combining equations 10.14b, 10.15 and 10.16, the total strain energy density per unit area of the Mindlin plate can then be written

$$\text{SE/area} = \tfrac{1}{2}\,(\boldsymbol{e}^{\mathrm{T}}\boldsymbol{D}\,\boldsymbol{e} + \kappa Gt\,\boldsymbol{g}^{\mathrm{T}}\boldsymbol{g}). \tag{10.18}$$

This is clearly a two-dimensional analogue of equation 9.25, the equivalent expression for a Timoshenko beam.

10.5.2 Mindlin plate elements, the general formulation

Universal expressions are now derived for the stiffness matrix of a Mindlin plate element. The first step in such a formulation is the selection of independent interpolations for w, ϕ and ψ. Suppose that the element we wish to formulate has m nodal values of w, and n nodal values of ϕ and ψ (translations and rotations are not necessarily defined at the same nodes in elements of this type). The interpolations for w, ϕ and ψ can then be written

$$w = \sum_{i=1}^{m} n_{\alpha i}\, w_i, \quad \phi = \sum_{j=1}^{n} n_{\beta j}\,\phi_j, \quad \psi = \sum_{j=1}^{n} n_{\beta j}\,\psi_j, \tag{10.19}$$

where $n_{\alpha i}(x, y)$, $(i = 1, \ldots, m)$ and $n_{\beta j}(x, y)$, $(j = 1, \ldots, n)$ are shape functions for displacement and rotation, respectively. Note that compatibility of displacement and rotation will be achieved between elements provided that both sets of shape functions preserve simple, inter-element continuity. It is convenient at this stage to define shape matrices $\boldsymbol{N}_\alpha$ and $\boldsymbol{N}_\beta$ as

$$\boldsymbol{N}_\alpha^e = [n_{\alpha 1}, \ldots, n_{\alpha m}] \text{ and } \boldsymbol{N}_\beta^e = [n_{\beta 1}, \ldots, n_{\beta n}]. \tag{10.20}$$

Discrete versions of the strain–displacement equations are now formed by substituting expressions 10.19 into equations 10.14c and 10.17. This gives

$$\boldsymbol{e} = \boldsymbol{B}_b^e \boldsymbol{d}^e \text{ and } \boldsymbol{g} = \boldsymbol{B}_s^e \boldsymbol{d}^e, \tag{10.21}$$

where the nodal displacement vector $\boldsymbol{d}^e$ has components

$$\boldsymbol{d}^e = [w_1, w_2, \ldots, w_m, \phi_1, \ldots, \phi_n, \psi_1, \ldots, \psi_n]^T,$$

and where the bending (subscript 'b') and shear (subscript 's') strain–displacement matrices $\boldsymbol{B}_b^e$ and $\boldsymbol{B}_s^e$ are given by

$$\boldsymbol{B}_b^e = \begin{bmatrix} \boldsymbol{0} & -\partial \boldsymbol{N}_\beta^e/\partial x & \boldsymbol{0} \\ \boldsymbol{0} & \boldsymbol{0} & -\partial \boldsymbol{N}_\beta^e/\partial y \\ \boldsymbol{0} & -\partial \boldsymbol{N}_\beta^e/\partial y & -\partial \boldsymbol{N}_\beta^e/\partial x \end{bmatrix}, \tag{10.22}$$

and

$$\boldsymbol{B}_s^e = \begin{bmatrix} \partial \boldsymbol{N}_\alpha^e/\partial x & -\boldsymbol{N}_\beta^e & \boldsymbol{0} \\ \partial \boldsymbol{N}_\alpha^e/\partial y & \boldsymbol{0} & -\boldsymbol{N}_\beta^e \end{bmatrix}. \tag{10.23}$$

Substitution into equation 10.18 then gives the strain energy of the element as the sum of bending and shear contributions,

$$U^e = U_b^e + U_s^e = \tfrac{1}{2}\boldsymbol{d}^{eT}\boldsymbol{K}_b^e\boldsymbol{d}^e + \tfrac{1}{2}\boldsymbol{d}^{eT}\boldsymbol{K}_s^e\boldsymbol{d}^e, \tag{10.24}$$

where the bending and shear stiffness matrices, $\boldsymbol{K}_b^e$ and $\boldsymbol{K}_s^e$, are given by

$$\boldsymbol{K}_b^e = \int_A \boldsymbol{B}_b^{eT}\boldsymbol{D}\boldsymbol{B}_b^e \,\mathrm{d}A, \text{ and } \boldsymbol{K}_s^e = (\kappa G t)\int_A \boldsymbol{B}_s^{eT}\boldsymbol{B}_s^e \,\mathrm{d}A. \tag{10.25}$$

A large family of plate elements can be created in this way using relatively simple shape functions. The compatibility problems associated with the direct Kirchhoff formulation are entirely removed provided that the interpolations for w, ϕ and ψ preserve simple continuity between elements. Moreover, element mappings can be used provided that they also maintain basic, inter-element continuity.

In practice, matters are not quite so straightforward. For a start, Mindlin formulations suffer from 'shear locking', a disturbing phenomenon whereby the shear component of deformation dominates the solution. Since this is generally much smaller than the bending component, the effect is to spectacularly underestimate the true deflection. Essentially, the discrete model is overly constrained by the integral constraint on shear strain energy (the second integral of expression 10.24). A standard remedy is to weaken the relative influence of this constraint by evaluating the integral of shear strain energy at reduced integration points while retaining full integration for the bending terms. This relaxation of the shear constraint can however introduce other problems, notably zero energy modes and 'rank deficiency' of the stiffness matrix which can permit spurious solutions to propagate throughout the mesh. These problems can however be overcome by careful selection both of shape functions and integration schemes. A comprehensive discussion of these aspects of the Mindlin formulation is to be found in Chapter 5 of [10]. They will be remarked upon again shortly when specific element topologies are discussed.

10.5.3 Mindlin plate elements, specific topologies

At first sight, the most straightforward, potential, Mindlin element is a three-noded triangle with linear interpolation for w, ϕ and ψ. Although simple to formulate on paper, the element is susceptible to shear locking even when reduced (single point) integration is used for the shear stiffness. There is no straightforward remedy for this and the element is never used. A more satisfactory, three-noded triangle which does not suffer from shear locking can be formed by including an additional (smoothing) interpolation for shearing strain. This approach goes somewhat beyond the basic formulation described above and is not presented here ([10] contains further details of this element).

The simplest, 'straightforward' Mindlin element which performs satisfactorily is a four-noded, twelve-degree of freedom quadrilateral shown as element (a) in Fig. 10.7. The element can be used with or without an isoparametric mapping. Its nodal variables are the values of w, ψ and ϕ ($-\phi$, if right-handed rotations are required) at the four corner nodes. The shape functions, for displacement and rotation, are those of a bilinear serendipity rectangle (section 8.3). When full (2×2) integration is used for $\boldsymbol{K}_b^e$ and $\boldsymbol{K}_s^e$, the element locks badly. Excellent results are obtained, however, when reduced (single point) integration is used for the shear stiffness. The element *does* suffer from rank deficiency but produces unsatisfactory solutions only in the somewhat unusual situation that w is constrained at *no* two adjacent nodes.

The problem of rank deficiency as a by-product of reduced integration can be dispelled entirely by using higher order elements of the 'heterosis' family, introduced by Hughes and Cohen [11]. These achieve high accuracy and correct rank and are in all regards paragons of plate behaviour. The simplest is a quadratic quadrilateral which uses an eight-noded serendipity representation for displacement and a nine-noded Lagrangian representation for rotation (Fig. 10.7(b)). When selective, reduced integration is used (2×2 for shear stiffness, 3×3 for bending stiffness) this element gives excellent results over a wide range of plate thicknesses. It also passes the patch test and generates no spurious, zero energy modes.

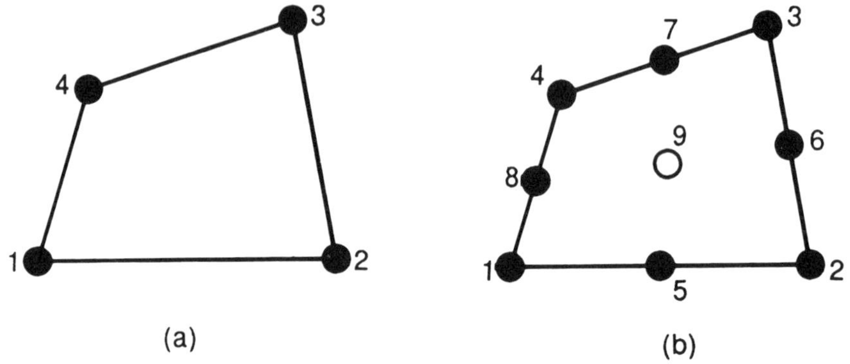

Fig. 10.7 Mindlin elements: (a) bilinear quadrilateral, (b) heterosis. Key: (•) ω_i, ψ_i, φ_i ;(○) ψ_i, φ_i.

10.6 DISCRETE KIRCHHOFF ELEMENTS

The 'discrete Kirchhoff' formulation for plate elements is similar in concept to the Mindlin formulation but applies in the limit as the plate thickness approaches zero. The original idea was first proposed by Wempner and others [12] and has been implemented successfully in a large number of elements. We will confine our attention here to a particularly simple element of this type, the three-noded 'discrete Kirchhoff triangle' (DKT), which was formulated by Stricklin in 1969 [5].

The discrete Kirchhoff formulation starts in the same way as the Mindlin formulation with the selection of independent interpolations for w, ϕ and ψ. In the case of the DKT element, the starting point is a six-noded triangle with nodes at its vertices and midsides (Fig 10.8(a)). Some ambiguity exists as to the required interpolation for w. As far as the DKT formulation is concerned, w is only required on the perimeter of the element. For completeness, however, let us assume that it is to be interpolated at all points inside the element and on its boundary. We will use a complete cubic polynomial for this purpose. This contains ten unknown coefficients which will be determined by ten degrees of freedom, taken to be the (nine) values of w θ_x ($=\partial w/\partial y$) and θ_y ($=-\partial w/\partial x$) at the vertices, plus an additional value of w at the centroid. In fact, this last nodal value is extraneous to the final formulation, since it plays no part in determining the values of w on the periphery of the element. To summarize, although the interpolation for w is technically determined by ten degrees of freedom, only nine of these, the values of w, θ_x and θ_y at the vertices, are actually used in the DKT formulation.

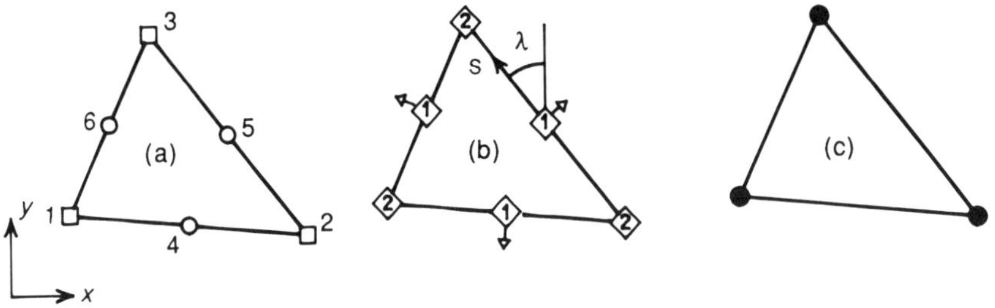

Fig. 10.8 *The discrete Kirchhoff triangle: (a) initial topology, (b) constraints, (c) final topology. Key: (□) $w, \theta_x, \theta_y, \phi, \psi$; (○) ϕ, ψ; (◈) Kirchhoff constraint (n equations); (•) w, θ_x, θ_y; (↗) rotational constraint.*

Complete quadratic interpolation is used for the rotations ϕ and ψ. Each involves six arbitrary constants and requires six degrees of freedom, the values of ψ or ϕ at each of the six nodes of the element. The initial, unconstrained DKT element therefore has 21 degrees of freedom, nine of these associated with w and six each with ϕ and ψ. Their disposition between the nodes is shown in Fig. 10.8(a).

Substitution of the shape functions for ϕ and ψ — those of a six-noded triangle, see equations 8.19 — into equation 10.23 yields a strain–displacement matrix $\boldsymbol{B}_b^e$, in the Mindlin terminology. The strain energy of the element, omitting shear contributions, is then

$$U^e = U_b^e = \tfrac{1}{2}\, \boldsymbol{d}_I^{eT}\, \boldsymbol{K}_b^e\, \boldsymbol{d}_I^e \tag{10.26}$$

where $\boldsymbol{d}_I^e$ is a nodal displacement vector containing the 21 initial degrees of freedom, and $\boldsymbol{K}_b^e$ is a bending stiffness matrix, given by the first integral of expression 10.25. This is of little practical use as it stands since $\boldsymbol{K}_b^e$ contains zeroes in all rows and columns which correspond to translational degrees of freedom.

Discrete constraints now provide the necessary coupling between the displacements and the rotations. Twelve constraint equations are required if we are to eliminate the twelve nodal values of ϕ and ψ in favour of the final, nine values of w_i, θ_{xi} and θ_{yi} ($i = 1, 2, 3$). The constraint equations are formed in the following way (see also Fig. 10.8(b)).

(i) At the vertices, the Kirchhoff constraint is applied to both components of shear strain. This yields six equations,

$$0 = \gamma_{xz} = -\theta_{yi} - \phi i \qquad (i = 1, 2, 3),$$

$$0 = \gamma_{yz} = \theta_{xi} - \psi i \qquad (i = 1, 2, 3).$$

(ii) At the midside nodes, the Kirchhoff constraint is applied to the tangential component of shear strain. This gives three equations,

$$0 = \gamma_{zs} \text{ at nodes 4, 5 and 6.}$$

Note: the shear strain γ_{zs} is evaluated from expressions of the form $(\partial w/\partial s)_i - \beta_i$, $i = 4, \ldots, 6$, where $\partial w/\partial s$ is obtained using cubic interpolation of w along the side of the element, and β_i is the rotation of a fibre about an axis normal to this direction and in the plane of the element. This is obtained by resolving the interpolated rotations ϕ and ψ perpendicular to the side. At point 5, for example, the required constraint is

$$\beta_5 = (\psi_5 \cos\lambda - \phi_5 \sin\lambda) = \left(\frac{\partial w}{\partial s}\right)_5 .$$

(iii) Along each side of the element, the rotation of a fibre about an axis parallel to that side is constrained to vary linearly. This is achieved by forcing the interpolated component of the rotations ϕ and ψ resolved along the side and evaluated at the midpoint, to be the mean of the corresponding, resolved components at each end. On the side 2–3 for example it reduces to the condition

$$\phi_5 \cos\lambda + \psi_5 \sin\lambda = \tfrac{1}{2}\,[(\phi_2 \cos\lambda + \psi_2 \sin\lambda) + (\phi_3 \cos\lambda + \psi_3 \sin\lambda)].$$

Similar constraints apply at nodes 4 and 6, giving a total of three equations.

These twelve constraints — six from (i) and three each from (ii) and (iii) — can now be combined with nine identities, $w_1 = w_1$, $\theta_{x1} = \theta_{x1}, \ldots, \theta_{y3} = \theta_{y3}$ to give twenty-one equations which relate the initial twenty-one degrees of freedom to the final, nine nodal values of w, θ_x and θ_y. Placing the latter in a displacement vector $\boldsymbol{d}^{\mathrm{e}}$, we obtain a matrix transformation of the form

$$\boldsymbol{d}_1^{\mathrm{e}} = \boldsymbol{T}^{\mathrm{e}}\boldsymbol{d}^{\mathrm{e}}, \tag{10.28}$$

where $\boldsymbol{T}^{\mathrm{e}}$ is a 21×9 matrix containing (suitably rearranged) coefficients from the twelve constraint equations and the nine identities. This is substituted into equation 10.26 to give

$$U^{\mathrm{e}} = \tfrac{1}{2}\,\boldsymbol{d}^{\mathrm{eT}}\,\boldsymbol{K}^{\mathrm{e}}\,\boldsymbol{d}^{\mathrm{e}}, \tag{10.29}$$

where $\boldsymbol{K}^{\mathrm{e}} = \boldsymbol{T}^{\mathrm{T}}\,\boldsymbol{K}_{\mathrm{b}}^{\mathrm{e}}\,\boldsymbol{T}$.

This defines a 9×9 stiffness matrix $\boldsymbol{K}^e$ for the final, nine-degree of freedom, DKT element. The element formed in this way passes the patch test and gives accurate and convergent results for most test problems. Further details of the formulation are given in the original article [5] and also in a somewhat simpler form in [13]. A discrete Kirchhoff quadrilateral [14] is formed along very similar lines (problem 13 at the end of this chapter).

A number of interesting variants exist on the discrete Kirchhoff approach. For example, discrete constraints can be supplemented by integral constraints over the area of the element or around its perimeter. A recent example of such an element is the twelve-degree of freedom triangle proposed by Meek and Tan [15]. The nodal variables of this element are displacements at the vertices, and rotations about each edge, evaluated at Gauss integration points along each side, termed 'loof' nodes. Discrete Kirchhoff constraints are applied at the vertices and at the loof nodes, and the transverse shear strain is constrained in an integral sense over the area of the element. A somewhat similar philosophy lies behind the popular 'semiloof' formulation of Irons. Here, integrated, shear strain constraints are applied around the perimeter of a quadrilateral to supplement discrete constraints at loof nodes along each side. Details of this extremely accurate element are to be found in [16].

10.7 ACCURACY AND CONVERGENCE

The elements discussed so far represent a very small sample of the many plate formulations in regular use. Assessments of the relative accuracy/efficiency of these elements appear regularly in the technical literature. No attempt is made here to provide a comprehensive summary of such information. The reader is directed towards review papers such as those by Batoz and others [13, 14] as a good starting point for an evaluation of the more popular formulations. Extensive performance studies for Mindlin-type elements appear also in Chapter 5 of Hughes' general text [10] and in Chapters 1 and 2 of the second volume of Zienkiewicz and Taylor's text [9]. In addition, contemporary journals, such as the *International Journal for Numerical Methods in Engineering*, contain a more or less continuous stream of such comparisons as new formulations are measured against existing elements.

We will confine our attention here to the performance of the specific elements which have been discussed in the preceding sections and will use simple test data to indicate the comparative levels of accuracy which can be expected from them. This data is not intended as a guide to element selection and should not be treated as such. It is to be emphasized that even quite thorough comparisons of this type such as those in [9] and [10] should be treated with caution since values obtained in regularly meshed test cases do not necessarily extend to irregular meshes or to more complex loadings and boundary conditions.

The test case which is used as a basis for comparison is that of a simply supported, square plate subject to a concentrated central load. Ideally, results should be presented for regular and irregular meshes ([10] contains some useful comparisons of this type for Mindlin type elements). In the present instance, however, we consider only a regular arrangement of elements. From

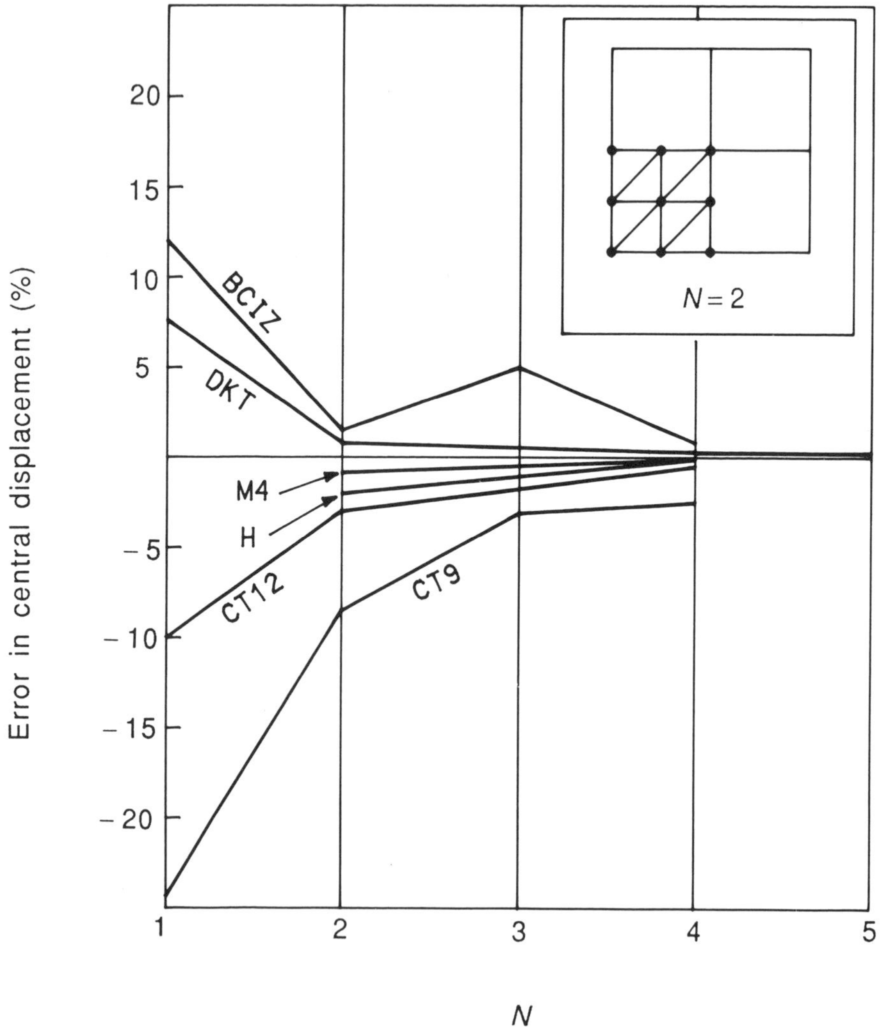

Fig. 10.9 *Error in the central displacement of a simply supported, centrally loaded plate. BCIZ — nonconforming cubic triangle [3], CT9 — Clough–Tocher conforming triangle, 9 degrees of freedom [2], CT12 — Clough–Tocher conforming triangle, 12 degrees of freedom [2], DKT — discrete Kirchhoff triangle [5], M4 — bilinear, Mindlin quadrilateral [4], H — heterosis rectangle [11].*

symmetry, only one quarter of the plate need be modelled. The meshes used are indicated as an inset in Fig. 10.9. The diagonal element dividers should be removed when results for quadrilateral elements are discussed. The variable plotted in each case is the percentage error in the central deflection, determined with respect to the exact Kirchhoff solution for a thin plate. This is plotted against N, the number of intervals between rows of equally spaced nodes in the x or y directions over one quadrant of the plate. In the case of the Mindlin elements (M4 and H) the computed solution is calculated for a plate thickness of the order of 0.01 of the width, so that results are directly comparable, for all intents and purposes, to the 'thin' plate solution.

Clearly all of the elements discussed so far, with the exception of the complete cubic triangle which is not included in these comparisons, converge to the exact solution as the mesh is successively refined. The BCIZ triangle does not in fact do so for *all* meshes, but the error is small, and the element is generally considered to be reliable, if not absolutely convergent. Comparing the two conventional Kirchhoff formulations, we observe that the performance of the fully conforming, Clough–Tocker triangle is generally less impressive than that of its BCIZ counterpart. This improves somewhat when it is used in its twelve-degree of freedom form (with midside slopes included as nodal variables) but there is clearly little to be gained in this particular instance by selecting a conforming rather than a nonconforming element.

Of the more advanced formulations, the DKT element produces results which are significantly better than those of the conventional Kirchhoff elements, especially as the mesh is refined. It also passes the patch test and must be regarded as an altogether superior element. The Mindlin rectangle, however, is clearly the star performer of the current selection, as is its close relative the Heterosis element of Fig. 10.7(b) which is also shown. Although the performance of both of these elements deteriorates somewhat when mappings are introduced (as required for irregular meshes) they are among the most accurate and elegant of the current formulations, and have the added advantage of being suitable for thick and thin plates since they include the effects of shear deformation.

10.8 SHELL ELEMENTS

10.8.1 Geometric coupling of membrane and bending effects

In the analysis of flat plates, we can uncouple bending and membrane deformations and treat them quite separately. In a finite element model of a flat sheet, for example, membrane stresses can be modelled using the plane stress elements of Chapters 6 and 8, and bending stresses using the plate elements discussed in the preceeding sections. This separation of membrane and bending analysis is no longer possible in folded plates or curved shells. Membrane and bending effects are then inextricably linked, as indicated in Fig. 10.10. This shows a local ('dashed') coordinate system at points P and Q on the

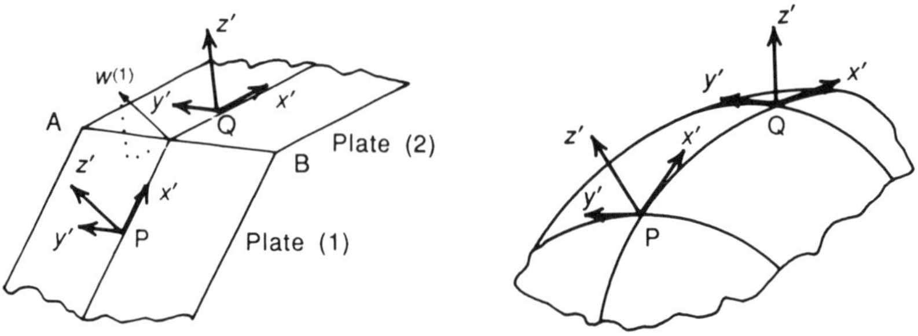

Fig. 10.10 *Local axes: (a) folded plate, (b) curved shell.*

surface of a folded plate and a curved shell, (Fig. 10.10(a) and (b), respectively). In both instances the z' coordinate is normal to the surface. In case (a) the orientation of this axis undergoes an abrupt change at the interface AB. At points on this line the bending displacement normal to the surface of plate 1, $w^{(1)}$ say, contributes a bending *and* an in-plane (stretching) component to the displacement in plate 2. This 'coupling' of bending and stretching is an essential characteristic of such structures.

The same effect is present in the curved shell but occurs as a continuous rather than a discrete process. In Fig. 10.10(b), for example, the direction of the normal axis z' does not change abruptly at 'folds' in the surface but varies continuously as we move from point P to point Q. A continuous coupling then occurs between membrane and bending displacements. This is reflected in theoretical models for curved shells by coupling of the normal and tangential equations of equilibrium [17].

In forming finite element models for curved shells several avenues can be followed. The most sophisticated formulations produce truly curved elements in which bending and stretching are coupled within the element itself. These approaches are based either on classical shell theory (with or without shear deformation) or on the notion of a shell as a degenerate form of an elastic solid. In the former category are elements of the Love–Kirchhoff and Mindlin types, based either on interpolation of displacement only, or on independent interpolation of displacement and rotation. The discrete Kirchhoff approach and associated formulations (semiloof for example) are also effective. No further discussion of curved elements is included here but Chapter 4 of [9] provides a useful introduction to such formulations as does Chapter 6 of [10]. An up to date review specifically of thin shell elements is to be found in [18].

A more straightforward approach to modelling curved shells is that of the 'facet shell' representation. This is based on the observation that the distinction between a folded plate and a continuously curved shell disappears for all intents and purposes when the curved surface is approximated by a series of facets. In the limit as the number of facets becomes infinite, such an approximation represents more and more closely the behaviour of the curved shell. Before this limit is reached the multifaceted representation often forms an acceptable physical approximation to the curved shell. Provided that we are prepared to accept this level of geometrical approximation prior to our formulation of a finite element model, the types of structure illustrated in Fig. 10.10(a) and (b) may both be treated in the same way, as assemblies of flat elements, one or more for each plane facet. The elements used for such analyses are obtained by superimposing membrane and plate elements in the same physical location in much the same way that beam and bar elements are superimposed to produce the frame element.

In selecting the membrane and plate components for a 'facet shell' element of this type, little is gained by using particularly complex interpolations. The argument that a few higher order elements are more effective than a larger number of more primitive ones is frustrated in the case of a curved shell by the purely geometrical consideration that as many facets as possible should be used to approximate the curved surface. Relatively simple bending and membrane elements are therefore favoured in most facet-shell formulations. The development of a simple triangular element of this type is outlined in the next section.

10.8.2 Facet shell triangle, local formulation

The first step in producing a triangular facet shell element is the formulation of a flat element (in the x–y plane) which incorporates bending and membrane effects. The formation of such an element is illustrated in Fig. 10.11. It involves little more than the selection and superposition of separate plane-stress and plate bending elements. In the case of a three-noded triangle, the obvious choice for the plane-stress component is the constant strain triangle of Chapter 6. The degrees of freedom of this element are the in-plane displacements u_i and v_i ($i = 1, 2, 3$) shown in Fig. 10.11.

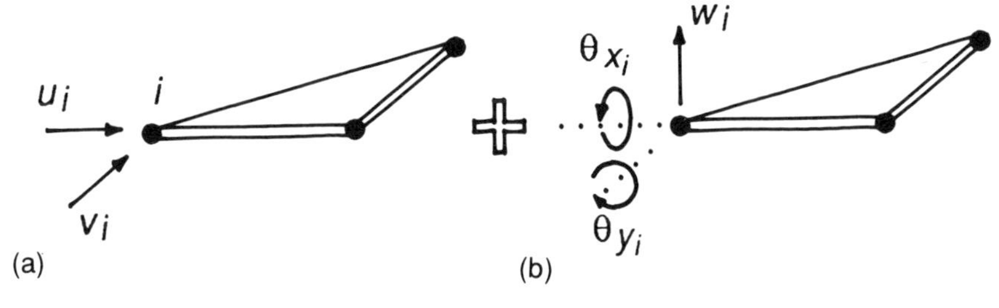

Fig. 10.11 *Superposition of membrane and plate elements to form a facet shell.*

The membrane strain energy (subscript m) of the element, $U_{\mathrm{m}}^{\mathrm{e}}$ say, is then given by

$$U_{\mathrm{m}}^{\mathrm{e}} = \tfrac{1}{2} \boldsymbol{d}_{\mathrm{m}}^{\mathrm{eT}} \boldsymbol{K}_{\mathrm{m}}^{\mathrm{e}} \boldsymbol{d}_{\mathrm{m}}^{\mathrm{e}}, \tag{10.30}$$

where $\boldsymbol{K}_{\mathrm{m}}^{\mathrm{e}}$ is the stiffness matrix given by equation 6.20, and $\boldsymbol{d}_{\mathrm{m}}^{\mathrm{e}}$ is the nodal displacement vector given by

$$\boldsymbol{d}_{\mathrm{m}}^{\mathrm{e}} = [u_1, v_1, u_2, v_2, u_3, v_3]^{\mathrm{T}}.$$

A three-noded plate element is then selected to model the bending strain energy. Any triangular element with the appropriate topology can be used for this purpose. Elements (a), (b) and (d) of Fig. 10.1, for example, are all suitable candidates. Whatever element is chosen, the strain energy due to bending, $U_{\mathrm{b}}^{\mathrm{e}}$, is given by

$$U_{\mathrm{b}}^{\mathrm{e}} = \tfrac{1}{2} \boldsymbol{d}_{\mathrm{b}}^{\mathrm{eT}} \boldsymbol{K}_{\mathrm{b}}^{\mathrm{e}} \boldsymbol{d}_{\mathrm{b}}^{\mathrm{e}}, \tag{10.31}$$

where $\boldsymbol{K}_{\mathrm{b}}^{\mathrm{e}}$ is the stiffness matrix of the plate element and $\boldsymbol{d}_{\mathrm{b}}^{\mathrm{e}}$ is the nodal displacement vector,

$$\boldsymbol{d}_{\mathrm{b}}^{\mathrm{e}} = [w_1, \theta_{x1}, \theta_{y1}, w_2, \ldots, \theta_{y3}]^{\mathrm{T}}.$$

The two elements, (a) and (b), are then superimposed, as indicated in Fig. 10.11, to give a composite element which incorporates both sets of degrees of freedom. In terms of strain energy, the process is equivalent to a summation of equations 10.30 and 10.31 to give

$$U^{\mathrm{e}} = U_{\mathrm{b}}^{\mathrm{e}} + U_{\mathrm{m}}^{\mathrm{e}} = \frac{1}{2}\left[\boldsymbol{d}_{\mathrm{b}}^{\mathrm{eT}}, \boldsymbol{d}_{\mathrm{m}}^{\mathrm{eT}}\right] \begin{bmatrix} \boldsymbol{K}_{\mathrm{b}}^{\mathrm{e}} & 0 \\ 0 & \boldsymbol{K}_{\mathrm{m}}^{\mathrm{e}} \end{bmatrix} \begin{bmatrix} \boldsymbol{d}_{\mathrm{b}}^{\mathrm{e}} \\ \boldsymbol{d}_{\mathrm{m}}^{\mathrm{e}} \end{bmatrix}. \tag{10.32}$$

The 15×15 matrix at the centre of this expression is the stiffness matrix for the fifteen-degree of freedom element. The next stage of the formulation is made simpler if the order of the variables in the nodal displacement vector of expression 10.32 is adjusted so that the displacements and rotations associated with a particular node are grouped together. The nodal displacement vector for the composite element is accordingly reordered so that it contains the fifteen nodal variables in the sequence

$$\boldsymbol{d}^{\mathrm{e}} = [u_1, v_1, w_1, \theta_{x1}, \theta_{y1}, u_2, \ldots, \theta_{y3}]^{\mathrm{T}}.$$

This involves a corresponding rearrangement of the rows and columns in the partitioned stiffness matrix on the right hand side of expression 10.32. In terms of the rearranged matrix, $\boldsymbol{K}^{\mathrm{e}}$ say, the strain energy of the element is then

$$U^{\mathrm{e}} = \tfrac{1}{2} \boldsymbol{d}^{\mathrm{eT}} \boldsymbol{K}^{\mathrm{e}} \boldsymbol{d}^{\mathrm{e}}. \tag{10.33}$$

Note that the degrees of freedom of the shell element formed in this way do not include a full complement of rotations at each node, since those about the z-axis receive no stiffness contribution either from the bending or the membrane element. This causes minor problems during the assembly of coplanar elements as noted in the next section.

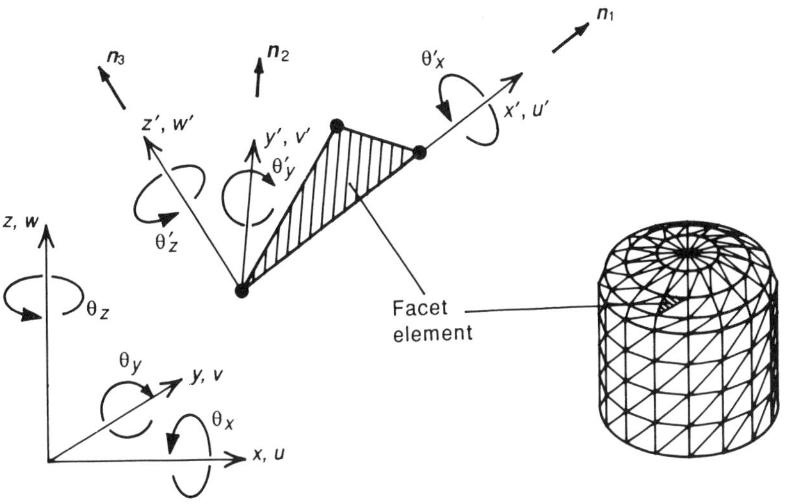

Fig. 10.12 *The facet shell element, local and global coordinates.*

10.8.3 Global stiffness matrix, assembly

The element of Fig. 10.11 can be rotated in three dimensions to produce a facet element which is arbitrarily orientated in space. Such an element is shown in Fig. 10.12. Its formulation is analogous to that which lead to the arbitrarily orientated frame element in Chapter 9.

Suppose that the triangular element of the previous section lies in the x'–y' plane of its own 'local' coordinate system (Fig. 10.12). The orientation of the ('dashed') local axes with respect to a set of global coordinates is defined by unit vectors $\boldsymbol{n}_i = (l_i, m_i, n_i)$, $(i = 1, 2, 3)$. If the displacements and rotations in the local system are denoted by the 'dashed' variables, u', v', w', θ'_x, θ'_y and θ'_z, we can write the strain energy of the element as

$$U^{\mathrm{e}} = \tfrac{1}{2}\boldsymbol{d}^{\mathrm{e}'\mathrm{T}}\boldsymbol{K}^{\mathrm{e}'}\boldsymbol{d}^{\mathrm{e}'}, \tag{10.34}$$

where $\boldsymbol{d}'_{\mathrm{e}} = [u'_1, v'_1, w'_1, \theta'_{x1}, \theta'_{y1}, u'_2, \ldots, \theta'_{y3}]^{\mathrm{T}}$ and where $\boldsymbol{K}^{\mathrm{e}'}$ is the local stiffness matrix of expression 10.33.. We now define a second nodal displacement vector $\boldsymbol{d}^{\mathrm{e}}$ with a full complement of *global* displacements and rotations, $u_i, v_i, \ldots, \theta_{xi}$, etc. This gives

$$\boldsymbol{d}^{\mathrm{e}} = [u_1, v_1, w_1, \theta_{x1}, \theta_{y1}, \theta_{z1}, u_2, v_2, w_2, \theta_{x2}, \theta_{y2}, \ldots, \theta_{z3}]^{\mathrm{T}}. \tag{10.35}$$

By resolving these components along the local axes of the element, we then obtain the transformation

$$\boldsymbol{d}^{\mathrm{e}'} = \boldsymbol{T}^{\mathrm{e}}\boldsymbol{d}^{\mathrm{e}} \tag{10.36a}$$

where

$$T^e = \begin{bmatrix} L & 0 & 0 \\ 0 & L & 0 \\ 0 & 0 & L \end{bmatrix}, \text{ and } L = \begin{bmatrix} l_1 & m_1 & n_1 & 0 & 0 & 0 \\ l_2 & m_2 & n_2 & 0 & 0 & 0 \\ l_3 & m_3 & n_3 & 0 & 0 & 0 \\ 0 & 0 & 0 & l_1 & m_1 & n_1 \\ 0 & 0 & 0 & l_2 & m_2 & n_2 \end{bmatrix}. \quad (10.36b)$$

Upon substitution into equation 10.34, this gives

$$U^e = \tfrac{1}{2} \boldsymbol{d}^{eT} \boldsymbol{K}^e \boldsymbol{d}^e,$$

where

$$\boldsymbol{K}^e = \boldsymbol{T}^{eT} \boldsymbol{K}^{e\prime} \boldsymbol{T}^e. \quad (10.37)$$

$\boldsymbol{K}^e$ is now a global stiffness matrix for the facet shell element, and can be assembled in the usual way to form multifaceted structures such as that shown in Fig. 10.12.

The above transformation increases the number of degrees of freedom associated with the original, 'local' element from 15 to 18. This is a consequence of the absence of rotational stiffness about the local z' axis. It causes no difficulty provided that all elements meet at oblique angles but can cause singularity of the assembled stiffness matrix when all elements at a particular node are coplanar. If this were to happen, for example, in the global x–y plane, the θ_{zi} row of the assembled stiffness matrix would then contain no terms at all, giving an indeterminate global equation of the form

$$0 \,.\, \theta_{zi} = 0. \quad (10.38)$$

This will cause an error in most equation solvers but is easily remedied either by constraining θ_{zi} to zero or by inserting a notional stiffness, k say, in the offending diagonal, which has the same effect since the indeterminate equation then becomes

$$k \,.\, \theta_{zi} = 0. \quad (10.39)$$

The situation is more complex when the coplanar elements lie at an oblique angle to the global axes. The rows of the assembled stiffness matrix which correspond to the global rotations at the coplanar node, then give non-trivial stiffness equations, none of which takes the form of equation 10.38, but which are nonetheless singular, all three equations being derived from two independent relationships in the local system. The most straightforward remedy is once again to insert a fictitious stiffness associated with the *local* z' rotation. This can be done either by modifying the local-global transformation 10.36 so that the degrees of freedom at 'coplanar' nodes remain in local components, or by inserting a rotational stiffness k in the element stiffness matrix prior to transformation and assembly (see problem 14 at the end of this chapter). A third alternative is to use a 'drilling' stiffness in all elements whether coplanar or not. Further discussion of all three remedies is to be found in section 3.5 of [9].

REFERENCES

[1] Melosh, R. J. (1963) Basis of derivation of matrices for direct stiffness method. *A.I.A.A. Journal*, **1**, 1631–7.

[2] Clough, R. W. and Tocher, J. L. (1965) Finite element stiffness matrices for analysis of plate bending. *Proceedings of the Conference on Matrix Methods in Structural Mechanics*, Air Force Institute of Technology, Wright Patterson Air Force Base, Dayton, Ohio.

[3] Bazeley, G. P., Cheung, Y. K., Irons, B. M., and Zienkiewicz, O. C. (1965) Triangular elements in bending — conforming and nonconforming solutions. *Proceedings of the Conference on Matrix Methods in Structural Mechanics*, Air Force Institute of Technology, Wright Patterson Air Force Base, Dayton, Ohio.

[4] Hughes, T. J. R., Taylor, R. L. and Kanoknukulchai, W. (1977) A simple and efficient finite element for plate bending, *International Journal for Numerical methods in Engineering*, **11** (10), 1529–43.

[5] Stricklin, J. A., Hasler, W., Tisdale, P., and Gunderson, R. (1969) A rapidly converging Triangular plate element. *A.I.A.A. Journal*, **7** (1), 180–1.

[6] Gallagher, R. H. (1974) *Finite Element Analysis, Fundamentals*, Prentice-Hall, Englewood-Cliffs, NJ, Chapter 6.

[7] Zienkiewicz, O. C. and Taylor, R. L. (1989) The Finite Element Method, 4th edn, Vol. I, *Basic Formulation and Linear Problems*. McGraw-Hill, London, Chapters 12 and 13.

[8] Argyris, J. H., Fried, I. and Scharpf, D. W. (1968) The TUBA family of plate bending elements for the matrix displacement method. *Journal of the Royal Aeronautical Society*, **72**, 701–9.

[9] Zienkiewicz, O. C. and Taylor, R. L. (1991) *The Finite Element Method*, 4th edn, Vol. 2, *Solid and Fluid Mechanics, Dynamics and Vibration*, McGraw-Hill, London, Chapters 1, 2 and 3.

[10] Hughes, T. J. R. (1987) *The Finite Element Method. Linear Static and Dynamic Finite Element Analysis*, Prentice-Hall, Englewood Cliffs, NJ, Chapters 5 and 6.

[11] Hughes, T. J. R. and Cohen, M. (1978) The 'Heterosis' element for plate bending. *Computers and Structures*, **9**, 445–50.

[12] Wempner, G., Oden J. T. and Kross, D. (1968) Finite element analysis of thin shells. *Proceedings of A.S.C.E., Journal of the Engineering Mechanics Division*, **94**, **EM6**, 1273–94.

[13] Batoz, J. L., Bathe, K. J. and Ho, L. W. (1980) A study of three node triangular plate bending elements. *International Journal for Numerical Methods in Engineering*, **15**, 1771–812.

[14] Batoz, J. L., and Tahar, M. B. (1982) Evaluation of a new quadrilateral thin plate bending element. *International Journal for Numerical Methods in Engineering*, **18**, 1655–77.

[15] Meek, J. L. and Tan, H. S. (1986) A faceted shell element with loof nodes. *International Journal for Numerical Methods in Engineering*, **23**, 49–67.

[16] Irons, B. M. (1976) The semiloof shell element. *Finite Elements for Thin Shells and Curved Members*, (eds D. G. Ashwell and R. H. Gallagher). Wiley, London, Chapter 11, 197–222.

[17] Donnell, L. H. (1976) *Beams, Plates and Shells*, McGraw Hill, New York, Chapter 6.

[18] Yang, H. T. Y., Saigal, S. and Liaw, D. G. (1990) Advances in thin shell elements Yand some applications — version I. *Computers and Structures*, **35** (4), 481–504.

[19] Morley, L. S. D. (1971) The constant-moment plate bending element. *Journal of Strain Analysis*, **6** (1), 20–24.

PROBLEMS

1. A Kirchhoff plate element of the type described in section 10.2 has m nodes at $(x_1, y_1), (x_2, y_2), \ldots, (x_m, y_m)$. The shape relationship for the element is

$$w(x, y) = n_1(x, y)w_1 + n_2(x, y)\theta_{x1} + n_3(x, y)\theta_{y1} + \ldots + n_{3m}(x, y)\theta_{ym},$$

where w_i, θ_{xi} and θ_{yi} are the displacement and rotations at node i. Show that n_1, n_2 and n_3 satisfy the following conditions:

(i) $n_1 = 1$ at node 1,
$n_1 = 0$ at nodes 2, . . . , p,
$\partial n_1/\partial x = 0$ and $\partial n_1/\partial y = 0$ at *all* nodes.

(ii) $\partial n_2/\partial y = 1$ at node 1,
$\partial n_2/\partial y = 0$ at nodes 2, . . . , p,
$n_2 = 0$ and $\partial n_2/\partial x = 0$ at *all* nodes.

(iii) $\partial n_3/\partial x = -1$ at node 1,
$\partial n_3/\partial x = 0$ at nodes 2, . . . , p,
$n_3 = 0$ and $\partial n_3/\partial y = 0$ at *all* nodes.

2. A rectangular plate element with nodes at $(\pm a, \pm b)$ as shown is formulated using the direct Kirchhoff approach. The displacement w_i and rotations, θ_{xi} and $\theta_{yi} (i = 1, \ldots, 4)$ are taken as nodal variables, and an interpolation is proposed of the form

$$w(x, y) = \alpha_1 + \alpha_2 x + \alpha_3 y + \alpha_4 x^2 + \alpha_5 xy + \alpha_6 y^2 + \alpha_7 x^3 + \alpha_8 x^2 y + \alpha_9 xy^2 + \alpha_{10} y^3 + \alpha_{11} x^3 y + \alpha_{12} xy^3 .$$

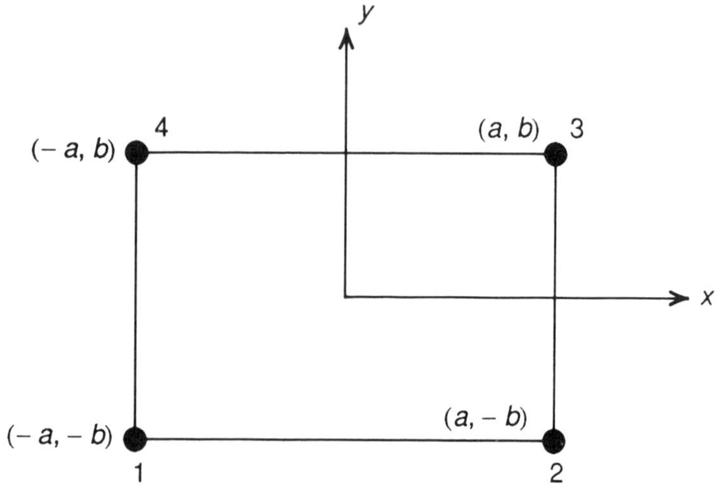

Problem 2.

Do not evaluate the constants $\alpha_i (i = 1, \ldots, 12)$, but confirm that continuity of displacement is guaranteed along a common edge between adjacent elements. Can the same assurance be given for continuity of the normal derivative of displacement?

3. The shape relationship for the element of problem 2 takes the form

$$w(x, y) = n_1(x, y)w_1 + n_2(x, y)\theta_{x1} + n_3(x, y)\theta_{y1} + \ldots + n_{12}(x, y)\theta_{y4}.$$

Confirm that suitable expressions for the first three shape functions are

$$n_1(x, y) = \frac{1}{8}\left(1 - \frac{x}{a}\right)\left(1 - \frac{y}{b}\right)\left(2 - \frac{x}{a} - \frac{y}{b} - \frac{x^2}{a^2} - \frac{y^2}{b^2}\right),$$

$$n_2(x, y) = \frac{1}{8} b \left(1 - \frac{x}{a}\right)\left(1 - \frac{y}{b}\right)^2 \left(1 + \frac{y}{b}\right),$$

$$n_3(x, y) = -\frac{1}{8} a \left(1 - \frac{x}{a}\right)^2 \left(1 - \frac{y}{b}\right)\left(1 + \frac{x}{a}\right).$$

(Do not attempt to calculate the unknown coefficients α_i, but use the properties established in problem 1 to confirm that n_1, n_2 and n_3 satisfy the required conditions.)

4. A concentrated load P acts at the centre of the rectangular element of problem 3. Show that the equivalent nodal loads which must be applied at

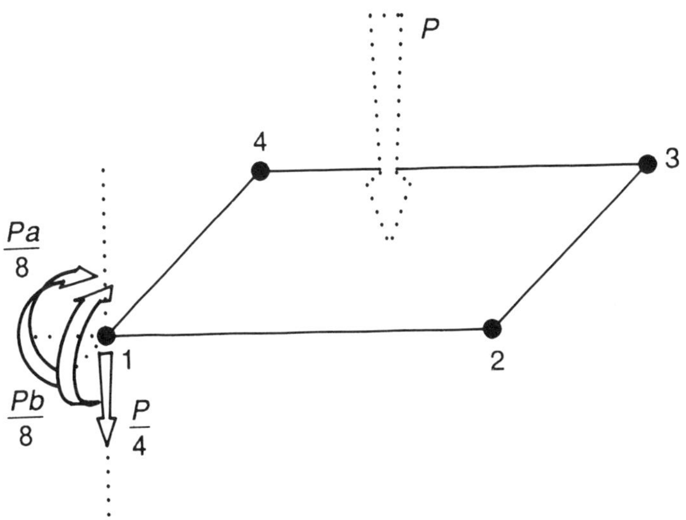

Problem 4.

node one to simulate this loading are: a nodal force $P/4$, and nodal moments $Pa/8$ and $Pb/8$ acting as shown.

5. What equivalent nodal forces and moments should be applied to the rectangular element of problem 3 to simulate a load P which is distributed uniformly over the upper surface of the element?

6. Calculate the terms in the first column of the strain–displacement matrix $\boldsymbol{B}^{\mathrm{e}}$ for the rectangular element of problem 3. Hence evaluate the 1–1 term in the matrix product $(\boldsymbol{B}^{\mathrm{eT}}\boldsymbol{D}\boldsymbol{B}^{\mathrm{e}})$ and show that the the 1–1 term in the element stiffness matrix is

$$K_{11}^{\mathrm{e}} = \frac{D}{ab}\left[\left(\frac{a}{b}\right)^2 + \left(\frac{b}{a}\right)^2 + \frac{7}{10} - \frac{\nu}{5}\right],$$

where D is the flexural rigidity of the plate.

7. A square plate of side $2a$ is clamped on all sides and subject to a concentrated load P at the centre. Model one quadrant of the plate using a single element of the type discussed in problems 2–6 and obtain an estimate for the the central deflection (Note: from symmetry, it may be assumed that θ_x and θ_y are zero at the central node).

8. A nonconforming BCIZ triangle of the type described in section 10.3 has nodes at $(a, 0)$, $(0, a)$ and $(0, 0)$ as shown. Confirm that the area coordinates at any point within the element are given by

$$L_1 = \xi,\ L_2 = \eta \text{ and } L_3 = 1 - \xi - \eta,$$

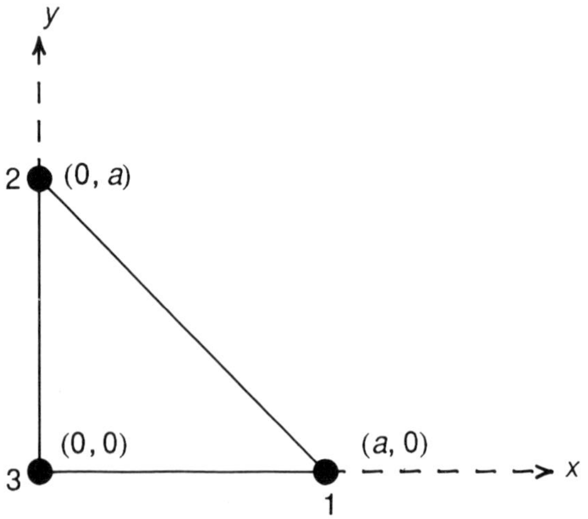

Problem 8.

where $\xi = x/a$ and $\eta = y/a$. Hence show that the shape function n_1 can be written

$$n_1 = 3\xi^2 + 2\xi\eta - 2\xi^3 - 2\eta^2\xi - 2\eta\xi^2.$$

Confirm that this explicitly satisfies condition (i) of problem 1. Obtain expressions for n_2 and n_3 in terms of ξ and η and y and show that they satisfy conditions (ii) and (iii).

9. Determine the equivalent nodal forces and moments which must be applied at node one to simulate a uniformly distributed load p_0 per unit area acting on the triangular element of problem 8.

10. Show that the terms in the first column of the strain–displacement matrix $\boldsymbol{B}^{\mathrm{e}}$ for the element of problem 8 are

$$\left(\frac{1}{a^2}\right)\begin{bmatrix} 12\xi + 4\eta - 6, & \dots \\ 4\xi, & \dots \\ 8\eta + 8\xi - 4, & \dots \end{bmatrix}.$$

Evaluate the 1–1 term in the integrand $(\boldsymbol{B}^{\mathrm{eT}}\boldsymbol{D}\boldsymbol{B}^{\mathrm{e}})$, and show that the 1–1 term of the stiffness matrix is

$$k_{11}^{\mathrm{e}} = \frac{6D}{a^2}$$

A mesh of eight triangular elements of this type models a square plate of side $2a$ as shown. The plate is clamped at the edges and loaded with a concen-

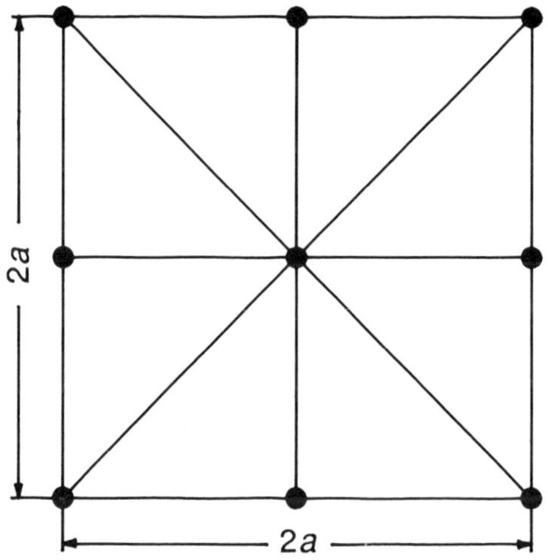

Problem 10.

trated load P at the centre. Obtain an estimate for the central deflection (assume from symmetry that θ_x and θ_y are zero at the central node).

11. A nonconforming six-noded Kirchhoff triangle is to be formed using a complete, quadratic interpolation of the form

$$w(x, y) = \alpha_1 + \alpha_2 x + \alpha_3 y + \alpha_4 x^2 + \alpha_5 xy + \alpha_6 y^2.$$

The nodal degrees of freedom are the values of w at the vertices and the values of the normal derivative, $\partial w/\partial n$, at the midpoints of the sides. What level of continuity, if any, exists at boundaries between adjacent elements of this type?

12. The shape relationship for the element of problem 11 takes the form

$$w(x, y) = n_1(x, y)w_1 + n_2(x, y)w_2 + n_3(x, y)w_3 + n_4(x, y)\left(\frac{\partial w}{\partial n}\right)_4 + n_5(x, y)\left(\frac{\partial w}{\partial n}\right)_5 + n_6(x, y)\left(\frac{\partial w}{\partial n}\right)_6$$

where nodes 1, 2 and 3 are at the vertices of the triangle and nodes 4, 5 and 6 at the midpoints of each side. Do not attempt to determine the shape functions explicitly, but confirm that $n_1(x, y)$ must satisfy the following conditions

(i) $n_1 = 1$ at node 1,
(ii) $n_1 = 0$ at nodes 2 and 3,
(iii) $\partial n_1/\partial n = 0$ at nodes 4, 5 and 6.

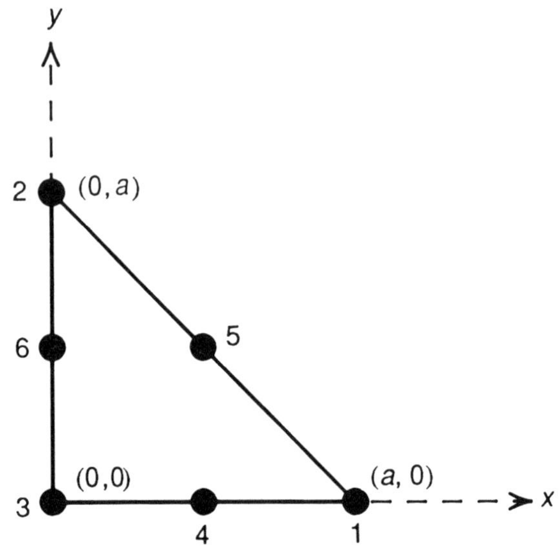

Problem 12.

A particular element of this type has vertices at $(a, 0)$, $(0, a)$ and $(0, 0)$ (as shown). Confirm that $n_1(x, y)$ is given by,

$$n_1(x, y) = \tfrac{1}{2}(\xi + \eta + \xi^2 - \eta^2 - 2\xi\eta), \quad (\xi = x/a, \eta = y/a).$$

Obtain expressions for the terms in the first column of the strain–displacement matrix $\boldsymbol{B}^e$ and determine the 1–1 term in the stiffness matrix. (Note: this unusual, nonconforming, element is described in detail in [19].)

13. An eight-noded discrete Kirchhoff rectangle has 28 initial degrees of freedom. In the notation of section 10.6, these are the values of ϕ and ψ at all nodes, and the values of w, $\theta_x(= \partial w/\partial y)$ and $\theta_y(= -\partial w/\partial x)$ at the four corner nodes (as shown). The displacement w is interpolated using a polynomial identical to that used in problem 2. The interpolation for ϕ and ψ is that of an eight-noded serendipity rectangle. The discrete Kirchhoff element is formed by applying constraints, both at the corners and at the midside nodes, which are identical to to those applied to the corresponding discrete Kirchhoff triangle (see section 10.6). Confirm that this procedure provides the correct number of relationships for the current element, and write explicit expressions for the constraints applied at nodes 1 and 5.

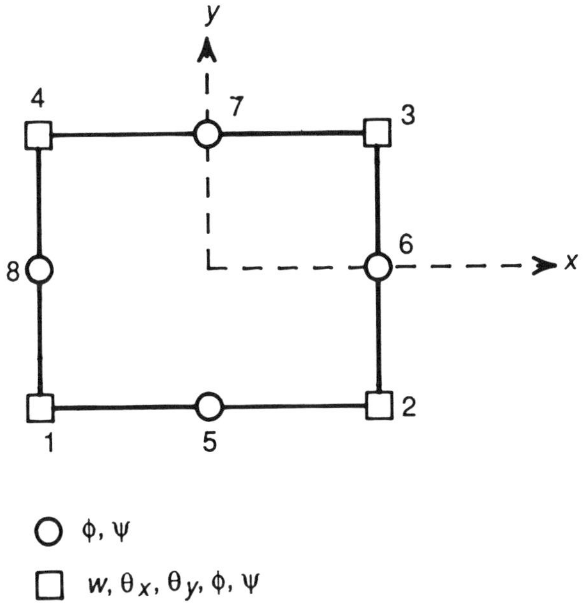

Problem 13.

14. Show that a strain energy contribution, $\frac{1}{2}k(\theta'_{zi})^2$, in the local coordinate system of Fig. 10.12 can be simulated by assembling an 'element' which has a nodal, displacement vector $\boldsymbol{d}^{\mathrm{e}} = [\theta_{xi}, \theta_{yi}, \theta_{zi}]^{\mathrm{T}}$, and a stiffness matrix,

$$\boldsymbol{K}^{\mathrm{e}} = k\begin{bmatrix} l_3^2 & l_3m_3 & l_3n_3 \\ m_3l_3 & m_3^2 & m_3n_3 \\ n_3l_3 & n_3m_3 & n_3^2 \end{bmatrix}.$$

Comment on the application of this element to coplanar assemblies of facet shell elements.

11

Dynamics and Vibration

11.1 INTRODUCTION

The finite element models of previous chapters have been based on the principle of minimum total energy as an alternative to a full statement of static equilibrium. This approach can accommodate only static (dead) loads. In practice the loads need not be absolutely unvarying but they must change slowly so that system behaves instantaneously as though they had always been present. A 'slowly' varying load in this context is one which does not change significantly during the time taken for a displacement or stress perturbation to reach the most distant point in the body. Such propagation times are often difficult to estimate and are strongly dependent on problem geometry. A flexural wave, for example, travelling as a 'ripple' in a plate progresses more slowly than an elastic wave propagating through a solid block of material. Fortunately, such estimates are seldom necessary since the timescales involved are characterized by the 'natural' frequencies of the system. These are frequencies at which it vibrates freely without external excitation. The calculation of the natural frequencies of an object and the determination of the associated modes of vibration is a preliminary to most dynamic analyses. In such cases a 'slowly varying' applied load can be defined as one which varies significantly only over a timescale which is large (two or three times greater say) compared to the period of the lowest natural frequency. If the loading takes place over a shorter timescale than this, a dynamic model must be used.

Dynamic analysis of a linear system is frequently performed in two parts. A 'normal mode' analysis is performed first, followed by a 'response' (or 'time history') analysis. A finite element, dynamic analysis generally follows the same sequence. A normal mode analysis is performed first to determine the natural frequencies of the system and the corresponding mode shapes. On occasions this may be all that is required. In most instances, however, a 'response' analysis is then performed to provide further details on the response of the system to a specific time-dependent loading (or base excitation).

In the discussion which follows, we confine our attention to those aspects of a dynamic analysis which are peculiar to the finite element idealization. The time-dependent, finite element equations are, in fact, identical to those of many other multi-degree of freedom, dynamical models. Standard techniques can be used to obtain solutions to such equations and these form the staple of many undergraduate texts on dynamics and vibration (see, for example, [1] and [2]). These methods are reviewed and applied where necessary but are not developed here in any detail.

11.2 DYNAMIC EQUATIONS

11.2.1 d'Alembert's principle

The application of finite element methodology to the dynamic case is presented here as an extension of the static formulation. This is one of several approaches which can be used to derive the dynamic equations. An obvious alternative is to reformulate the discrete problem from first principles by including kinetic energy in the total energy functional. This produces Hamilton's principle which replaces the principle of minimum total energy as the basis for a Rayleigh–Ritz model and subsequent finite element discretization.

The dynamic equations can be derived more simply in the current instance by using d'Alembert's principle. This converts a dynamic problem into a static one subject to fictitious 'inertial' forces. The underlying logic is demonstrated by the following argument. Consider a single particle of mass m which is subject to a time dependent, external force $\boldsymbol{F}(t)$ and experiences an instantaneous acceleration $\boldsymbol{a}(t)$ (Fig. 11.1(a)). Force and acceleration are related by Newton's second law

$$\boldsymbol{F} = m\boldsymbol{a}. \tag{11.1}$$

However, by moving the accelerative term from the right hand side of this equation to the left we obtain

$$\boldsymbol{F} + \boldsymbol{F}_{\text{I}} = 0, \tag{11.2}$$

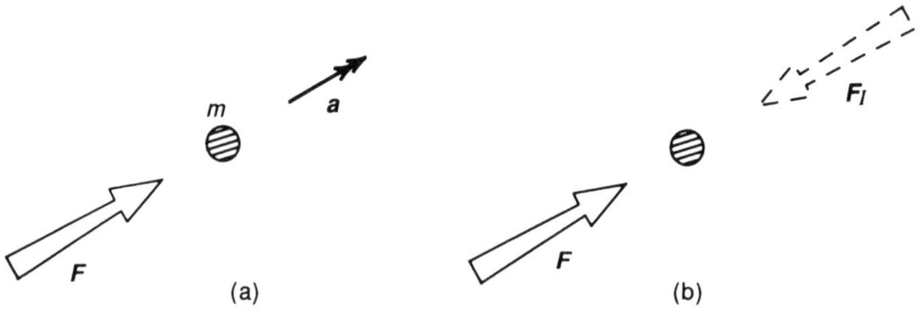

Fig. 11.1 *d'Alembert's principle: (a) dynamic equilibrium of a particle, (b) equivalent static forces.*

where $\boldsymbol{F}_{\mathrm{I}} = -m\boldsymbol{a}$. This is equivalent to a statement of *static* equilibrium for the same particle provided that an 'inertial' load $\boldsymbol{F}_{\mathrm{I}}$ is applied in a direction opposite to that of the acceleration (Fig. 11.1(b)). Equation 11.2 is a mathematical statement of d'Alembert's principle for a single particle. It states that the forces required for dynamic equilibrium are instantaneously identical to those required for static equilibrium provided that an 'inertial' force is applied to the system.

The same logic extends to a continuum. This is easily demonstrated by reformulating the static equations of stress equilibrium (derived in section 2.2.3) for the unsteady case. Dynamic effects are included by equating the net force on the differential element of Fig. 2.7 not to zero but to the mass of the element multiplied by its acceleration. Equations 2.4 and 2.5 then become

$$\begin{aligned} \frac{\partial \sigma_x}{\partial x} + \frac{\partial \tau_{yx}}{\partial y} + f_x &= \rho a_x, \\ \frac{\partial \tau_{xy}}{\partial x} + \frac{\partial \sigma_y}{\partial y} + f_y &= \rho a_y, \end{aligned} \tag{11.3}$$

where a_x and a_y are the accelerations in the x and y directions, and ρ is the density of the material. By moving the accelerative terms from the right-hand side of each equation to the left, we obtain

$$\begin{aligned} \frac{\partial \sigma_x}{\partial x} + \frac{\partial \tau_{yx}}{\partial y} + (f_x - \rho a_x) &= 0, \\ \frac{\partial \tau_{xy}}{\partial x} + \frac{\partial \sigma_y}{\partial y} + (f_y - \rho a_y) &= 0. \end{aligned} \tag{11.4}$$

In other words, the stress field required for dynamic equilibrium is identical at any instant to that required for static equilibrium provided that an inertial body force, $(-\rho\boldsymbol{a})$ per unit volume, is applied throughout the body. The finite element formulations of previous chapters can now be modified to include dynamic effects simply by incorporating additional, distributed, body forces proportional to the instantaneous acceleration.

11.2.2 Spatial interpolation in the dynamic case

In a dynamic, finite element model the displacement and stress fields vary both with space and time. The same spatial discretization for displacement can however be used as in previous chapters. That is to say, the body can be divided into discrete elements (fixed in space) and within each of these the *instantaneous* displacement field can be interpolated using standard element shape functions. The only difference in the dynamic case is that the nodal degrees of freedom are no longer constants but vary with time. The shape relationship within an element can therefore be written

$$\boldsymbol{u} = \boldsymbol{N}^{\mathrm{e}}(\boldsymbol{x})\,\boldsymbol{d}^{\mathrm{e}}(t), \tag{11.5}$$

where $\boldsymbol{N}^{\mathrm{e}}$ is the standard element shape matrix — written as an explicit function of the position vector $\boldsymbol{x}$ to emphasize its dependence on position only — and $\boldsymbol{d}^{\mathrm{e}}(t)$ is a vector of time-dependent, nodal displacements. In an element with p degrees of freedom, for example, $\boldsymbol{d}^{\mathrm{e}}$ has components,

$$\boldsymbol{d}^{\mathrm{e}} = [\delta_1(t), \delta_2(t), \ldots, \delta_p(t)]^{\mathrm{T}}. \tag{11.6}$$

11.2.3 Inertial loads, the element mass matrix

d'Alembert's principle tells us that the stiffness equations for a dynamic model are equivalent to those of a static model subject to an inertial body force $(-\rho\boldsymbol{a})$ per unit volume. The assembled finite element equations are, therefore,

$$\boldsymbol{K}\,\boldsymbol{d}(t) = \boldsymbol{f}(t) + \boldsymbol{f}_{\mathrm{I}}(t) \tag{11.7}$$

where $\boldsymbol{K}$ is the (static) stiffness matrix, $\boldsymbol{d}(t)$ is the vector of instantaneous nodal displacements, $\boldsymbol{f}(t)$ is the vector of applied nodal forces and $\boldsymbol{f}_{\mathrm{I}}(t)$ is the vector of equivalent inertial forces. The components of $\boldsymbol{f}_{\mathrm{I}}$ derive from element contributions, $\boldsymbol{f}_{\mathrm{I}}^{\mathrm{e}}$, obtained by treating the inertial load, $-\rho\boldsymbol{a}$, as a distributed body force over each element. $\boldsymbol{f}_{\mathrm{I}}^{\mathrm{e}}$ is therefore given by (equations 4.39)

$$\boldsymbol{f}_{\mathrm{I}}^{\mathrm{e}} = \int_V \boldsymbol{N}^{\mathrm{eT}}(-\rho\boldsymbol{a})\,\mathrm{d}V. \tag{11.8}$$

However, the acceleration within each element is consistently approximated by the second derivative of the interpolated displacement with respect to time, that is

$$\boldsymbol{a} = \frac{\partial^2 \boldsymbol{u}}{\partial t^2}, \tag{11.9}$$

where $\boldsymbol{u}$ is given by equation 11.5. Substituting, we obtain

$$\boldsymbol{a} = \boldsymbol{N}^{\mathrm{e}}\ddot{\boldsymbol{d}}^{\mathrm{e}} \tag{11.10}$$

where $\ddot{\boldsymbol{d}}^{\mathrm{e}}$ is the nodal acceleration vector with components,

$$\ddot{\boldsymbol{d}}^{\mathrm{e}} = \left[\frac{\partial^2\delta_1}{\partial t^2}, \frac{\partial^2\delta_2}{\partial t^2}, \ldots, \frac{\partial^2\delta_p}{\partial t^2}\right]^{\mathrm{T}}. \tag{11.11}$$

Substituting equations 11.10 into equation 11.8, we finally obtain

$$\boldsymbol{F}_{\mathrm{I}}^{\mathrm{e}} = -\int_V \rho[\boldsymbol{N}^{\mathrm{eT}}\boldsymbol{N}^{\mathrm{e}}]\,\ddot{\boldsymbol{d}}^{\mathrm{e}}\mathrm{d}V = -\boldsymbol{M}^{\mathrm{e}}\ddot{\boldsymbol{d}}^{\mathrm{e}}. \tag{11.12}$$

The square matrix $\boldsymbol{M}^e$ is the 'mass matrix' of the element. It is defined by the integral

$$\boldsymbol{M}^e = \int_V \rho \boldsymbol{N}^{eT} \boldsymbol{N}^e \, dV. \tag{11.13}$$

Although not apparent from the above derivation, $\boldsymbol{M}^e$ is intimately related to the kinetic energy of the element (see problem 1 at the end of this chapter).

Mass matrices can be formed using expression 11.13 for any of the elements derived in previous chapters. By way of illustration let us calculate explicit matrices for some simple cases.

(i) *The uniaxial bar element*

The bar element discussed in section 4.2 is shown again in Fig. 11.2(a). It has a nodal displacement vector $\boldsymbol{d}^e = [u_1, u_2]^T$ and a shape matrix (see equation 4.7)

$$\boldsymbol{N}^e = \left[1 - \frac{x'}{L}, \frac{x'}{L}\right],$$

where x' is a local axial coordinate. Using expression 11.13 with $dV = A dx'$, $\boldsymbol{M}^e$ is therefore given by

$$\boldsymbol{M}^e = \int_0^L \rho \begin{bmatrix} (1 - x/L) \\ (x/L) \end{bmatrix} \left[1 - \frac{x'}{L}, \frac{x'}{L}\right] A dx',$$

where A is the cross-sectional area of the bar. Integrating, we obtain

$$\boldsymbol{M}^e = \begin{bmatrix} m/3 & m/6 \\ m/6 & m/3 \end{bmatrix}, \tag{11.14}$$

where m ($= \rho AL$) is the mass of the bar.

(ii) *The constant strain triangle*

The constant strain triangle of Chapter 6 is shown again in Fig. 11.2(b). It has a nodal displacement vector $\boldsymbol{d}^e = [u_1, v_1, u_2, v_2, u_3, v_3]^T$, and a shape matrix

$$\boldsymbol{N}^e = \begin{bmatrix} L_1 & 0 & L_2 & 0 & L_3 & 0 \\ 0 & L_1 & 0 & L_2 & 0 & L_3 \end{bmatrix},$$

where L_1, L_2 and L_3 are area coordinates in the usual notation. Equation 11.13 then gives

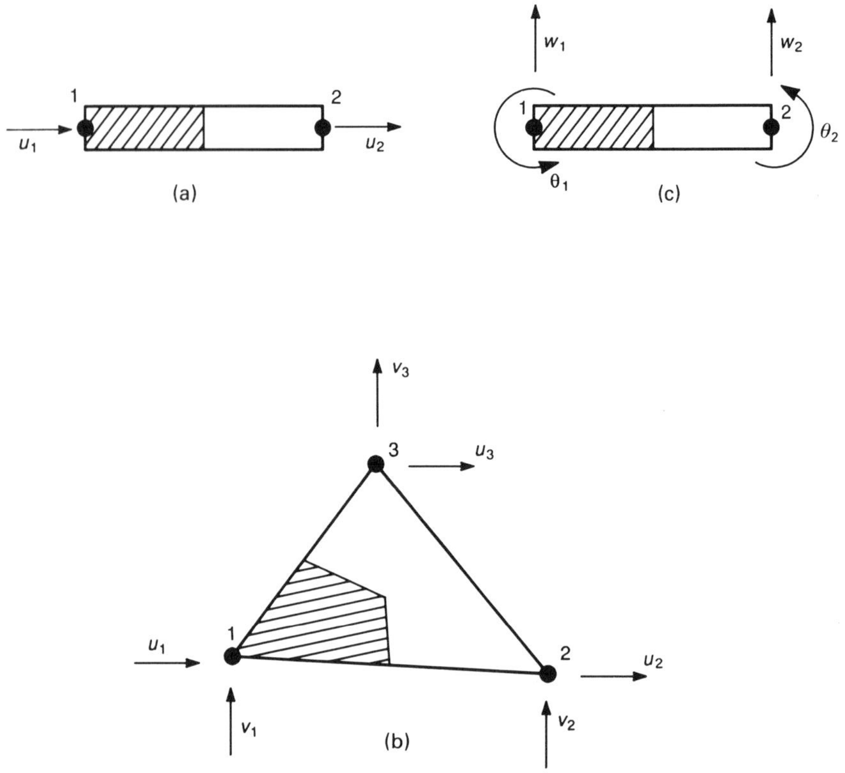

Fig. 11.2 *(a) Bar element, (b) beam element, (c) linear triangle. The shaded area represents the mass associated with node 1 in lumped formulation.*

$$M^{\mathrm{e}} = \int_A \rho \begin{bmatrix} L_1^2 & 0 & L_1L_2 & 0 & L_1L_3 & 0 \\ 0 & L_1^2 & 0 & L_1L_2 & 0 & L_1L_3 \\ L_2L_1 & 0 & L_2^2 & 0 & L_2L_3 & 0 \\ \vdots & \vdots & \vdots & \vdots & \vdots & \vdots \\ 0 & L_3L_1 & 0 & L_3L_2 & 0 & L_3^2 \end{bmatrix} h\mathrm{d}A,$$

where h is the element thickness. Using the identity in equation 8.22 to evaluate these integral terms, we obtain

$$M^{\mathrm{e}} = \frac{m}{12}\begin{bmatrix} 2 & 0 & 1 & 0 & 1 & 0 \\ 0 & 2 & 0 & 1 & 0 & 1 \\ 1 & 0 & 2 & 0 & 1 & 0 \\ 0 & 1 & 0 & 2 & 0 & 1 \\ 1 & 0 & 1 & 0 & 2 & 0 \\ 0 & 1 & 0 & 1 & 0 & 2 \end{bmatrix}, \qquad (11.15)$$

where m (= ρAh) is the mass of element.

(iii) *The simple beam element*

The Bernoulli–Euler beam element of section 9.2, is shown again in Fig. 11.2(c). It has a nodal displacement vector $\boldsymbol{d}^e = [w_1, \theta_1, w_2, \theta_2]^T$ and a shape matrix (equations 9.6 and 9.8)

$$\boldsymbol{N}^e = [1 - 3\xi^2 + 2\xi^3, \xi L(1 - 2\xi + \xi^2), 3\xi^2 - 2\xi^3, \xi L(\xi^2 - \xi)],$$

where $\xi = x'/L$ and x' is a local axial coordinate. Substitution into equation 11.13 gives

$$\boldsymbol{M}^e = \int_0^L \rho AL \begin{bmatrix} 1 - 3\xi^2 + 2\xi^3 \\ \xi L(1 - 2\xi + \xi^2) \\ 3\xi^2 - 2\xi^3 \\ \xi L(\xi^2 - \xi) \end{bmatrix} [1 - 3\xi^2 + 2\xi^3, \ldots, \xi L(\xi^2 - \xi)]\mathrm{d}\xi,$$

which can be integrated (with some effort) to give

$$\boldsymbol{M}^e = \left(\frac{m}{420}\right) \begin{bmatrix} 156, & 22L, & 54, & -13L \\ 22L, & 4L^2, & 13L, & -3L^2 \\ 54, & 13L, & 156, & -22L \\ -13L, & -3L^2, & -22L, & 4L^2 \end{bmatrix}, \tag{11.16}$$

where m (= ρAL) is the mass of element.

11.2.4 Lumped and diagonal masses

Matrices derived using expression 11.13 are properly termed 'consistent' mass matrices. Cruder representations of inertial loads are obtained by concentrating (or 'lumping') the mass of an element at its nodes. In the case of simple elements such as those discussed above, the mass is distributed equally between the nodes. In a 'lumped' bar element, for example, each node is subject to an inertial force equivalent to that experienced by a particle of mass $m/2$. The resulting inertial load vector is therefore

$$\mathbf{f}_I^e = -\begin{bmatrix} \frac{1}{2}m\ddot{u}_1 \\ \frac{1}{2}m\ddot{u}_2 \end{bmatrix} = -\begin{bmatrix} \frac{1}{2}m & 0 \\ 0 & \frac{1}{2}m \end{bmatrix} \begin{bmatrix} \ddot{u}_1 \\ \ddot{u}_2 \end{bmatrix},$$

giving a 'lumped' (subscript 'L') mass matrix, (*cf.* equation 11.14)

$$M_L^e = \begin{bmatrix} \frac{1}{2}m & 0 \\ 0 & \frac{1}{2}m \end{bmatrix}. \tag{11.17}$$

By a similar process, the lumped mass matrix for the constant strain triangle is obtained by concentrating one third of the mass at each vertex to give (*cf.* equation 11.15)

$$M_L^e = \left(\frac{m}{3}\right) \begin{bmatrix} 1 & 0 & 0 & 0 & 0 & 0 \\ 0 & 1 & 0 & 0 & 0 & 0 \\ 0 & 0 & 1 & 0 & 0 & 0 \\ 0 & 0 & 0 & 1 & 0 & 0 \\ 0 & 0 & 0 & 0 & 1 & 0 \\ 0 & 0 & 0 & 0 & 0 & 1 \end{bmatrix}. \tag{11.18}$$

The lumped representation of a beam element places one half of the mass at each end, giving (*cf.* equation 11.16)

$$M_L^e = \begin{bmatrix} \frac{1}{2}m & 0 & 0 & 0 \\ 0 & 0 & 0 & 0 \\ 0 & 0 & \frac{1}{2}m & 0 \\ 0 & 0 & 0 & 0 \end{bmatrix}. \tag{11.19a}$$

Lumped representations for elements with more complex topologies are less straightforward since it is not then obvious that an equal proportion of the total mass should be assigned to each node and indeed this is not generally the case. It is more satisfactory in such cases to distribute the total mass of the element among its translational displacements in the same proportion as the corresponding terms on the principal diagonal of the *consistent* mass matrix. The algorithm can be summarized as follows [3]:

(i) The diagonal terms m_{ii} of the consistent mass matrix are summed for the translational degrees of freedom δ_i. Denote the summed quantity by m'.

(ii) The lumped matrix is formed by placing a term $m_{ii}\,(\alpha m/m')$ in the ith position on the principal diagonal of an initially zero matrix. Here, m is the total mass of the element, i is a translational degree of freedom number and $\alpha\,(= 1, 2, 3)$ is the number of translations at each node.

A practical difficulty arises in implementing the lumped formulation in elements which have rotational degrees of freedom (beam and plate elements for example). By definition, a concentrated mass has no rotational inertia. Zeros therefore occur on the diagonal of any lumped matrix in rows which correspond to rotational degrees of freedom. In the case of the beam element for example (equation 11.19a) zero terms occur in rows two and four, which correspond to rotational degrees of freedom θ_1 and θ_2. The lumped mass matrix is then singular. This can generate spurious 'infinite' frequencies of vibration in any model formed from such elements. A simple but effective remedy is to add rotational inertia to the diagonal terms of the lumped matrix in the same proportion as in the consistent formulation. Step (ii) above is then modified so that *i* refers to rotational as well as translational degrees of freedom. In the case of the beam element, for example, this gives a modified lumped matrix, $\boldsymbol{M}_{\mathrm{L}}^{\mathrm{e}\prime}$, with components

$$\boldsymbol{M}_{\mathrm{L}}^{\mathrm{e}\prime}=\begin{bmatrix} m/2 & 0 & 0 & 0 \\ 0 & mL^2/78 & 0 & 0 \\ 0 & 0 & m/2 & 0 \\ 0 & 0 & 0 & mL^2/78 \end{bmatrix}. \tag{11.19b}$$

Confirmation of the above result is left to the reader as an exercise.

11.2.5 The assembled equations, inclusion of damping

The element contributions $\boldsymbol{f}_{\mathrm{I}}^{\mathrm{e}}$ can be assembled in the usual way to form an inertial force vector for the whole system. Given that each contribution is of the form $-\boldsymbol{M}^{\mathrm{e}}\ddot{\boldsymbol{d}}^{\mathrm{e}}$, the assembled vector $\boldsymbol{f}_{\mathrm{I}}$ is given by

$$\boldsymbol{f}_{\mathrm{I}}=-\boldsymbol{M}\ddot{\boldsymbol{d}},$$

where $\ddot{\boldsymbol{d}}$ is a vector of nodal accelerations and $\boldsymbol{M}$ is a mass matrix obtained by assembling the element mass matrices $\boldsymbol{M}^{\mathrm{e}}$ in the same way that the element stiffness matrices are assembled to form $\boldsymbol{K}$. Since each element mass matrix is symmetric, the assembled matrix is also symmetric.

The dynamic stiffness relationship for the assembled system, equation 11.7, then becomes

$$\boldsymbol{M}\ddot{\boldsymbol{d}}+\boldsymbol{K}\boldsymbol{d}=\boldsymbol{f}(t). \tag{11.20}$$

In scalar terms, this represents a system of coupled, second order, ordinary differential equations which must be solved for the nodal parameters $\delta_i(t)$. As currently formulated, equation 11.20 applies only to an undamped system. In practice some measure of damping is invariably present due to hysteretic loss or friction at joints. Although the damping forces can seldom be modelled with precision, they are generally assumed to be viscous in nature, that is to say, linearly proportional to the nodal velocities. Equation 11.20 therefore becomes

$$\boldsymbol{M}\ddot{\boldsymbol{d}}+\boldsymbol{C}\dot{\boldsymbol{d}}+\boldsymbol{K}\boldsymbol{d}=\boldsymbol{f}(t). \tag{11.21}$$

Here $\boldsymbol{C}$ is a constant 'damping' matrix. Unlike $\boldsymbol{K}$ and $\boldsymbol{M}$, the damping matrix cannot readily be assembled from element contributions since the physics of the damping mechanism is generally unclear at an element level. $\boldsymbol{C}$ must therefore be modelled by using empirical global models, or by estimating the damping in each mode of vibration. (Section 11.4.5 contains further discussion of both these approaches)

11.3 NORMAL MODE ANALYSIS, FREE VIBRATIONS

Consider now solutions of the undamped dynamic equations in the absence of external loading. In particular, let us look at discrete solutions which vary harmonically with time, that is, solutions of the form

$$\boldsymbol{d} = \bar{\boldsymbol{d}} \cos(\omega t), \tag{11.22}$$

where ω is a circular frequency in radians per second and $\boldsymbol{d}$ is a vector of nodal amplitudes. Nodal accelerations are then given by

$$\ddot{\boldsymbol{d}} = \partial^2 \boldsymbol{d}/\partial t^2 = -\omega^2 \bar{\boldsymbol{d}} \cos(\omega t), \tag{11.23}$$

and equation 11.21 becomes (after removal of the damping matrix $\boldsymbol{C}$ and the forcing vector $\boldsymbol{f}$)

$$[\boldsymbol{K} - \lambda \boldsymbol{M}]\bar{\boldsymbol{d}} = 0, \tag{11.24}$$

where $\lambda = \omega^2$.

This poses an eigenvalue problem in λ and has non-trivial solutions only if the coefficient matrix $[\boldsymbol{K} - \lambda \boldsymbol{M}]$ is singular. If $\boldsymbol{M}$ and $\boldsymbol{K}$ are of order n, and if $\boldsymbol{M}$ itself is non-singular (this is certainly true for consistent mass formulations) there are n eigenvalues, λ_i $(i = 1, 2, \ldots, n)$. These are the roots of the scalar equation

$$\det | \boldsymbol{K} - \lambda \boldsymbol{M} | = 0. \tag{11.25}$$

The eigenvalues themselves define physical frequencies, $\omega_i = \sqrt{\lambda_i}$, at which the system vibrates freely. Associated with each is an 'eigenvector' $\bar{\boldsymbol{d}}_i$ which satisfies equation 11.24 and which contains the nodal amplitudes for that particular mode of vibration.

In normal mode analysis, unlike static analysis, it is quite common for a model to be constrained in such a way that rigid body displacements are possible (an airframe in flight, for example). In such cases the stiffness matrix is singular and rigid body modes appear as non-trivial, 'zero frequency' solutions to the eigenvalue problem. These are valid physical solutions but can cause computational problems. In such situations it is sometimes convenient to modify the coefficient matrix of equation 11.24 by adding and subtracting a multiple of the mass matrix. A 'shift' of this type yields a new but equivalent eigenproblem

$$[(\boldsymbol{K} + \alpha\boldsymbol{M}) - (\lambda + \alpha)\boldsymbol{M}]\,\bar{\boldsymbol{d}} = 0. \tag{11.26}$$

The stiffness matrix $[\boldsymbol{K} + \alpha\boldsymbol{M}]$ is then non-singular, and although the eigenvalues have been increased by α, the eigenvectors are unaltered.

11.3.1 Free vibration of bars and beams, simple solutions

The solution of the eigenproblem in equation 11.25 for realistic finite element problems generally requires extensive computation. The algorithms used are discussed briefly in the next section. First, however, some simple manual eigensolutions are demonstrated for systems with one- and two-degrees of freedom.

Example (i): Longitudinal vibration of a uniform bar

An exact solution for the axial modes of vibration of a prismatic bar can readily be derived using analytic techniques. Details are given in [1]. In the case of a bar of length L, fixed at one end and free at the other, the eigenfrequencies are

$$\omega_i = (2i - 1)\frac{\pi}{2}\sqrt{\frac{k}{m}}, \quad i = 1, 2, 3, \ldots, \infty,$$

where k is the axial stiffness of the bar $(= EA/L)$ and m is its mass $(= \rho AL)$. The corresponding eigenmodes are

$$u_i(x) = \sin\,(2i - 1)\frac{\pi x}{2L}, \quad i = 1, 2 \ldots$$

where x is an axial coordinate whose origin is at the fixed end. The first two mode shapes are illustrated in Fig. 11.3. Approximate solutions for this problem are presented using the two finite element models shown in Fig. 11.3. Model (a) consists of a single element with one unconstrained degree of freedom, δ_1. Model (b) is formed from two elements and has two degrees of freedom, δ_1 and δ_2.

The partitioned stiffness and mass matrices for system (a), using the 'consistent' formulation, are

$$\boldsymbol{K} = [k] \text{ and } \boldsymbol{M} = \left[\frac{m}{3}\right].$$

The determinant of $[\boldsymbol{K} - \lambda\boldsymbol{M}]$ is then simply $(k - \lambda(m/3))$, and equation 11.25 yields a single eigenvalue $\lambda_1 = (3k/m)$, giving an eigenfrequency $\omega_1 = 1.732\sqrt{(k/m)}$. This is quite close to the exact result $1.5708\sqrt{(k/m)}$, surprisingly so perhaps given the crudeness of the physical model. The corresponding eigenvector $\bar{\boldsymbol{d}}_1$ is simply

$$\bar{\boldsymbol{d}}_1 = \{1\}\,.$$

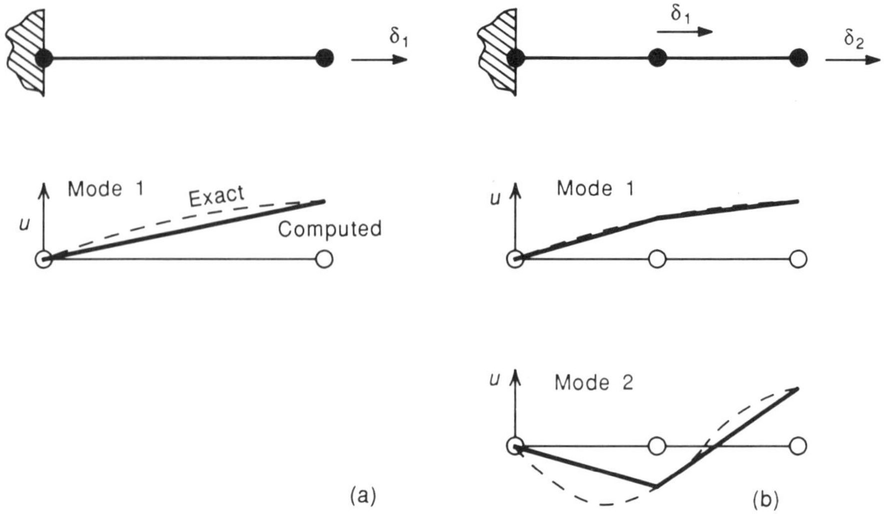

Fig. 11.3 *Axial vibration of a uniform bar: (a) one-degree of freedom model, mesh and first eigenmode, (b) two-degree of freedom model, mesh and eigenmodes.*

The resulting eigenmode — obtained using linear interpolation within the element — is shown in Fig. 11.3(a) and forms a primitive approximation to the half sine wave of the exact solution.

The two-element model is less trivial. The matrix eigenproblem 11.24 becomes

$$\left[\begin{bmatrix} 4k & -2k \\ -2k & 2k \end{bmatrix} - \lambda \begin{bmatrix} m/3 & m/12 \\ m/12 & m/6 \end{bmatrix}\right] \begin{bmatrix} \delta_1 \\ \delta_2 \end{bmatrix} = \begin{bmatrix} 0 \\ 0 \end{bmatrix},$$

and the eigenvalues are then the roots of

$$\det |\boldsymbol{K} - \lambda \boldsymbol{M}| = \left(4k - \frac{\lambda m}{3}\right)\left(2k - \frac{\lambda m}{6}\right) - \left(2k + \frac{\lambda m}{12}\right)^2 = 0.$$

These are $\lambda_1 = 2.597(k/m)$ and $\lambda_2 = 31.69(k/m)$, giving $\omega_1 = 1.611\sqrt{(k/m)}$ and $\omega_2 = 5.629\sqrt{(k/m)}$. The first is within 3% of the exact solution, the second some 20% in error. The corresponding eigenmodes, normalized with respect to displacement at the free end, are

$$\bar{\boldsymbol{d}}_1 = \begin{bmatrix} 1/\sqrt{2} \\ 1 \end{bmatrix} \text{ and } \bar{\boldsymbol{d}}_2 = \begin{bmatrix} -1/\sqrt{2} \\ 1 \end{bmatrix}.$$

The ratio of nodal values in each eigenvector is then correct when compared to the exact solution, and the interpolated eigenmodes are primitive but recognizable representations of the exact (sinusoidal) eigensolutions (Fig. 11.3(b)).

We proceed no further here in refining the mesh, but the accuracy with which each eigenfrequency is resolved improves steadily as the number of elements increases (see problem 10 at the end of this chapter).

Example (ii). Lateral vibration of a cantilever beam

The natural frequencies of lateral vibration of an Euler beam of length L, fixed at one end and free at the other (Fig. 11.4), can be calculated analytically without difficulty (as detailed in [1] and [2]). The first two eigenfrequencies are

$$\omega_1 = 3.516\sqrt{\frac{EI}{mL^3}} \text{ and } \omega_2 = 22.03\sqrt{\frac{EI}{mL^3}},$$

A finite element model of the beam is constructed using a single Euler beam element. The model has two unconstrained degrees of freedom, a displacement δ_1 and a rotation δ_2 at the free end (Fig. 11.4). The discrete eigenproblem is then formed by assembling and partitioning the stiffness and (consistent) mass matrices to give

$$\left[\begin{bmatrix} 12EI/L^3 & -6EI/L^2 \\ -6EI/L^2 & 4EI/L \end{bmatrix} - \lambda \begin{bmatrix} 156m/420 & -22mL/420 \\ -22mL/420 & 4mL^2/420 \end{bmatrix}\right]\begin{bmatrix} \delta_1 \\ \delta_2 \end{bmatrix} = \begin{bmatrix} 0 \\ 0 \end{bmatrix}.$$

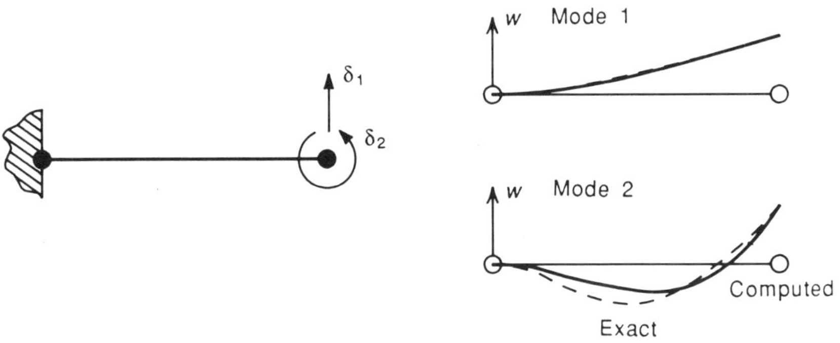

Fig. 11.4 *Flexural vibration of a beam, mesh and eigenmodes.*

Equating the determinant of the coefficient matrix to zero, we deduce that the eigenvalues are roots of

$$140\beta^2 - 408\beta + 12 = 0,$$

where $\beta = \lambda mL^3/420EI$, thus giving: $\beta_1 = 0.02971$ and $\beta_2 = 2.885$, or

$$\omega_1 = 3.533\sqrt{\frac{EI}{mL^3}} \text{ and } \omega_2 = 34.81\sqrt{\frac{EI}{mL^3}}.$$

The corresponding eigenvectors, normalized with respect to the displacement of the free end, are

$$\bar{\boldsymbol{d}}_1 = \begin{bmatrix} 1.0 \\ 1.378/L \end{bmatrix}, \text{ and } \bar{\boldsymbol{d}}_2 = \begin{bmatrix} 1.0 \\ 7.622/L \end{bmatrix}.$$

Interpolated eigenmodes based on these nodal values are plotted in Fig. 11.4 along with their exact counterparts. The first mode is clearly a very close approximation to the exact solution (the error in the eigenfrequency is less than 0.5%) whereas the second, although clearly recognizable, gives a frequency which is some 60% in error. As the mesh is refined, this mode also is more accurately resolved, the error dropping to 1% with the addition of a second element.

11.3.2 Condensation of the eigenvalue problem, Guyan reduction

The proportion of reasonably accurate eigenvalues and recognizable eigenmodes produced in the simple, manual calculations of the previous section is quite high. Both of the two-degree of freedom models, for example, resolve the lowest eigenfrequency to within a few percent of its exact value and give at least a crude approximation to the second. Such performance is seldom observed proportionally in larger and more complex models. The finite element method is, in fact, quite profligate in the number of degrees of freedom which must be used to resolve a prescribed number of modes. Commonly, one quarter or less of the computed eigenmodes are predicted with useful accuracy. The disproportion between the number of reliably resolved modes and the order of the eigenproblem is exacerbated in many instances by purely geometrical considerations which dictate the number of elements which are needed to model the basic shape of the object irrespective of its dynamic behaviour. A typical model of this sort is illustrated in Fig. 11.5. This shows a doubly curved impeller blade. A portion of the conical hub to which it is fixed is also shown, but not modelled. The blade in Fig. 11.5(a) is represented by 28 rectangular plate elements. The first four eigenmodes and their computed modeshapes are shown in Fig. 11.5(b)–(e). The mesh used, although quite crude in terms of the curved geometry of the blade, involves 192 degrees of freedom. It is, however, marginal for an accurate representation of the third and fourth modes. In effect we have been forced to solve an eigenproblem of order 192 to obtain *four* reliable eigenmodes. This order of disparity between the dimensionality of the eigenmatrix and the number of useful eigensolutions is not uncommon.

A number of 'condensation' methods can be used to improve this ratio while retaining accuracy in the lower order modes. The most common of these is 'Guyan reduction' (also termed 'mass condensation' and 'eigenvalue economization'). Here, the nodal variables of the full model are divided into 'masters'

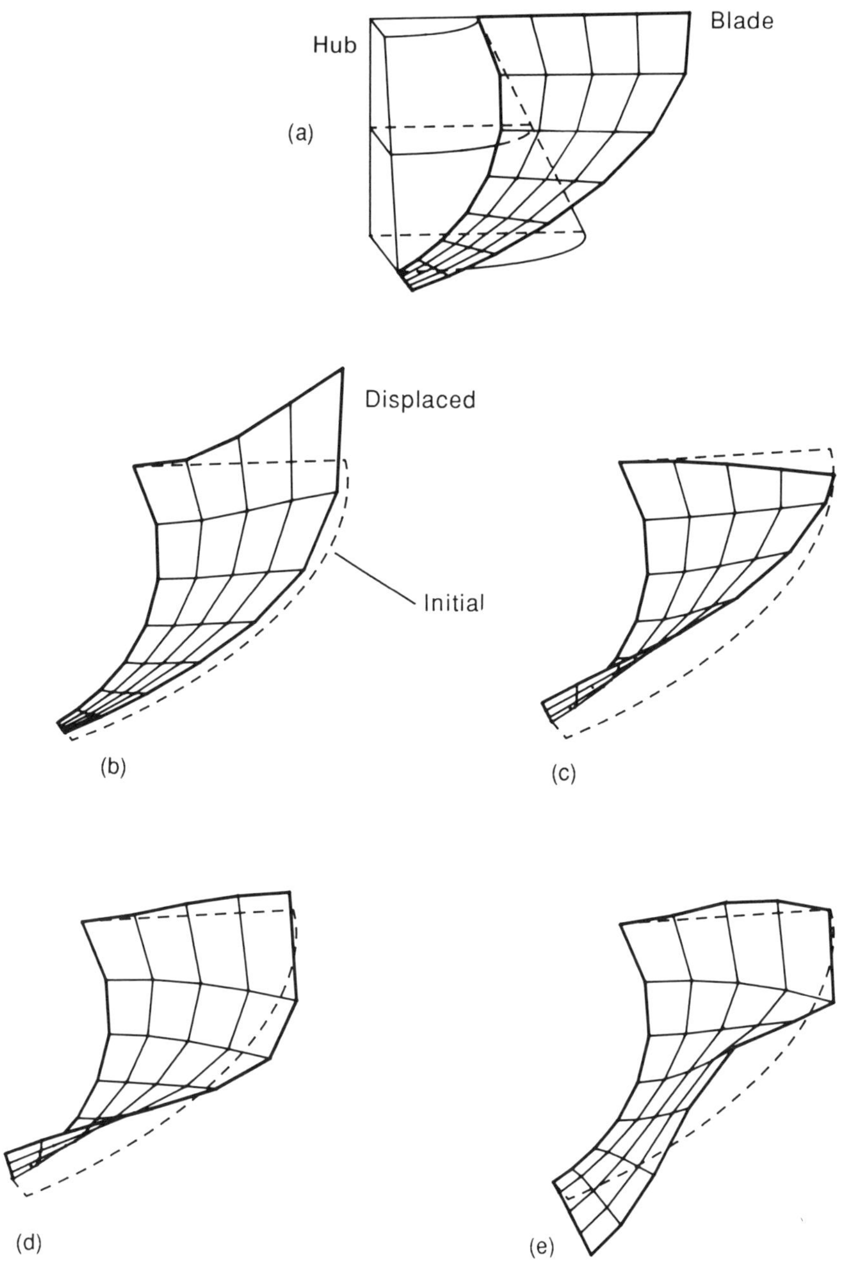

Fig. 11.5 *Vibration of a helical blade, mesh and first four eigenmodes. (a) plate elements, (b) mode 1:* $\omega_1 = 190$Hz, *(c) mode 2:* $\omega_2 = 265$Hz, *(d) mode 3:* $\omega_3 = 326$Hz, *(e) mode 4:* $\omega_4 = 428$Hz.

and 'slaves'. The masters are placed in the solution vector of a modified eigenproblem and the 'slaves' are tied to them through the discrete equations of static equilibrium. The method is most easily visualized by re-ordering the degrees of freedom of the full problem so that the masters, $\bar{\boldsymbol{d}}_{\mathrm{m}}$, come first and the slaves, $\bar{\boldsymbol{d}}_{\mathrm{s}}$, last. In partitioned form, equation 11.24 becomes,

$$\left[\begin{bmatrix} \boldsymbol{k}_{\mathrm{mm}} & \boldsymbol{K}_{\mathrm{ms}} \\ \boldsymbol{K}_{\mathrm{sm}} & \boldsymbol{K}_{\mathrm{ss}} \end{bmatrix} - \lambda \begin{bmatrix} \boldsymbol{M}_{\mathrm{mm}} & \boldsymbol{M}_{\mathrm{ms}} \\ \boldsymbol{M}_{\mathrm{sm}} & \boldsymbol{M}_{\mathrm{ss}} \end{bmatrix}\right] \begin{bmatrix} \bar{\boldsymbol{d}}_{\mathrm{m}} \\ \bar{\boldsymbol{d}}_{\mathrm{s}} \end{bmatrix} = \begin{bmatrix} 0 \\ 0 \end{bmatrix} \tag{11.27}$$

By neglecting entirely the mass terms in the bottom partitioned row, we obtain an approximate relationship between $\bar{\boldsymbol{d}}_{\mathrm{m}}$ and $\bar{\boldsymbol{d}}_{\mathrm{s}}$ based on stiffness considerations only, that is

$$\boldsymbol{K}_{\mathrm{sm}}\bar{\boldsymbol{d}}_{\mathrm{m}} + \boldsymbol{K}_{\mathrm{ss}}\bar{\boldsymbol{d}}_{\mathrm{s}} = 0, \text{ or } \bar{\boldsymbol{d}}_{\mathrm{s}} = -\boldsymbol{K}_{\mathrm{ss}}^{-1}\boldsymbol{K}_{\mathrm{sm}}\bar{\boldsymbol{d}}_{\mathrm{m}}.$$

This may be incorporated in the transformation,

$$\begin{bmatrix} \bar{\boldsymbol{d}}_{\mathrm{m}} \\ \bar{\boldsymbol{d}}_{\mathrm{s}} \end{bmatrix} = \begin{bmatrix} \boldsymbol{I} \\ -\boldsymbol{K}_{\mathrm{ss}}^{-1}\boldsymbol{K}_{\mathrm{sm}} \end{bmatrix} [\bar{\boldsymbol{d}}_{\mathrm{m}}] \text{ or } \bar{\boldsymbol{d}} = \boldsymbol{T}\bar{\boldsymbol{d}}_{\mathrm{m}}, \tag{11.28}$$

which relates the nodal displacements of the entire system back to the components of $\bar{\boldsymbol{d}}_{\mathrm{m}}$. Substituting into equation 11.27 and premultiplying by $\boldsymbol{T}^{\mathrm{T}}$, we obtain the 'condensed' eigenvalue problem,

$$[\boldsymbol{K}_{\mathrm{c}} - \lambda\boldsymbol{M}_{\mathrm{c}}]\bar{\boldsymbol{d}}_{\mathrm{m}} = 0, \text{ where } \boldsymbol{K}_{\mathrm{c}} = \boldsymbol{T}^{\mathrm{T}}\boldsymbol{K}\boldsymbol{T} \text{ and } \boldsymbol{M}_{\mathrm{c}} = \boldsymbol{T}^{\mathrm{T}}\boldsymbol{M}\boldsymbol{T}. \tag{11.29}$$

The condensed (subscript c) stiffness matrix $\boldsymbol{K}_{\mathrm{c}}$ and mass matrix $\boldsymbol{M}_{\mathrm{c}}$ then contain much of the information of the original model, and the lower order eigenmodes can be resolved to a much greater accuracy than would be possible using a coarser, uncondensed model of the same dimensionality. The only obvious disadvantage of the procedure is that the condensed matrices are no longer banded and do not preserve sparse characteristics in the originals. Large computational savings can however be made, depending on the type of eigensolver used (see further comments in next section).

The effectiveness of the method is demonstrated by the test case illustrated in Fig. 11.6. Here, 50 triangular plate elements are used to model a square cantilever plate of side a and flexural rigidity D fixed along one edge. The displacement of each node is represented by three degrees of freedom, a lateral displacement and two rotations. The full model therefore has 90 unconstrained degrees of freedom. The eigenfrequencies of the plate were computed [6] using the full model and also using a condensed model obtained by taking six lateral displacements as master degrees of freedom (these are indicated in Fig. 11.6(b)). Values of the computed, nondimensional eigenfrequencies $\Omega_i = \omega_i/\sqrt{(D/m)}$ (m is the mass of the plate, D its flexural rigidity) for the first four modes are tabulated below each model. Although the condensed problem is only of order one fifteenth that of the original, the fundamental eigenfrequency is resolved to within 0.12% of the value obtained from the full model. Corresponding discrepancies in modes two, three and four are 1%, 6% and 9%. Much of the accuracy of the original model is thus preserved in the resolution of lower order modes while the size of the eigenproblem is dramatically reduced.

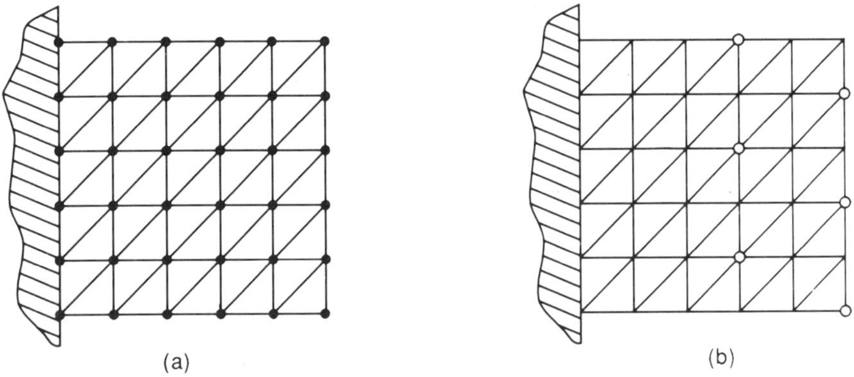

$\Omega_1 = 3.469$, $\Omega_2 = 8.535$, $\Omega_3 = 21.450$, $\Omega_4 = 27.059$.

$\Omega_1 = 3.473$, $\Omega_2 = 8.604$, $\Omega_3 = 22.690$, $\Omega_4 = 29.490$.

Fig. 11.6 *Vibration of a cantilever plate [6]: (a) full model, 90 degrees of freedom, (b) condensed model, 6 degrees of freedom.* • *— convention node (three degrees of freedom).* ○ *— master node (one degree of freedom).*

The selection of master degrees of freedom can be performed 'manually' or 'automatically'. When selected manually, the master displacements must be chosen so that they are capable of representing and distinguishing between the lower order modes. There is little point, for example, in choosing master degrees of freedom which are clustered together in a small portion of the mesh, since the lower modes will generally exhibit large scale variations of displacement over the whole model. Automatic selection of masters removes such decisions from the finite element user. The simplest and most popular of these automatic procedures simply selects those degrees of freedom which have the largest ratio of diagonal mass to stiffness in successively condensed versions of the eigenproblem. Further details are to be found in [7].

11.3.3 Computational eigensolvers

Whether or not the full problem is condensed, a relatively large matrix eigenproblem must ultimately be solved in most normal mode analyses of practical interest. This is done using a variety of computational methods. The effort involved depends very much on the nature of the eigenproblem, the order and sparseness of the eigenmatrices, whether they are banded, how many eigenvalues are to be found, where they lie in the total spectrum of eigenvalues and so on. Some observations are now made on the interactions which occur between some of these factors and computational eigenvalue solvers in common use.

All eigensolvers are iterative and differ in this regard from many of the linear equation solvers discussed in Chapter 5. The solution of an eigenvalue problem is, in fact, equivalent in most instances to the solution of *many* sets of linear equations of the same order. The computational effort is consequently much greater, particularly when a full set of eigenvalues is required. The use of efficient and appropriate algorithms is of the utmost importance.

We make no attempt here to explain in detail the mathematical bases and practical workings of the eigensolvers to be discussed. The most effective solvers are quite complex on both counts and details must be sought in more specialized texts such as [10]. Discussion is limited here to a brief description of the various categories of solver with comments on their applicability and efficiency in the context of finite element calculations. Readers seeking further information on particular methods should, in the first instance, consult general treatments such as those of [8] and [9]. More detailed discussion specifically orientated towards finite element application can be found in [10] and [11] and a useful up to date review in [12]. The methods used to solve eigenproblems include the following:

(i) *Vector iteration.* Methods of this type — referred to also as 'power' methods — extract the highest (forward iteration) or lowest (inverse iteration) eigenvalue by iterative improvement of the corresponding eigenvector. The eigenproblem can then be 'deflated' and the next largest (smallest) root sought. Vector iteration methods are simple to program and can be coded to exploit bandedness and sparsity. They are computationally inefficient however and are seldom used for eigenvalue extraction in large problems. When used with a 'shift', however, they form an efficient method for generating eigenvectors for known eigenvalues and are used in this way as an adjunct to more efficient extraction algorithms.

(ii) *Subspace iteration.* Subspace iteration is one of the most effective algorithms for finite element eigenvalue extraction. It is particularly efficient in situations where a relatively small number of eigenvalues must be extracted from a large sparse system. The method is based on iteration of a vector subspace rather than iteration of individual vectors. A comprehensive description (and FORTRAN listing) is presented in Chapter 12 of [10]. Extraction of the lowest p eigenvectors using this method, requires a subspace of q vectors where q is larger than p but not greatly so (a recommended value is $2p$ or $(p+8)$ whichever is the least). The choice of the initial vectors is important, but can be automated using logic similar to that employed in the automatic selection of dynamic masters for Guyan reduction. In this regard, the use of subspace iteration generally replaces any requirement for prior condensation of the eigenproblem. A Sturm sequence check is usually incorporated to ensure that no eigenvectors have been 'missed'. The whole algorithm can be coded to utilize bandedness and sparsity.

(iii) *Determinant search.* These methods search directly for roots of the implicit polynomial $f(\lambda) = \det|\boldsymbol{K} - \lambda\boldsymbol{M}|$ using standard techniques (Newton–Rhapson, false position, repeated bisection etc). The determinant itself is generally calculated by factorizing $|\boldsymbol{K} - \lambda\boldsymbol{M}|$ into lower and upper matrices, a process which is expedited by first reducing the eigenproblem to tridiagonal form (see (iv) below). Once again the Sturm sequence property can be used to ensure that no root has been missed, and indeed can be used in its own right as part of the search algorithm. Solvers based on the determinant search approach can generally be coded to preserve bandedness and sparsity.

(iv) *Transformation methods.* These generally require the eigenproblem to be written in 'standard' form as

$$[\boldsymbol{A} - \lambda\boldsymbol{I}]\boldsymbol{y} = 0.$$

This is accomplished while preserving symmetry in the matrix $\boldsymbol{A}$ by pre and post multiplying the original eigenproblem (equation 11.24) by $\boldsymbol{L}^{-1}$ and $\boldsymbol{L}^{-\mathrm{T}}$ where $\boldsymbol{M} = \boldsymbol{L}\boldsymbol{L}^{\mathrm{T}}$, giving $\boldsymbol{A} = \boldsymbol{L}^{-1}\boldsymbol{K}\boldsymbol{L}^{-\mathrm{T}}$ and $\boldsymbol{y} = \boldsymbol{L}^{\mathrm{T}}\boldsymbol{d}$ (this factorization is greatly simplified when $\boldsymbol{M}$ is a diagonal matrix). Transformation methods then rely on the property that $\boldsymbol{A}^* = \boldsymbol{P}^{\mathrm{T}}\boldsymbol{A}\boldsymbol{P}$ has the same eigenvalues as $\boldsymbol{A}$ provided that $\boldsymbol{P}$ possesses the property, $\boldsymbol{P}^{\mathrm{T}}\boldsymbol{P} = \boldsymbol{I}$. A sequence of orthogonal transformations of this type can be applied to produce a matrix $\boldsymbol{A}^*$ whose eigenvalues are more easily extracted than those of the original matrix. In particular, when $\boldsymbol{A}^*$ is a triangular matrix, its eigenvalues lie on the principal diagonal. The most straightforward method of this type, Jacobi iteration, iterates progressively to reduce or eliminate the largest off-diagonal term at each iteration. When a specified threshold value is reached for all of the off-diagonal terms, the eigenvalues are simply read from the diagonal. More efficient, transformation methods extract the eigenvalues in two distinct phases. First, the eigenmatrix $\boldsymbol{A}$ is reduced to tridiagonal form using a finite number of orthogonal transformations (Givens or Householder transformations are commonly used). Once in tridiagonal form, a final iterative reduction is performed. This is typically a QR iteration but any of the methods already mentioned (determinant search for example) may also be used. The eigenvectors, if required, are obtained using inverse iteration. The most effective algorithm of this general description is the Householder–QR–Inverse iteration (HQRI) method. It is a feature of the HQRI approach, and transformation methods in general, that they produce a complete set of eigenvalues. They do this with extreme efficiency, when compared to most other methods, but cannot be used to selectively extract a few eigenvalues from a large system. They tend therefore to be most effective when the eigenmatrix is full or has a large bandwidth, when all or most of the eigenvectors and eigenvalues are required, and when Guyan reduction has been used.

(v) *The Lanczos algorithm*. This is the most recent of the eigenvalue extraction algorithms and does not fit conveniently into any of the previous categories. The 'classical' Lanczos approach was introduced in the early 1950s as a method for calculating a few extreme eigenvalues. It was also used, with limited success, to reduce standard eigenproblems to tridiagonal form. Recent developments, however, have extended the scope of the method well beyond these original applications and have overcome many of the early difficulties associated with 'round off' errors. More recent versions of the Lanczos algorithm share with the subspace iteration method the ability to selectively extract a limited set of eigenfrequencies for a large problem. This is an extremely useful feature in the finite element context. Moreover although the Lanczos algorithm is complex (details and program listings are given in [11] and [12]) it can be programmed to respect banded and skyline storage in a way that most transformation methods cannot. It is not widely used in commercial programs at the present time, but is likely to become more so and is perhaps 'destined to supplant other eigenproblem algorithms in the finite element community' in the near future [4].

11.4 DYNAMIC RESPONSE, MODAL METHODS

Consider now the second phase of a dynamic analysis, the calculation of the response of a system to a given time-dependent loading. In such calculations, the forcing vector $\boldsymbol{f}(t)$ of equation 11.21 is a known function of time and the dynamic equations are integrated to give the displacement history for given initial values of velocity and displacement. To summarize, the equations

$$\boldsymbol{M}\ddot{\boldsymbol{d}} + \boldsymbol{C}\dot{\boldsymbol{d}} + \boldsymbol{K}\boldsymbol{d} = \boldsymbol{f}(t), \qquad (11.30)$$

are solved subject to initial conditions

$$\boldsymbol{d}(0) = \boldsymbol{d}_0 \text{ and } \dot{\boldsymbol{d}}(0) = \boldsymbol{v}_0, \qquad (11.31)$$

where $\boldsymbol{d}_0$ and $\boldsymbol{v}_0$ are nodal displacements and velocities at time $t = 0$. In the case of a system with n degrees of freedom, we must solve a set of n second order, differential equations for the nodal parameters $\delta_i(t)$ $(i = 1, \ldots, n)$. The equations are coupled through the off-diagonal components of $\boldsymbol{M}$, $\boldsymbol{K}$ and $\boldsymbol{C}$ and must be solved simultaneously.

If a normal mode analysis is performed first, an equivalent set of *uncoupled* equations can readily be formed by using modal amplitudes, rather than nodal variables, as the time-dependent parameters of the system. These 'modal' equations can be solved more easily than their 'nodal' equivalents and form the basis for the 'modal' approaches to the calculation of dynamic response. A direct numerical assault on equation 11.26 is also possible, and will be discussed in the following section.

11.4.1 Orthogonality of eigenmodes, modal mass, damping and stiffness

Since $\boldsymbol{K}$ and $\boldsymbol{M}$ are symmetric, eigenvectors $\bar{\boldsymbol{d}}_i$ and $\bar{\boldsymbol{d}}_j$ which satisfy the undamped eigenproblem 11.24 are orthogonal to both matrices, in other words

$$\bar{\boldsymbol{d}}_i^{\mathrm{T}} \boldsymbol{M} \bar{\boldsymbol{d}}_j = 0 \text{ and } \bar{\boldsymbol{d}}_i^{\mathrm{T}} \boldsymbol{K} \bar{\boldsymbol{d}}_j = 0 \qquad (i \neq j). \tag{11.32a}$$

This result is readily established for distinct eigenvalues, but holds also for repeated eigenvalues provided that the eigenmodes are chosen appropriately.

There is no physical reason why the damping matrix should be governed by the same orthogonality condition but it is often assumed that this is the case (this will be discussed further in section 11.4.5). Equations 11.32a are then supplemented by an additional orthogonality condition

$$\bar{\boldsymbol{d}}_i^{\mathrm{T}} \boldsymbol{C} \bar{\boldsymbol{d}}_j = 0, \qquad (i \neq j). \tag{11.32b}$$

In the case $i = j$, the matrix products 11.32a and 11.32b define nonzero scalar quantities termed the 'modal' mass, stiffness and damping. These are given by

$$m_i = \bar{\boldsymbol{d}}_i^{\mathrm{T}} \boldsymbol{M} \bar{\boldsymbol{d}}_i, \quad k_i = \bar{\boldsymbol{d}}_i^{\mathrm{T}} \boldsymbol{K} \bar{\boldsymbol{d}}_i \text{ and } c_i = \bar{\boldsymbol{d}}_i^{\mathrm{T}} \boldsymbol{C} \bar{\boldsymbol{d}}_i. \tag{11.33}$$

By substituting λ_i and $\bar{\boldsymbol{d}}_i$ into equation 11.24 and premultiplying by $\bar{\boldsymbol{d}}_i^{\mathrm{T}}$, we note for future reference that the modal stiffnesses and masses are related by the identity

$$k_i = \lambda_i m_i = \omega_i^2 m_i. \tag{11.34}$$

11.4.2 Transformation to modal coordinates

A solution to equation 11.30 is now sought as an expansion of undamped eigenvectors and time dependent coefficients. In other words, the time-dependent solution vector $\boldsymbol{d}(t)$ is written

$$\boldsymbol{d}(t) = \sum_{j=1}^{n} a_j(t) \bar{\boldsymbol{d}}_j = \boldsymbol{E}\, \boldsymbol{a}(t), \tag{11.35}$$

where $\boldsymbol{a}(t)$ is a column vector of the modal amplitudes, $a_1(t), a_2(t), \ldots, a_n(t)$, and $\boldsymbol{E}$ is a matrix whose columns are the eigenmodes of equation 11.24. That is,

$$\boldsymbol{E} = [\bar{\boldsymbol{d}}_1, \bar{\boldsymbol{d}}_2, \ldots, \bar{\boldsymbol{d}}_n]. \tag{11.36}$$

By substituting equation 11.35 into equation 11.30 and premultiplying by $\boldsymbol{E}^{\mathrm{T}}$ we obtain the 'modal' equations of dynamic equilibrium

$$\boldsymbol{M}^* \ddot{\boldsymbol{a}} + \boldsymbol{C}^* \dot{\boldsymbol{a}} + \boldsymbol{K}^* \boldsymbol{a} = \boldsymbol{f}^*(t), \tag{11.37}$$

where

$$\boldsymbol{M}^* = \boldsymbol{E}^{\mathrm{T}}\boldsymbol{M}\boldsymbol{E}, \quad \boldsymbol{K}^* = \boldsymbol{E}^{\mathrm{T}}\boldsymbol{K}\boldsymbol{E}, \quad \boldsymbol{C}^* = \boldsymbol{E}^{\mathrm{T}}\boldsymbol{C}\boldsymbol{E} \text{ and } \boldsymbol{f}^*(t) = \boldsymbol{E}^{\mathrm{T}}\boldsymbol{f}(t). \tag{11.38}$$

11.4.3 Mode-displacement method

Orthogonality of the eigenmodes with respect to $\boldsymbol{M}$, $\boldsymbol{K}$ and $\boldsymbol{C}$ — assuming in the latter case that equation 11.32b holds — ensures that $\boldsymbol{M}^*, \boldsymbol{K}^*$ and $\boldsymbol{C}^*$ are diagonal matrices with components m_i, k_i and c_i on their principal diagonals. Equations 11.37 can therefore be decoupled to give n independent, scalar equations,

$$m_i \ddot{a}_i + c_i \dot{a}_i + k_i a_i = f_i^*(t), \qquad (i = 1, \ldots, n). \tag{11.39}$$

The initial conditions can also be transformed to the modal representation by substituting equation 11.35 into equation 11.31 and premultiplying by $\boldsymbol{E}^{\mathrm{T}}\boldsymbol{M}$. After some manipulation — and further use of the orthogonality condition — we obtain

$$a_i(0) = \left(\frac{1}{m_i}\right)\bar{\boldsymbol{d}}_i^{\mathrm{T}}\boldsymbol{M}\,\boldsymbol{d}_0, \text{ and } \dot{a}_i(0) = \left(\frac{1}{m_i}\right)\bar{\boldsymbol{d}}_i^{\mathrm{T}}\boldsymbol{M}\,\boldsymbol{v}_0. \tag{11.40}$$

The solution of equations 11.39 subject to initial conditions in equations 11.40 is now a much more straightforward exercise, the original problem having been reduced to the solution of n uncoupled, single-degree of freedom systems. Procedures for solving equation 11.39 are well established and are described in many elementary texts on dynamics and vibration (for example, see Chapter 6 of [1] or Chapter 7 of [2]), the standard solutions being written in terms of convolution integrals. In applying such methods to equation 11.39, it is convenient to express the modal damping component c_i in terms of a critical damping coefficient $\xi_i = c_i/(2m_i\omega_i)$. A typical member of equation set 11.39 can then be written

$$m_i(\ddot{a}_i + 2\xi_i\omega_i\dot{a}_i + \omega_i^2 a_i) = f_i^*(t). \tag{11.41}$$

This has the general solution (equation 6.7 of [1])

$$a_i(t) = a_i(0)\exp(-\xi_i\omega_i t)\cos(\omega_i^{\mathrm{d}}t)$$

$$+\frac{1}{\omega_i^{\mathrm{d}}}\{\dot{a}_i(0)+\xi_i\omega_i a_i(0)\}\exp(-\xi_i\omega_i t)\sin(\omega_i^{\mathrm{d}}t)$$

$$+\left(\frac{1}{m_i\omega_i^{\mathrm{d}}}\right)\int_0^t\{f_i^*(\tau)\exp(-\xi_i\omega_i(t-\tau))\sin(\omega_i^{\mathrm{d}}(t-\tau))\}\,\mathrm{d}\tau. \quad (11.42)$$

where ω_i^{d} is the 'damped' natural frequency ($=\omega_i\sqrt{(1-\xi_i^2)}$). The convolution integral on the right hand side is usually evaluated numerically but can be calculated analytically in simple cases. In either event the physical solution is reconstructed by substituting the calculated values for $a_i(t)$ into equation 11.35. This constitutes the 'mode-displacement' approach. A simple application of the method to an undamped, two-degree of freedom finite element model is included as an exercise at the end of this chapter (problem 21).

11.4.4 Mode-acceleration method

We have assumed so far that *all* of the eigenmodes of the discrete problem are present in expansion 11.35. In practice, the series is truncated after a relatively small number of terms, m say where $m < n$. The amplitudes of the modes retained are still governed by expression 11.41 but in the reconstruction of the solution, only m terms are included. This introduces an element of approximation into expression 11.35 as a representation of the full, discrete solution. One might reasonably expect that this approach will give a good approximation to the overall response, provided that those modes whose eigenfrequencies lie within the spectrum of the exciting force are included in the truncated series. Broadly speaking this is correct, and in many finite element models higher frequency eigenmodes can be neglected entirely without adversely affecting the accuracy of the solution. There are exceptions however. In impact problems, for example, high frequency modes make a significant contribution and cannot necessarily be omitted, and at the other extreme of the loading spectrum, difficulties arise for problems in which the quasistatic portion of the response is significant. In the latter case, many modes are required to represent accurately the (almost) 'steady' portion of the solution. Paradoxically therefore, as the loading more closely approaches the static case, the number of terms required in the modal expansion increases. An alternative reconstruction of the solution — the 'mode-acceleration' approach — can significantly improve the accuracy of the modal representation in such situations. The mode-acceleration solution is obtained by rewriting the original untransformed dynamic equation 11.30 as

$$\boldsymbol{d} = \boldsymbol{K}^{-1}(\boldsymbol{f}(t) - \boldsymbol{M}\ddot{\boldsymbol{d}} + \boldsymbol{C}\dot{\boldsymbol{d}}). \quad (11.43)$$

Modal amplitudes are again obtained from equation 11.42, but the truncated modal expansion is then substituted into the mass and damping terms only. After some manipulation, this gives

$$\boldsymbol{d}(t) = \boldsymbol{K}^{-1}\boldsymbol{f}(t) - \sum_{i=1}^{m} \left\{ \left(\frac{2\xi_i}{\omega_i}\right) \dot{a}_i(t)\, \boldsymbol{d}_i + \left(\frac{1}{\omega_i}\right)^2 \ddot{a}_i(t)\, \boldsymbol{d}_i \right\}. \tag{11.44}$$

The first term now models exactly the 'steady' or quasistatic portion of the response and it is found that fewer eigenmodes are required to provide an adequate representation of the truly transient portion of the solution (problem 22 at the end of this chapter).

11.4.5 Further comments on damping

It is assumed in the mode-displacement and mode-acceleration solutions that the modal damping ratio ξ_i is known for each mode. Values may of course be defined explicitly mode by mode, based on experience and engineering judgement. More commonly, however, they are derived from more general models for damping of the system as a whole. The most common of these models is Rayleigh or 'proportional' damping, in which the damping matrix $\boldsymbol{C}$ is assumed to be a linear combination of $\boldsymbol{M}$ and $\boldsymbol{K}$, that is

$$\boldsymbol{C} = \alpha \boldsymbol{M} + \beta \boldsymbol{K}. \tag{11.45}$$

This not only guarantees orthogonality of the eigenmodes with respect to $\boldsymbol{C}$ (by virtue of their orthogonality with respect to $\boldsymbol{M}$ and $\boldsymbol{K}$) but implicitly defines a damping coefficient ξ_i for each mode, where

$$\begin{aligned} \xi_i &= \frac{c_i}{2m_i\,\omega_i} \\ &= \frac{\alpha m_i + \beta k_i}{2m_i\,\omega_i} \\ &= \frac{\alpha}{2\omega_i} + \frac{\beta\,\omega_i}{2}. \end{aligned} \tag{11.46}$$

Note also that since the Rayleigh damping matrix is defined without direct reference to the modal representation, it can be used in the direct integration models to be discussed shortly. In both cases however, appropriate values must first be assigned to the constants α and β. This is usually done by specifying damping ratios ξ_a and ξ_b for two characteristic frequencies ω_a and ω_b (Fig. 11.7). The lower frequency ω_a is usually taken close to the fundamental frequency of the system, for which a damping ratio can be estimated with reasonable accuracy (values of 0–5% are common for hysteretic loss, dependent on the level of stress, and somewhat higher values are used in jointed structures). The upper frequency ω_b is chosen so that the frequency range $\omega_a < \omega < \omega_b$ contains those modes which are likely to contribute significantly to the response. Since the damping associated with the stiffness — the $\frac{1}{2}\beta\omega_i$ term in 11.46 — increases linearly with frequency, higher frequency modes are then progressively damped from the solution, a desirable feature as we will see shortly.

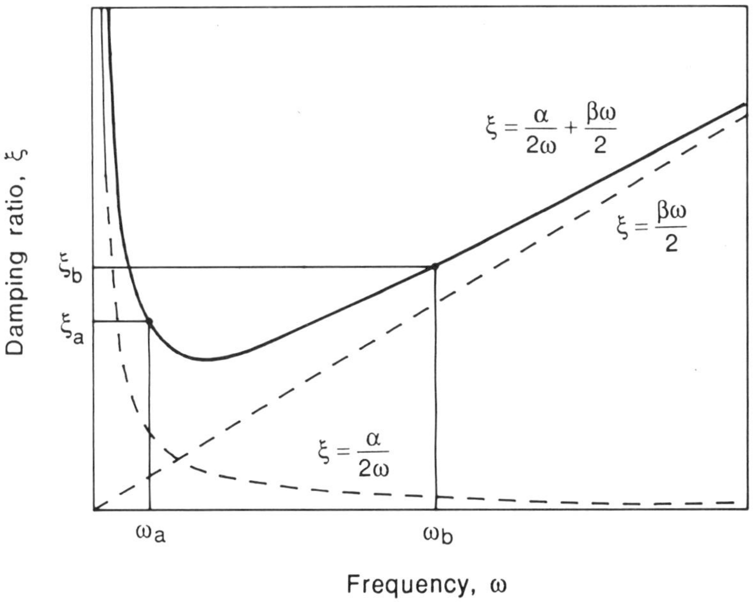

Fig. 11.7 *Rayleigh damping, variation of damping coefficients with frequency.*

11.5 DYNAMIC RESPONSE, DIRECT INTEGRATION

In the 'direct integration' or 'step-by-step' approach to dynamic response, the time derivatives of the nodal displacements are approximated explicitly by finite difference expressions involving displacements at adjacent instants in time, usually at a fixed interval, Δt, apart. Dynamic equilibrium is then enforced at discrete instants rather than in a continuous sense. Methods of this type are well established in the solution of ordinary differential equations and may be extended with little difficulty to simultaneous systems of equations such as those represented by equation 11.30. Although many of these methods have been modified or extended for finite element use, we confine our attention here to two traditional algorithms which have been imported without modification, but which demonstrate many of the general characteristics of more sophisticated schemes.

Before looking at specific algorithms, however, a shorthand notation is needed to define the values of a variable at discrete time intervals, that is, at times, t_0 (=0), t_1, t_2, . . . , where $t_n = n(\Delta t)$. The superscript (n) is used for this purpose. The value of the variable $\delta(t)$ at time t_n is therefore denoted by $\delta^{(n)}$. The same notation applies to vector quantities $\boldsymbol{d}^{(n-1)}$, $\boldsymbol{d}^{(n)}$ and $\boldsymbol{d}^{(n+1)}$ representing values of the vector $\boldsymbol{d}(t)$ at times t_{n-1}, t_n and t_{n+1}, respectively.

11.5.1 Central difference method, an explicit scheme

The 'exact' solution of equations 11.30 gives a nodal displacement vector $\boldsymbol{d}(t)$ which varies continuously with time. In a direct integration scheme, we content ourselves with a solution which is defined only at times t_0, t_1, t_2, . . . and which satisfies equations 11.30 in some sense at those instants. Such solutions are obtained by approximating the nodal velocities and accelerations ($\dot{\boldsymbol{d}}$ and $\ddot{\boldsymbol{d}}$) by difference formulae involving nodal displacements at adjacent time steps. Most schemes of this type can be categorized as 'explicit' or 'implicit' although some formulations, not discussed here, incorporate elements of both. An *explicit* scheme is one in which the solution at time t_{n+1} is obtained by imposing dynamic equilibrium at the previous time step, that is at time t_n. The new estimate of displacement is therefore derived entirely from 'historical' information about acceleration and velocity. An *implicit* method, on the other hand, uses 'current' information, at time t_{n+1}, to predict the new values of displacement. The two methods have very different characteristics.

We look first at a widely used explicit method, the 'central difference' scheme. This derives from traditional central difference approximations for velocity and acceleration. It is an explicit scheme in the sense that the estimate for $\boldsymbol{d}^{(n+1)}$ is obtained from conditions of dynamic equilibrium at time t_n. In other words, the scheme takes as its starting point the equilibrium statement

$$\boldsymbol{M}\,\ddot{\boldsymbol{d}}^{(n)} + \boldsymbol{C}\,\dot{\boldsymbol{d}}^{(n)} + \boldsymbol{K}\,\boldsymbol{d}^{(n)} = \boldsymbol{f}^{(n)}. \tag{11.47}$$

The nodal accelerations and velocities at time t_n are then expressed in terms of nodal displacements at t_{n-1}, t_n and t_{n+1} by using the central difference formulae,

$$\dot{\boldsymbol{d}}^{(n)} = \frac{1}{2\Delta t}\,[\boldsymbol{d}^{(n+1)} - \boldsymbol{d}^{(n-1)}], \tag{11.48}$$

and

$$\ddot{\boldsymbol{d}}^{(n)} = \frac{1}{\Delta t^2}\,[\boldsymbol{d}^{(n+1)} - 2\boldsymbol{d}^{(n)} + \boldsymbol{d}^{(n-1)}]. \tag{11.49}$$

These derive from an assumption that $\boldsymbol{d}$ varies quadratically with time over the interval $[t_{n-1}, t_{n+1}]$. After substituting equations 11.48 and 11.49 into equations 11.47 and rearranging terms so that those involving $\boldsymbol{d}^{(n+1)}$ appear on the left hand side, we obtain

$$\left[\frac{1}{\Delta t^2}\boldsymbol{M} + \frac{1}{2\Delta t}C\right]\boldsymbol{d}^{(n+1)} = \boldsymbol{f}^{(n)} - \boldsymbol{K}\boldsymbol{d}^{(n)} + \frac{1}{\Delta t^2}M\,\{2\boldsymbol{d}^{(n)} - \boldsymbol{d}^{(n-1)}\} + \frac{1}{2\Delta t}\boldsymbol{C}\boldsymbol{d}^{(n-1)}. \tag{11.50}$$

The above equation defines a two-step, implicit scheme; 'two-step' because it involves values of $\boldsymbol{d}$ at two previous time steps (t_n and t_{n-1}), and 'explicit' because the calculation is based on equilibrium considerations at time t_n. It is an important characteristic not just of the central difference scheme but of all explicit solutions, that the stiffness matrix $\boldsymbol{K}$ appears only on the right hand side of the equation.

Since the values of d at two previous time steps are required for each new prediction a difficulty arises at the first time step where only one value (the initial displacement $\boldsymbol{d}^{(0)}$) is available. A special starting procedure is therefore required, a problem common to all two-step schemes. In practice, a fictitious displacement $\boldsymbol{d}^{(-1)}$ at time t_{-1} $(=(-1)\Delta t)$ is formed by writing equations 11.47 and 11.49 with $n = 0$ and eliminating $\boldsymbol{d}^{(1)}$ to give

$$\boldsymbol{d}^{(-1)} = \boldsymbol{d}^{(0)} - \Delta t\,\dot{\boldsymbol{d}}^{(0)} + \frac{1}{2}\Delta t^2\,\ddot{\boldsymbol{d}}^{(0)},$$

where $\boldsymbol{d}^{(0)}$ and $\dot{\boldsymbol{d}}^{(0)}$ are specified as initial conditions (equations 11.31) and where the initial acceleration $\ddot{\boldsymbol{d}}^{(0)}$ is obtained from equation 11.47 by writing

$$\ddot{\boldsymbol{d}}^{(0)} = \boldsymbol{M}^{-1}\,[\boldsymbol{f}^{(0)} - \boldsymbol{C}\,\dot{\boldsymbol{d}}^{(0)} - \boldsymbol{K}\,\boldsymbol{d}^{(0)}]. \tag{11.51}$$

With this adjustment, equations 11.50 form a set of simultaneous algebraic equations which can be solved at each time increment to give $\boldsymbol{d}^{(n+1)}$ in terms of $\boldsymbol{d}^{(n)}$ and $\boldsymbol{d}^{(n-1)}$. In the implementation of this scheme, the following points should be noted:

(i) In the undamped case, the mass matrix $\boldsymbol{M}$ can be made diagonal by using a lumped or diagonal representation of element mass. The coefficient matrix on the left hand side is then diagonal and the solution for $\boldsymbol{d}^{(n+1)}$ is trivial. When damping is present, the coefficient matrix is no longer diagonal since even the simplest damping matrices seldom possess this property. In such situations, however, the velocity can be lagged by one half time step to give a modified scheme of somewhat diminished accuracy in which the coefficient matrix is once again diagonal. It must be emphasized that the achievement of a diagonal coefficient matrix is an important objective in an explicit schemes, and is generally essential if it is to be competitive in terms of computational effort with the implicit schemes to be discussed shortly.

(ii) The central difference scheme is only 'conditionally stable'. This means that when the time step exceeds a particular value, errors grow exponentially and the solution 'diverges' from the correct, time dependent solution of the discrete equations. The critical time step depends upon the *spatial* discretization of the original finite element model, the smaller the size of the elements, the smaller the time step. This is a characteristic of most explicit schemes and indeed of some implicit schemes also. The critical time step for the central difference scheme is

$$\Delta t_{cr} = \frac{2.0}{\omega_{max}}, \tag{11.52}$$

where ω_{max} is the *highest*, natural frequency of the undamped system. In physical terms, the critical time step is related to the time taken for a stress wave to propagate across the smallest element in the mesh and can indeed be estimated in this way. In practice, condition 11.52 places a severe restriction on time intervals which might otherwise be used if accuracy rather than stability were the prime consideration.

Notes (i) and (ii) summarize many of the advantages and disadvantages not just of the central difference scheme but of explicit schemes in general. They involve minimal computational effort at each time step but require a large number of steps because of stability limitations. If they are to be competitive in computational terms with unconditionally stable 'implicit' schemes (to be discussed shortly) they must, in effect, be implemented with lumped, rather than consistent, masses. The solution at each time step is then extremely fast — the coefficient matrix being diagonal — and this compensates in some measure for the excessive number of steps which are usually required to satisfy the stability condition. On the other hand, if an implicit scheme is stable, accuracy is generally excellent, since time steps which are small enough to satisfy the stability criteria are extremely small compared to the overall timescale of the response. Moreover, the time increment can be taken very close to the stability limit without appreciable loss of accuracy. There is in fact little to be gained by using time increments which are appreciably smaller than this limit, since accuracy is thereby improved only in the higher frequency modes which contribute little to the overall response and are seldom accurately resolved.

11.5.2 Implicit schemes, average acceleration method

Implicit schemes exhibit behaviour which is roughly the inverse of that described above. That is to say, each time step requires substantial computational effort, but stability is generally unconditional and fewer time steps are therefore required.

A straightforward implicit scheme which is widely used in the solution of single degree of freedom systems is the 'average acceleration' method, known also as the 'trapezoidal' scheme for reasons which will become apparent shortly. It takes as its starting point, the equations of dynamic equilibrium evaluated at time t_{n+1}. We start therefore with the statement that

$$\boldsymbol{M}\,\ddot{\boldsymbol{d}}^{(n+1)} + \boldsymbol{C}\,\dot{\boldsymbol{d}}^{(n+1)} + \boldsymbol{K}\,\boldsymbol{d}^{(n+1)} = \boldsymbol{f}^{(n+1)}. \tag{11.53}$$

Difference relationships for $\dot{\boldsymbol{d}}$ and $\ddot{\boldsymbol{d}}$ are obtained by integrating these quantities over the preceeding time step using the trapezoidal rule of numerical integration. In the case of acceleration, the integral approximation is illustrated (for a scalar variable $\ddot{d}$) in Fig. 11.8. Equating the integral of $\ddot{d}$ to the shaded area, gives

$$\dot{d}^{(n+1)} - \dot{d}^{(n)} = \int_{n\Delta t}^{(n+1)\Delta t} \ddot{d}(t)\,\mathrm{d}t \simeq \frac{1}{2}\Delta t\,(\ddot{d}^{(n+1)} + \ddot{d}^{(n)}).$$

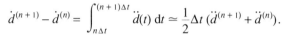

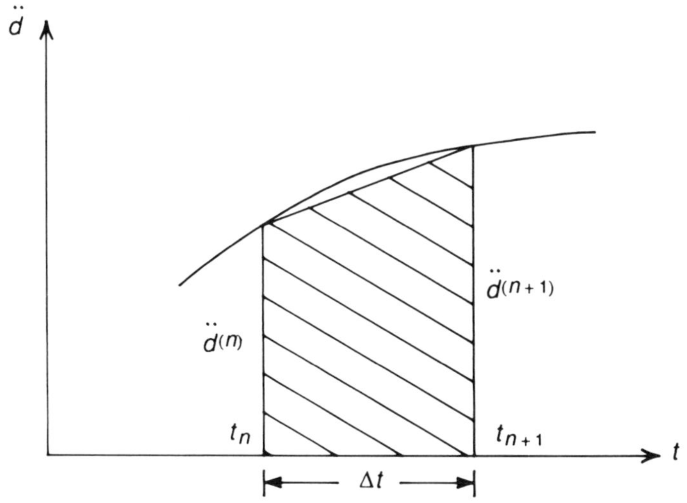

Fig. 11.8 *Trapezoidal rule for integrating $\ddot{d}(t)$.*

After some rearrangement of terms, and treating $\ddot{\boldsymbol{d}}$ as a vector rather than a scalar, we obtain

$$\ddot{\boldsymbol{d}}^{(n+1)} = \left(\frac{2}{\Delta t}\right)[\dot{\boldsymbol{d}}^{(n+1)} - \dot{\boldsymbol{d}}^{(n)}] - \ddot{\boldsymbol{d}}^{(n)}. \tag{11.54}$$

Applying the same procedure to $\dot{\boldsymbol{d}}$ gives

$$\dot{\boldsymbol{d}}^{(n+1)} = \frac{2}{\Delta t}\,[\boldsymbol{d}^{(n+1)} - \boldsymbol{d}^{(n)}] - \dot{\boldsymbol{d}}^{(n)}, \tag{11.55}$$

and by combining 11.54 and 11.55 we obtain

$$\ddot{\boldsymbol{d}}^{(n+1)} = \frac{4}{\Delta t^2}\left[\boldsymbol{d}^{(n+1)} - \boldsymbol{d}^{(n)}\right] - \frac{4}{\Delta t}\dot{\boldsymbol{d}}^{(n)} - \ddot{\boldsymbol{d}}^{(n)}. \tag{11.56}$$

Expressions 11.55 and 11.56 can now be used to replace $\dot{\boldsymbol{d}}^{(n+1)}$ and $\ddot{\boldsymbol{d}}^{(n+1)}$ in equation 11.53. After collecting all terms involving $\boldsymbol{d}^{(n+1)}$ to the left hand side, we finally obtain

$$\left[\left(\frac{4}{\Delta t^2}\right)M + \left(\frac{2}{\Delta t}\right)C + K\right]\boldsymbol{d}^{(n+1)} = \boldsymbol{f}^{(n+1)} + \boldsymbol{g}^{(n)}, \tag{11.57}$$

where $\boldsymbol{g}^{(n)}$ is a vector derived from nodal velocities, accelerations and displacements at time t_n and is given by

$$\boldsymbol{g}^{(n)} = M\left[\left(\frac{4}{\Delta t^2}\right)\boldsymbol{d}^{(n)} + \left(\frac{4}{\Delta t}\right)\dot{\boldsymbol{d}}^{(n)} + \ddot{\boldsymbol{d}}^{(n)}\right] + C\left[\left(\frac{2}{\Delta t}\right)\boldsymbol{d}^{(n)} + \dot{\boldsymbol{d}}^{(n)}\right]. \tag{11.58}$$

For the algorithm for stepping forward by a one time increment, assuming that $\boldsymbol{d}^{(n)}, \dot{\boldsymbol{d}}^{(n)}$ and $\ddot{\boldsymbol{d}}^{(n)}$ are known at time t_n, proceed as follows:

1. Equation 11.57 is solved to give $\boldsymbol{d}^{(n+1)}$ in terms of $\boldsymbol{f}^{(n+1)}, \boldsymbol{d}^{(n)}, \dot{\boldsymbol{d}}^{(n)}$ and $\ddot{\boldsymbol{d}}^{(n)}$.
2. Equations 11.55 and 11.56 are evaluated to give $\dot{\boldsymbol{d}}^{(n+1)}$ and $\ddot{\boldsymbol{d}}^{(n+1)}$ in terms of $\boldsymbol{d}^{(n+1)}$ (newly calculated), $\boldsymbol{d}^{(n)}, \dot{\boldsymbol{d}}^{(n)}$, and $\ddot{\boldsymbol{d}}^{(n)}$.
3. n is incremented by 1 and the cycle is repeated.

The algorithm is repeated as many times as are required, starting at time $t = 0$ with known values of $\boldsymbol{d}$, $\dot{\boldsymbol{d}}$ and $\ddot{\boldsymbol{d}}$, the former supplied by the initial conditions in equation 11.31 and the latter ($\ddot{\boldsymbol{d}}$) obtained using equation 11.51 at $t = 0$.

An important characteristic of the above scheme is the presence of the stiffness matrix $\boldsymbol{K}$ within the coefficient matrix on the left hand side of 11.57. The individual equations are therefore coupled by off-diagonal stiffness terms even when the mass matrix is diagonal. Their solution then requires a time-consuming factorization or inversion of the coefficient matrix. This is a much more demanding computational procedure than in the analogous phase of the explicit formulation where the coefficient matrix is diagonal and its inversion trivial.

11.5.3 Accuracy and stability, artificial damping

The average acceleration method and most other implicit schemes are unconditionally stable. In other words, they produce bounded — though not necessarily accurate — solutions, irrespective of the length of the time step. Stability considerations which limit the length of the time increment in explicit calculations are therefore replaced in implicit schemes by considerations of accuracy. In systems with many degrees of freedom, these are somewhat more subtle than might appear at first sight.

The accuracy and stability of time stepping methods in general — both implicit and explicit — can be deduced from single-degree of freedom models. In particular, the performance of algorithms such as that in equations 11.50 and 11.57 can be assessed by applying them to an uncoupled, undamped system governed by the scalar equation,

$$m\ddot{d} + kd = 0. \tag{11.59}$$

The exact solution has the general form

$$d(t) = A\cos(\omega t) + B\sin(\omega t), \tag{11.60}$$

where $\omega = \sqrt{(k/m)}$ and where A and B are constants determined by the initial conditions. The difference equations obtained by applying the 'average acceleration', 'central difference' — or any other two-step scheme — to this equation, yields an *analytic* difference solution of the form,

$$d^{(n)} = \exp(-\alpha' t_n)\,[A'\cos(\omega' t_n) + B'\sin(\omega' t_n)]. \tag{11.61a}$$

The analysis which leads to such difference solutions is not presented here but can be found at a relatively elementary level in [4] and in greater detail in [11] (see also problems 23–25 at the end of this chapter). Technically, such solutions exist only at discrete times $t = t_n$, but can be interpreted as discrete values of a continuous solution. The 'equivalent' continuous solution is then

$$d(t) = \exp(-\alpha' t)\,[A'\cos(\omega' t) + B'\sin(\omega')], \tag{11.61b}$$

and can be compared directly to the exact result (equation 11.60). If we neglect specific values of A, B, A' and B' (these are determined by the initial conditions) the extent to which expression 11.61b can approximate the exact result in equation 11.60 is determined by the extent to which ω' approximates ω and (in the undamped case) α' approaches zero. Comparisons of this type form the basis of much formal analysis of stability and accuracy. No further details are given here but a comprehensive discussion is to be found in Chapter 9 of [10] and [11].

A convenient measure of the correspondence between the numerical solution and the exact difference solution is provided by the 'amplitude decay' ε_a and 'period elongation' ε_T over a single cycle. These quantities are illustrated in Fig. 11.9. In terms of α', ω and ω', they are given by

$$\varepsilon_a = 1.0 - \exp(-2\alpha'\pi), \tag{11.62a}$$

and

$$\varepsilon_T = \frac{T' - T}{T} = \frac{\omega - \omega'}{\omega'}, \tag{11.62b}$$

and represent fractional differences in amplitude and period between the exact and difference solutions.

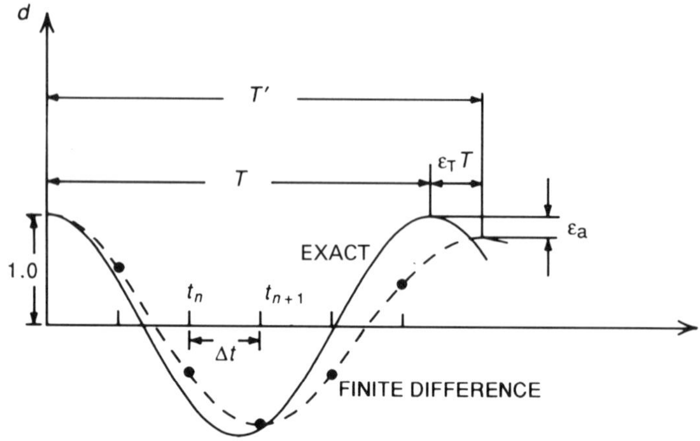

Fig. 11.9 *Physical representation of period and amplitude errors ε_T and ε_a.*

In calculating ε_a and ε_T for the average acceleration scheme, it is found that ε_a is identically zero for all values of Δt, giving zero amplitude error irrespective of the length of the time step. The central difference scheme gives the same result provided that $\Delta t < (2/\omega)$. For larger values, a negative root exists for α' and the difference solution then grows exponentially with time. This is a manifestation of the instability noted in section 11.5.1. In fact, since ω is the largest (and smallest) natural frequency of the one-degree of freedom system, this limitation on the size of the time step is simply a restatement of equation 11.49.

The period error of the central difference and average acceleration schemes is nonzero and increases with the time increment. It is plotted against $(\Delta t/T)$ in Fig. 11.10(a). The values for the two schemes are broadly comparable, although the central difference method gives a period contraction whereas the average acceleration scheme gives an elongation (the curve for the central difference scheme terminates at the onset of instability, $\Delta t/T = 1/\pi$)). In both schemes, the error reduces to a few percent when twenty or more time steps are used for each period of the solution (say $\Delta t/T < 0.05$).

As we have noted already, both the average acceleration and the central difference scheme have zero amplitude error for all (stable) time steps. Although admirable in response calculations for single-degree of freedom systems, such behaviour is less than ideal in most finite element models. The reasons are as follows. We noted in section 11.3 that the higher frequency modes in a finite element model are seldom represented with any accuracy. This is not a disadvantage in most instances since the overall response of the system is dominated by the lower order modes. In a modal response calculation, the inaccurate higher modes are simply eliminated from the solution by truncating the modal expansion after a reduced number of terms. In a direct numerical scheme, this is not possible since the individual modes are inextriceably coupled through the nodal variables. The same effect can be achieved by 'filtering' the higher order modes from the solution by using a step-by-step algorithm which artificially damps higher frequency disturbances. Such schemes are characterized by an amplitude decay which increases with $\Delta t/T$. The time increment Δt can then be chosen so that the amplitude decay is small for disturbances of time scale T where T is characteristic of the overall response of the system, but large for shorter time scales corresponding to higher frequency modes. This has the desired effect of damping the high frequency contributions after a few oscillations while accurately representing the lower order modes.

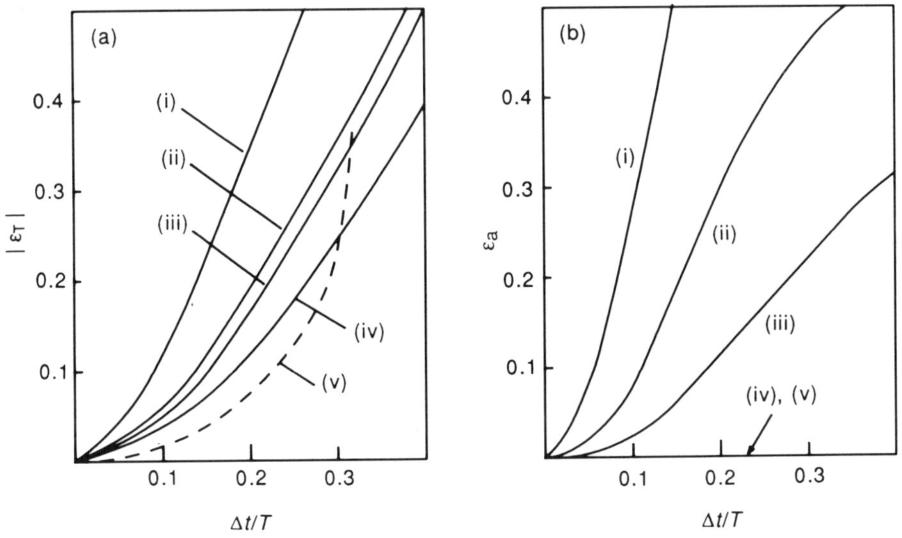

Fig. 11.10 *Variation of (a) period error (ε_T) and (b) amplitude error (ε_a) with step size: (i) Houbolt [14], (ii) Wilson–θ, $\theta = 1.4$ [15], (iii) α–method, $\alpha = 0.3$ [16], (iv) average acceleration, and (v) central difference. Key: (———) period elongation, (- - - -) period contraction.*

Much effort has gone into the development of step-by-step integration schemes which possess these characteristics, schemes in other words which exhibit substantial damping at high frequencies while maintaining good accuracy at low frequencies. Ideally they should also give low period error. A number of multi-step schemes have been proposed which more or less satisfy these requirements. The behaviour of some of these is illustrated in Fig. 11.10(a) and (b). These show the period and amplitude errors for a number of these methods, including the 'central difference' and 'average acceleration' schemes already discussed. The methods chosen for comparison — the Houbolt, Wilson–θ and Hughes–Hilber–Taylor–α methods — are detailed in [14–16]. Further discussion lies beyond the scope of the current treatment, but is to be found in comprehensive form in Chapter 9 of [11].

REFERENCES

[1] Craig, R. R. (1981) *Structural Dynamics, an Introduction to Computer Methods*, John Wiley, New York.

[2] Clough, R. W. and Penzien, J. (1975) *Dynamics of Structures*, McGraw-Hill, New York.

[3) Hinton, E., Rock, T. and Zienkiewicz, O. C. (1976) A note on mass lumping and related processes in the finite element method. *International Journal of Earthquake Engineering and Structural Dynamics*, **4** (3), 245–9.

[4] Cook, R. D., Malkus, D. S. and Plesha, M. E. (1989) *Concepts and Applications of Finite Element Analysis*, 3rd edn, John Wiley & Sons, New York, Chapter 13.

[5] Surana, K. S. (1978) Lumped mass matrices with nonzero inertia for general shell and axisymmetric shell elements. *International Journal for Numerical Methods in Engineering*, **12** (11), 1635–50.

[6) Anderson R. G., Irons, B. M. and Zienkiewicz, O. C. (1968) Vibration and Stability of plates using finite elements, *International Journal of Solids and Structures*, **4** (10), 1031–55.

[7] Henshell, R. D. and Ong, J. H. (1975) Automatic masters for eigenvalue economisation. *Earthquake Engineering and Structural Dynamics*, **3** (4), 701–3.

[8] Griffiths, D. V. and Smith, I. M. (1991) *Numerical Methods for Engineers*, Blackwell Scientific Publications, Oxford, Chapter 4.

[9] Al-Khaffaji, A. W. and Tooley, J. R. (1986) *Numerical Methods in Engineering Practice*, CBS publishing, New York, Chapter 8.

[10] Bathe, K. J. and Wilson, E. L. (1976) *Numerical Methods in Finite Element Analysis*, Prentice hall, Englewood Cliffs, NJ, Chapters 10, 11 and 12.

[11] Hughes, T. J. R. (1987) *The Finite Element Method: Linear Static and Dynamic Finite Element Analysis*, Prentice Hall, Englewood Cliffs, NJ, Chapters 9 and 10.

[12] Kardestuncer, H. and Norrie, D. H. (1987) *Finite Element Handbook*, Part 4, McGraw-Hill, New York, Chapter 1.

[13] Zienkiewicz, O. C. (1977) The Finite Element Method, 3rd edn, McGraw-Hill, London, Chapter 21, pp. 288–301.

[14] Houbolt, J. C. (1950) A recurrence matrix solution for the dynamic response of elastic aircraft. *Journal of the Aeronautical Sciences*, **17**, 540–50.

[15] Wilson, E. L., L. Farhoomand and Bathe, K. J. (1973) Nonlinear dynamic analysis of complex structures. *International Journal of Earthquake Engineering and Structural Dynamics*, **1**, 241–52.

[16] Hilber, M. M., Hughes T. J. R. and Taylor, R. L. (1977) Improved numerical dissipation for time integration algorithms in structural dynamics. *International Journal of Earthquake Engineering and Structural Dynamics*, **5**, 283–92.

PROBLEMS

1. The kinetic energy per unit volume of an elastic continuum is $\frac{1}{2}\rho\boldsymbol{v}\,.\,\boldsymbol{v}$ where $\boldsymbol{v}(\boldsymbol{x}, t)$ is the velocity field. If the continuum is modelled using a finite element discretization in which the spatial interpolation of velocity is identical to that of displacement, show that the kinetic energy $\boldsymbol{T}^{\mathrm{e}}$ of an element is given by

$$\boldsymbol{T}^{\mathrm{e}} = \dot{\boldsymbol{d}}^{\mathrm{eT}}\boldsymbol{M}^{\mathrm{e}}\dot{\boldsymbol{d}}^{\mathrm{e}}, \qquad \text{(i)}$$

where $\boldsymbol{M}^{\mathrm{e}}$ is the mass matrix of the element — defined by equation 11.13 — and $\dot{\boldsymbol{d}}^{\mathrm{e}}$ is a vector of nodal velocities. If the degrees of freedom comprise nodal displacements only (rather than displacements and rotations as in the case of a structural element) show that the matrix $\boldsymbol{M}^{\mathrm{e}}$ has the property that

$$\boldsymbol{x}^{\mathrm{T}}\boldsymbol{M}^{\mathrm{e}}\boldsymbol{x} = \alpha m, \qquad \text{(ii)}$$

where m is the mass of the element, α is the number of displacement components at each node (1, 2 or 3) and $\boldsymbol{x}$ is a vector with 1.0 in every place. (Suggestion: evaluate expression (i) for the particular case when $\boldsymbol{v}(\boldsymbol{x}, t)$ is a constant vector with one, two or three cartesian components, each of which takes the value 1.0).

2. The shape relationship, $\boldsymbol{u} = \boldsymbol{N}^{\mathrm{e}}\boldsymbol{d}^{\mathrm{e}}$, of a particular element has the property that $\boldsymbol{u}(\boldsymbol{x}) = 0$ at all points if and only if the nodal displacement vector $\boldsymbol{d}^{\mathrm{e}}$ is identically zero. Show that the mass matrix $\boldsymbol{M}^{\mathrm{e}}$ is then 'positive definite' in the sense that $\boldsymbol{x}^{\mathrm{T}}\boldsymbol{M}^{\mathrm{e}}\boldsymbol{x}$ is strictly positive for *all* non-trivial vectors $\boldsymbol{x}$ (use the definition of $\boldsymbol{M}^{\mathrm{e}}$ derived in problem 1).

3. A three-noded, quadratic bar element of length L (as shown) has degrees of freedom u_1, u_2, and u_3, and a shape matrix

$$\boldsymbol{N}^{\mathrm{e}} = \left[\frac{1}{2}\left(\frac{2x'}{L}\right)\left(\frac{2x'}{L} - 1\right),\; 1 - \left(\frac{2x'}{L}\right)^2,\; \frac{1}{2}\left(\frac{2x'}{L}\right)\left(1 + \frac{2x'}{L}\right)\right],$$

where x' is a local axial coordinate whose origin is at the central node. Show that the consistent mass matrix for this element is

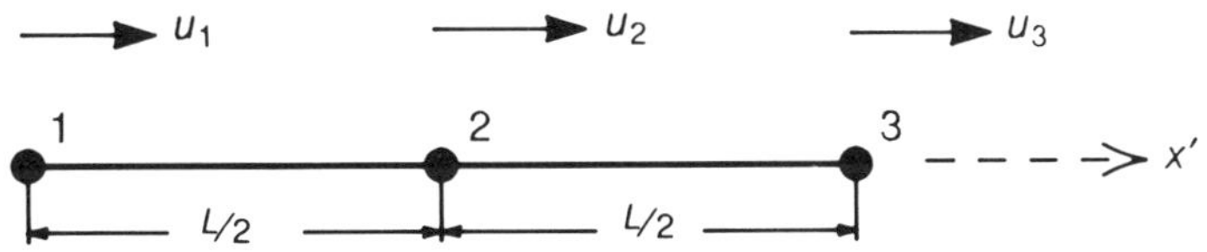

Problem 2.

$$\boldsymbol{M}^{e} = \left(\frac{m}{30}\right)\begin{bmatrix} 4 & 2 & -1 \\ 2 & 16 & 2 \\ -1 & 2 & 4 \end{bmatrix},$$

where m is the mass of the element. Confirm that this satisfies condition (ii) of problem 1 with $\alpha = 1$.

4. Show that the consistent mass matrix for the four-noded serendipity rectangle shown as element (i) of Fig. 8.4 is

$$\boldsymbol{M}^{e} = \left(\frac{m}{36}\right)\begin{bmatrix} 4 & 0 & 2 & 0 & 1 & 0 & 2 & 0 \\ 0 & 4 & 0 & 2 & 0 & 1 & 0 & 2 \\ 2 & 0 & 4 & 0 & 2 & 0 & 1 & 0 \\ 0 & 2 & 0 & 4 & 0 & 2 & 0 & 1 \\ 1 & 0 & 2 & 0 & 4 & 0 & 2 & 0 \\ 0 & 1 & 0 & 2 & 0 & 4 & 0 & 2 \\ 2 & 0 & 1 & 0 & 2 & 0 & 4 & 0 \\ 0 & 2 & 0 & 1 & 0 & 2 & 0 & 4 \end{bmatrix},$$

where m is the mass of the element. Confirm that this satisfies condition (ii) of problem 1 with $\alpha = 2$.

5. A bar element with degrees of freedom u_1 and u_2 (as shown) has a cross-sectional area A which varies linearly from A_1 at the left hand end to A_2 at the right. If linear interpolation is used within the element, show that the consistent mass matrix has components

$$\boldsymbol{M}^{e} = \left(\frac{\rho L}{12}\right)\begin{bmatrix} (3A_1 + A_2) & (A_1 + A_2) \\ (A_1 + A_2) & (A_1 + 3A_2) \end{bmatrix}.$$

(Note: at any point along the element, the cross-sectional area $A(x')$, is given by $A(x') = A_1 + (A_2 - A_1)(x'/L)$ where x' is a local axial coordinate.)

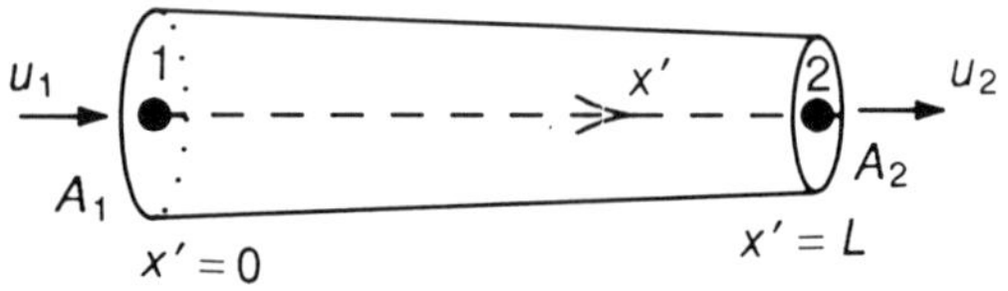

Problem 5.

6. Show that the lumped mass matrix for the three-noded bar element of problem 3 obtained using the algorithm of section 11.2.4 is

$$\boldsymbol{M}_{L}^{e} = \left(\frac{m}{6}\right)\begin{bmatrix} 1 & 0 & 0 \\ 0 & 4 & 0 \\ 0 & 0 & 1 \end{bmatrix}$$

where m is the mass of the element.

7. A six-noded, quadratic triangle has the shape functions given by equations 8.19. Obtain expressions for the consistent and lumped mass matrices for the element. In the case of the lumped matrix use the procedure outlined in section 11.2.4.

8. In what proportion should the mass of the two-noded bar element of problem 5 be distributed among the nodes in a lumped formulation? (Use the algorithm of section 11.2.4).

9. In what proportion should the mass of an eight-noded serendipity rectangle be distributed among the nodes in a lumped formulation? (Use the algorithm of section 11.2.4).

10. A uniform bar of length L, Young's modulus E, density ρ and cross sectional area A is fixed at the left-hand end and free at the right. It is modelled by three equal elements (as shown). Show that the natural modes of axial vibration of the model are solutions of the eigenproblem

$$\left[\begin{bmatrix} 3k & -3k & 0 \\ -3k & 6k & -3k \\ 0 & -3k & 6k \end{bmatrix} - \left(\frac{\omega^2}{18}\right)\begin{bmatrix} 2m & m & 0 \\ m & 4m & m \\ 0 & m & 4m \end{bmatrix}\right]\begin{bmatrix} \delta_1 \\ \delta_2 \\ \delta_3 \end{bmatrix} = \begin{bmatrix} 0 \\ 0 \\ 0 \end{bmatrix},$$

where $m = \rho AL$ and $k = EA/L$. Solve to obtain the natural frequencies in terms of m and k and compare with the exact values. (See example (i) of section 11.3.1).

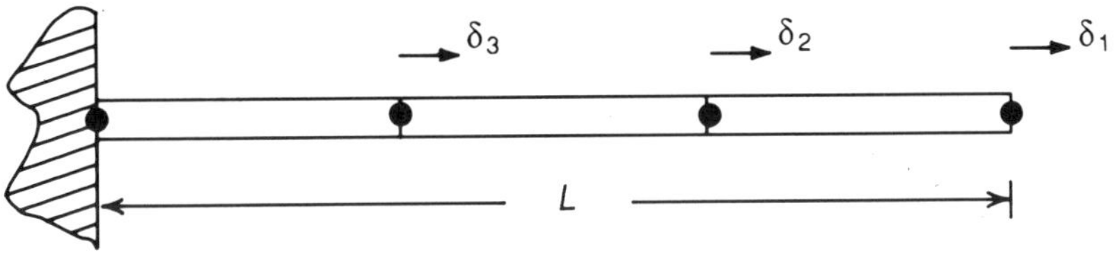

Problem 10.

11. A uniform bar of Young's modulus E, density ρ, cross sectional area A and length L is fixed at each end. Estimate the natural frequencies of axial vibration of the bar using a finite element model formed from three equal bar elements and compare with the exact values,

$$\omega_n = n\pi\sqrt{\frac{E}{\rho L^2}}, \ (n = 1, 2, 3, \dots).$$

Use a consistent mass formulation.

12. Repeat problem 11 using a lumped mass formulation.

13. Repeat example (i) of section 11.3.1 using a three-noded quadratic element in place of the two elements of Fig. 11.3(b). Use the consistent mass formulation (the mass matrix is calculated in problem 3) and confirm that the eigenfrequencies are more accurately determined than in the model of section 11.3.1. (Note: the stiffness matrix for the quadratic element is given in problem 2 of Chapter 4).

14. Repeat problem 13 using a lumped mass representation for the quadratic element (see problem 6 above) and compare the accuracy of the resulting eigenfrequencies with those obtained using the consistent formulation.

15. A beam of flexural rigidity *EI* and mass *m* is constrained so that it cannot displace at one end and cannot rotate at the other (as shown). The beam is modelled by a single beam element with unconstrained degrees of freedom δ_1 and δ_2. Show that the resulting eigenproblem for the free modes of vibration of the system, using a consistent mass formulation, is

$$\left[\left(\frac{EI}{L^3}\right)\begin{bmatrix} 12 & 6L \\ 6L & 4L^2 \end{bmatrix} - \frac{m\omega^2}{420}\begin{bmatrix} 156 & -13L \\ -13L & 4L^2 \end{bmatrix}\right]\begin{bmatrix} \delta_1 \\ \delta_2 \end{bmatrix} = \begin{bmatrix} 0 \\ 0 \end{bmatrix}.$$

Solve for the first two eigenfrequencies and compare with the exact values, $\omega_1 = 2.467\sqrt{(EI/mL^3)}$, and $\omega_2 = 22.21\sqrt{(EI/mL^3)}$.

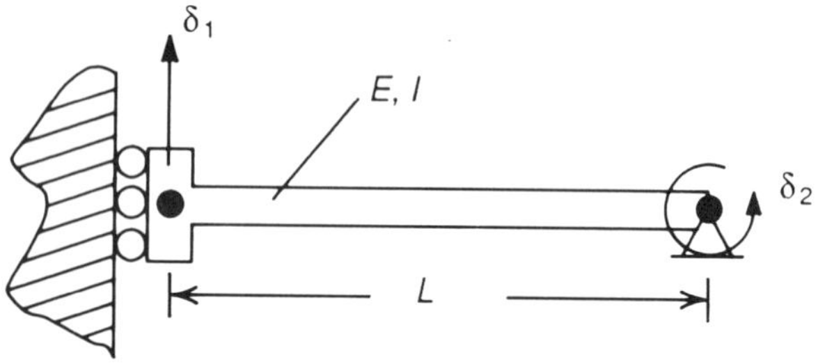

Problem 15.

16. Repeat problem 15 using the modified, lumped mass formulation of expression (11.19b).

17. A beam of length $2L$ and flexural rigidity *EI* is rigidly fixed at each end and is modelled using two beam elements of equal length. Calculate the eigenfrequencies of the system by using: (a) a consistent mass formulation; and (b) a modified lumped mass formulation as per expression (11.19b). Compare with the exact solution in each case and comment

(the first two exact eigenfrequencies are $\omega_1 = 7.910\sqrt{(EI/mL^3)}$ and $\omega_2 = 21.80\sqrt{(EI/mL^3)}$, where m is the mass of the beam).

18. Guyan reduction is used to eliminate degree of freedom δ_3 from the 3×3 eigenproblem of problem 10. Show that the transformation matrix $\boldsymbol{T}$ (see equation 11.28) is given by

$$\boldsymbol{T} = \begin{bmatrix} 1 & 0 \\ 0 & 1 \\ 0 & \frac{1}{2} \end{bmatrix}.$$

Obtain expressions for the condensed matrices $\boldsymbol{M}_c$ and $\boldsymbol{K}_c$ and calculate the eigenfrequencies of the reduced problem. Compare with exact values and with those obtained from the two-degree of freedom model of section 11.3.1.

19. Repeat problem 18 eliminating δ_2 and δ_3 retaining only δ_1 as a master.

20. Use Guyan reduction to eliminate the rotational degree of freedom from example (ii) of section 11.3.1 and calculate the fundamental frequency of the system. Compare with the exact value and comment.

21. A bar of unit axial stiffness ($EA/L = 1$), unit mass ($m = 1$) and unit length ($L = 1$) is modelled by two bar elements as shown in Fig. 11.3(b). The system is initially at rest and unstrained at time $t = 0$. A unit axial load is then applied to the free end. Use the mode displacement method to obtain a solution for the nodal displacements of the system at all subsequent times (the free modes of vibration have been calculated in example (ii) of section 11.3.1). If the modal expansion is truncated after the first term show that the resulting solution is

$$\begin{bmatrix} \delta_1(t) \\ \delta_2(t) \end{bmatrix} = \left(\frac{1}{m_1 \omega_1^2}\right)(1 - \cos(\omega_1 t))\begin{bmatrix} 1/\sqrt{2} \\ 1 \end{bmatrix},$$

where $\omega_1 = 1.611$ and $m_1 = 0.4512$.

22. Repeat problem 21 using the mode-acceleration method and show that the analogous solution, truncating the modal series after the first term, is,

$$\begin{bmatrix} \delta_1(t) \\ \delta_2(t) \end{bmatrix} = \begin{bmatrix} 1/2 \\ 1 \end{bmatrix} - \left(\frac{1}{m_1 \omega_1^2}\right)\cos\omega_1 t \begin{bmatrix} 1/\sqrt{2} \\ 1 \end{bmatrix}.$$

Confirm that if *both* modes are included, the mode-acceleration solution and the mode-displacement solution yield identical results.

23. Confirm that the difference equation obtained by applying the central difference scheme to equation 11.59 is

$$d^{(n+1)} + [(\omega \Delta t)^2 - 2]\, d^{(n)} + d^{(n-1)} = 0, \qquad \text{(i)}$$

where $\omega = \sqrt{(k/m)}$. Show by substitution, that

$$d^{(n)} = A' \cos \omega' t_n + B' \sin \omega' t_n$$

is an exact solution of equation (i) provided that ω' satisfies the relationship

$$\cos \omega' \Delta t = [1 - \tfrac{1}{2}(\omega \Delta t)^2].$$

Show that this condition can only be satisfied for real values of ω' if $(\omega \Delta t) < 2.0$. Calculate (ω'/ω) for $\omega \Delta t = 0.5$, 1.0, 1.5 and 2.0, and show that these values are consistent with the values of of ε_T plotted in Fig. 11.11(a).

24. An exact solution of the form $d^{(n)} = A' \lambda^n$ is proposed for difference equation (i) of the previous problem. Show that such a solution exists provided that λ is a root of

$$\lambda^2 + [(\omega \Delta t)^2 - 2]\lambda + 1 = 0.$$

Confirm that real roots exist for the above equation only if $\omega(\Delta t) > 2.0$ and that one of these has an absolute value greater than 1.0. Comment on the significance of this result in the context of criterion 11.53.

25. Confirm that the difference equation obtained by applying the average acceleration scheme to equation 11.59 is

$$(1 + \beta)\, d^{(n+1)} + 2\,(\beta - 1)\, d^{(n)} + (1 + \beta)\, d^{(n-1)} = 0, \qquad \text{(i)}$$

where $\beta = (\omega \Delta t)^2/4$ and $\omega = \sqrt{(k/m)}$. Show by substitution, that

$$d^{(n)} = A' \cos \omega' t_n + B' \sin \omega' t_n$$

is an exact solution provided that ω' satisfies

$$\cos \omega' \Delta t = \frac{1 - \beta}{1 + \beta}. \qquad \text{(ii)}$$

Determine values of ω' for $\omega \Delta t =$ 0.5, 1.0, 1.5 and 2.0, and show that these are consistent with the curve for ε_T plotted in Fig. 11.11(a).

Appendix A: List of symbols

The main symbols used in this book are listed below. In cases where a symbol is used for more than one variable its principal use is listed first.

Bold 'lower case' characters are used for vector quantities, both physical and algebraic. Bold 'upper case' characters are used for matrices. The superscript 'e' is used throughout to denote a scalar or matrix quantity which is associated with a particular element rather than with the system as a whole.

GREEK

α	Coefficient of thermal expansion
α_i, β_i	Interpolation shape coefficients
δ_i	Nodal displacement or rotation
δ_{ij}	Displacement δ_i due to a nodal force f_j
$\varepsilon_{i'}$	True strain
ε_a	Amplitude error
ε_T	Period error
γ_{ij}	Shear strain component
ν	Poissons ratio
σ, σ_i	Normal stress
τ, τ_{ij}	Shear stress
σ^0, σ_i^0, τ_{ij}^0	Components of membrane stress (in a beam or plate)
σ^b, σ_i^b, τ_{ij}^b	Components of bending stress (in a beam or plate)
θ_x, θ_y	Rotations of the midplane of a plate
ϕ, ψ	Rotations of a plane normal to the midsurface of a plate
ξ, ζ, η	Parent coordinates in a mapped element
ξ_i	Modal damping ratio
ω_i	I'th undamped natural frequency
ω_i^d	I'th damped natural frequency
ρ	Mass density
χ, χ^e	Total energy

LATIN

a_i	Rayleigh–Ritz coefficient
A	Cross-sectional area of a beam or bar *or* any plane area
$\mathbf{B}^e$	Strain–displacement matrix

$\mathbf{B}_b^e$, $\mathbf{B}_s^e$	Strain–displacement matrices (Mindlin plate element)
$\mathbf{C}$	Flexibility matrix
c_i	Modal damping
D	Flexural rigidity of a plate
$\mathbf{D}$	Stress–strain matrix
$\mathbf{d}$	Vector of nodal displacements (assembled)
$\mathbf{d}'$	Vector of unconstrained nodal displacements
$\mathbf{d}_0$	Vector of constrained nodal displacements
$\mathbf{d}^e$	Element displacement vector (global)
$\mathbf{d}^{e\prime}$	Element displacement vector (local)
E	Young's modulus
$\mathbf{e}$	Vector of strain components at a point
e, e_i	Direct strain
e^T, e_i^T	Thermal strain
e^0, e_i^0, γ_{ij}^0	Membrane strains (in a beam or plate)
e^b, e_i^b, γ_{ij}^b	Bending strains (in a beam or plate)
f_i	Nodal force
$\mathbf{f}$	Vector of nodal forces
$\mathbf{f}_g$, $\mathbf{f}_g^e$	Equivalent nodal forces due to a distributed body force
$\mathbf{f}_T$, $\mathbf{f}_T^e$	Equivalent nodal forces due to thermal effects
$\mathbf{f}_p$	Concentrated nodal loads
$\mathbf{g}$	Body force vector *or* vector of shear strains (Mindlin plate formulation)
I, I_Y, I_Z	Second moments of area of a beam
$\mathbf{J}$	Jacobian matrix
k	Axial stiffness of a bar
$\mathbf{K}^e$	Element stiffness matrix (global)
$\mathbf{K}^{e\prime}$	Element stiffness matrix (local)
$\mathbf{K}_b^e$, $\mathbf{K}_s^e$	Bending and shear stiffness matrices (Mindlin element)
$\mathbf{K}$	Stiffness matrix (assembled)
k_i	Modal stiffness
L_1, L_2, L_3	Natural coordinates
m	Mass of an element
M_x, M_y, M_{xy}	Distributed bending moments in a plate
$\mathbf{M}^e$	Element mass matrix (consistent)
$\mathbf{M}_L^e$	Element mass matrix (lumped)
$\mathbf{M}$	Mass matrix (assembled)
m_i	Modal mass
$n_i(x)$	Element shape function
$\mathbf{N}^e$	Element shape matrix
$n_{\alpha i}(x)$, $n_{\beta i}(x)$	Element shape functions (Mindlin element)
$\mathbf{N}_\alpha^e$, $\mathbf{N}_\beta^e$	Element shape matrices (Mindlin element)

s	Vector of stress components at a point *or* vector stress on an oblique plane
t	Surface traction
T	Temperature
u, v, w	Displacement components (cartesian)
u	Physical displacement vector *or* column vector of displacement components at a point
u_r, u_θ, u_z	Displacement components (cylindrical polar)
u_0, v_0, w_0	Displacements of the midplane of a beam or plate
U, U^e	Strain energy
V, V^e	Potential energy
W, ΔW	Virtual work
W_i	Weighting factor for numerical integration
x, y, z	Cartesian coordinates (global)
x′, y′, z′	Cartesian coordinates (local to an element)
x	Position vector

Index